Impact of Generative AI on Food Supply Chain Management

Bhupinder Pal Singh Chahal
Yorkville University, Canada

Arokiaraj David
SBS Swiss Business School, RAK, UAE

Amrinder Singh
Jain University, India

Geetika Madaan
Marwadi University, India

Gurmeet Singh
The University of the South Pacific, Fiji

IGI Global
Scientific Publishing
Publishing Tomorrow's Research Today

Vice President of Editorial Melissa Wagner
Director of Acquisitions Mikaela Felty
Director of Book Development Jocelynn Hessler
Production Manager Mike Brehm
Cover Design Jose Rosado

Published in the United States of America by
IGI Global Scientific Publishing
701 East Chocolate Avenue
Hershey, PA, 17033, USA
Tel: 717-533-8845
Fax: 717-533-7115
Website: https://www.igi-global.com E-mail: cust@igi-global.com

Library of Congress Cataloging-in-Publication Data

Names: Chahal, Bhupinder, 1980- editor. | David, Arokiaraj, 1986- editor. | Singh, Amrinder, 1983- editor. | Madaan, Geetika, 1983- editor.
Title: Impact of generative AI on food supply chain management / edited by Bhupinder Chahal, Arokiaraj David, Amrinder Singh, Geetika Madaan, Gurmeet Singh.
Description: Hershey, PA : IGI Global Scientific Publishing, [2026] | Includes bibliographical references and index. | Summary: "The book aims to examine how generative AI technologies can transform and optimize the food supply chain, addressing critical issues such as efficiency, sustainability, and waste reduction. By exploring AI's potential in decision-making, predictive analytics, and process automation, the book aims to bridge the gap between technological innovation and real-world application in the food industry"-- Provided by publisher.
Identifiers: LCCN 2024057011 (print) | LCCN 2024057012 (ebook) | ISBN 9798369398562 (hardcover) | ISBN 9798369398579 (paperback) | ISBN 9798369398586 (ebook)
Subjects: LCSH: Food supply. | Food industry and trade. | Supply chain management--Technological innovations. | Artificial intelligence.
Classification: LCC HD9000.5 .I4545 2025 (print) | LCC HD9000.5 (ebook) | DDC 664.0068/7--dc23/eng/20250202
LC record available at https://lccn.loc.gov/2024057011
LC ebook record available at https://lccn.loc.gov/2024057012

British Cataloguing in Publication Data
A Cataloguing in Publication record for this book is available from the British Library.

Table of Contents

Detailed Table of Contents

Artificial intelligence (AI) and Industry 4.0 are transforming food supply chain (FSC) by providing leading-edge solutions to enduring problems. This transformation uses latest technologies like blockchain, Internet of Things (IoT), AI-driven automation, and big data analytics to improve supply chain resilience, product and process quality along with significant improvement of operational efficiency. AI will help contribute towards waste reduction, improve inventory management while enabling predictive analytics for demand forecasting. The integration of AI and Industry 4.0 technologies in frozen FSC management is examined in this study, along with its advantages, disadvantages, and future prospects supporting the assumption that together AI and Industry 4.0 have the distinct ability to bring about significant transformation in the frozen FSC management through more effective decisions and collaboration. The study reflects on real benefits being waste reduction, cost reduction, and huge impact on the ecology.

The chapter explores how AI technologies can be applied to overcome some of the challenges facing quality assurance, compliance with regulation, and disruption of supply chains in the global food supply chain. This chapter further highlights that AI-powered systems, machine learning, and predictive analytics enhance supply chain resilience. The present chapter takes into consideration various other aspects of

used as data analysis tools, and hypothesis testing involved basic bivariate correlation coefficients and bootstrapping to make the results conservative and replicable. The study points out how applying AI enhances decision-making, cuts costs and wastes, and can also enhance logistics. It also provides valuable information on developing sustainable and flexible food supply chains in complex contexts.

Sustainable logistics pivotal focus in modern, driven by the urgent need to address environmental, social, and economic challenges. As global trade expands the sector faces significant pressure to reduce its carbon footprint while maintaining efficiency and competitiveness. Traditional logistics practices have often prioritized speed and cost over sustainability, leading to increased Consequently, the logistics industry is at a crossroads, compelled to rethink its operational frameworks and embrace sustainable practices.One of the most pressing challenges in sustainable logistics is the reliance on fossil fuels, climate change. Transitioning to alternative fuels and electric vehicles is essential but fraught with obstacles, including high upfront costs, limited infrastructure, and technological readiness. Additionally, among various stakeholders, including suppliers, manufacturers, and retailers, each of whom may have differing priorities and capabilities.Another critical challenge is the efficient management of resources.

The modern logistics sector, crucial to worldwide trade, is also a major source of carbon emissions. The environmental impact is significant, from transporting goods over long distances to the energy-intensive activities in warehouses. With the pressing necessity to cut down greenhouse gas emissions and address climate change, the logistics industry is at a crucial point. Introducing artificial intelligence (AI), an impressive technology that has the ability to transform logistics and greatly reduce its environmental impact.

Warehousing is an integral component of supply chain management, serving as a link between procurement and distribution. After witnessing a downfall during the COVID-19 pandemic, the 3PL market has experienced a rebound due to the adoption of digital solutions and technology. Gone are the days when organisations were using traditional warehouse management systems to manage and track the activities of a warehouse. Modern warehouse managing systems are software-based and are designed to be cost-effective, improve order fulfillment and delivery times, optimise inventory levels, improve lead times, reduce logistics costs and increase competitive advantage. This chapter focuses on introducing and analyzing the concept of smart warehousing. This chapter aims to enable the students to understand how the supply chain industry is embracing digital transformation by using Artificial Intelligence, IoT and analytics, advanced robotics and automation, RFID, automated guided vehicles (AGV) and Blockchain Technology.

The role of generative AI in enhancing inventory control and reducing holding costs in supply chain management. Traditional inventory management methods often struggle to balance stock levels, leading to overstocking or understocking, which incurs additional costs. Generative AI, through advanced predictive models, allows businesses to simulate various demand scenarios, improving forecasting accuracy and optimizing stock levels. By incorporating real-time data and automating decision-making, generative AI helps reduce overstocking, minimizes holding costs, and improves operational efficiency. This paper discusses key applications of generative AI, including demand forecasting, stock optimization, and automated reordering processes, while also addressing challenges such as data quality and system integration.

Preface

INTRODUCTION

This book explores the transformative role of Generative Artificial Intelligence (Generative AI) in the domain of Food Supply Chain Management (FSCM). It investigates how generative AI technologies—such as large language models, generative design tools, and predictive systems—can enhance forecasting, demand planning, inventory optimization, food traceability, and sustainability across all stages of the supply chain. The book provides theoretical frameworks, technological insights, and practical case studies to illustrate how generative AI is being deployed and its potential to disrupt traditional practices in food systems.

The food supply chain is under immense pressure due to global challenges such as climate change, pandemics, geopolitical conflicts, labor shortages, and consumer demands for transparency and sustainability. At the same time, digital transformation is accelerating, with AI at its forefront. Generative AI, in particular, has moved beyond experimentation into real-world implementation, offering new capabilities for data synthesis, simulation, and decision automation. As the global food system seeks to become more resilient and future-ready, the integration of generative AI represents a critical frontier in innovation and sustainability.

This book is intended for a multidisciplinary audience, including:

- **Academics and researchers** in artificial intelligence, food systems, supply chain management, and sustainability science.
- **Industry professionals and managers** working in agriculture, food processing, logistics, and retail who are exploring AI-driven solutions.
- **Technology developers and entrepreneurs** building AI platforms and tools for agri-food applications.

- **Students** pursuing studies in agribusiness, AI, operations management, or related fields.
- **Policy makers and sustainability leaders** interested in fostering ethical and efficient food systems through digital innovation.

CHAPTER OVERVIEW

Chapter 1: Industry 4.0 and AI Transformation in Food SCM

Artificial intelligence (AI) and Industry 4.0 are transforming food supply chain (FSC) by providing leading-edge solutions to enduring problems. This transformation uses latest technologies like blockchain, Internet of Things (IoT), AI-driven automation, and big data analytics to improve supply chain resilience, product and process quality along with significant improvement of operational efficiency. AI will help contribute towards waste reduction, improve inventory management while enabling predictive analytics for demand forecasting. The integration of AI and Industry 4.0 technologies in frozen FSC management is examined in this study, along with its advantages, disadvantages, and future prospects supporting the assumption that together AI and Industry 4.0 have the distinct ability to bring about significant transformation in the frozen FSC management through more effective decisions and collaboration. The study reflects on real benefits being waste reduction, cost reduction, and huge impact on the ecology.

Chapter 2: Integrating AI-Enabled Food Safety, Compliance, and Future Innovations for Resilient Food Supply Chain

The chapter explores how AI technologies can be applied to overcome some of the challenges facing quality assurance, compliance with regulation, and disruption of supply chains in the global food supply chain. This chapter further highlights that AI-powered systems, machine learning, and predictive analytics enhance supply chain resilience. The present chapter takes into consideration various other aspects of food safety, such as AI-based hazard detection, automated quality control, predictive maintenance in food processing equipment, AI-enhancing traceability of food items, compliance monitoring, real-time reporting, food safety-related decision support, incorporation of AI into standards and regulations on food safety, automation in compliance, real-time monitoring for regulatory compliance, AI-driven audit systems, risk management, smart contracts on food safety, and automated regulatory updates. The chapter emphasizes that AI plays a crucial role in shaping the future of food safety compliance, improving global food security and earning consumer trust.

Chapter 3: Generative AI Effect on Circular Economy Sustainability by Mediation of Food Supply Chain Agility

In order to comprehend how these elements support sustainable supply chain practices, this study examines the interactions between Food Supply Chain Agility (FSCA), Generative AI (GAI), and Circular Economy Sustainability (CES). Agile supply chains are better positioned to adopt sustainable practices, according to the research, which uses a mixed-methods approach and finds a substantial positive association between FSCA and CES. The results also demonstrate how important GAI is in promoting supply chain agility, which helps businesses become more responsive and efficient. In order to accomplish sustainable goals, the study emphasizes the significance of using cutting-edge technology and encouraging agility, providing insightful information for both managers and researchers. In the conclusion, the study advances the theoretical knowledge of sustainable supply chains and offers useful suggestions for businesses.

Chapter 4: Enhancing Traceability and Transparency in Food Supply Chains Through Generative AI

This raises the need for traceability while ensuring its responsible use, a daunting task. Emerging technologies like blockchain and AI, including GPT-2, offer potential solutions. GPT-2, with its capabilities in generation synthesis, machine learning, and document automation, addresses practical challenges such as synthetic data generation, product simulation, and document processing. By applying these technologies, traceability and visibility in food supply chains can be significantly improved. Virtual product models can replicate physical characteristics and provenance, enhancing food safety, sustainability, and consumer confidence. GPT-2's ability to automate documents and generate realistic simulations strengthens supply chain transparency, ensuring secure and sustainable operations while building trust among consumers.

Chapter 5: Training and Upskilling Employees for Generative AI-Powered Industry 4.0 Supply Chains

The advent of Industry 4.0 has revolutionized supply chains by integrating advanced technologies such as IoT, Big Data, AI and Generative AI, in particular, automates and enhances key supply chain functions like demand forecasting, inventory management, production scheduling, and logistics. This book proposal highlights the necessity for a skilled workforce to leverage these AI systems for optimal efficiency. It emphasizes the need for substantial investment in training and upskilling programs to prepare employees for AI-enabled roles. Addressing the existing skills gap, the proposal outlines strategies to bridge this gap through targeted educational initiatives and industry partnerships. It also discusses the challenges in fostering AI talent, stresses the importance of collaborative efforts among stakeholders to ensure successful AI integration. This comprehensive approach aims to equip the workforce with essential digital skills.

Chapter 6: AI-Driven Route Optimization and Sustainable Logistics: Transforming Stakeholder Perspectives for a Greener Future

This chapter examines how generative AI helps the food SCM evolve and aims to improve its efficiency, robustness, and sustainability performance. Using generative AI models, predictive analytics, and digital technologies, the paper identifies the AI's usage, data availability, digital supply chain, and the accuracy of the demand forecast. Therefore, the structured quantitative approach was adopted, and responses were collected on a seven-point Likert scale from 215 participants: financial analysts, market specialists, and other supply chain practitioners. Simple descriptive statistics such as frequency distributions and measures of central tendencies and variability were used as data analysis tools, and hypothesis testing involved basic bivariate correlation coefficients and bootstrapping to make the results conservative and replicable. The study points out how applying AI enhances decision-making, cuts costs and wastes, and can also enhance logistics. It also provides valuable information on developing sustainable and flexible food supply chains in complex contexts.

Chapter 7: Reducing Carbon Footprint with AI-Driven Logistics Planning

Sustainable logistics pivotal focus in modern, driven by the urgent need to address environmental, social, and economic challenges. As global trade expands the sector faces significant pressure to reduce its carbon footprint while maintaining efficiency and competitiveness. Traditional logistics practices have often prioritized speed and cost over sustainability, leading to increased Consequently, the logistics industry is at a crossroads, compelled to rethink its operational frameworks and embrace sustainable practices. One of the most pressing challenges in sustainable logistics is the reliance on fossil fuels, climate change. Transitioning to alternative fuels and electric vehicles is essential but fraught with obstacles, including high upfront costs, limited infrastructure, and technological readiness. Additionally, among various stakeholders, including suppliers, manufacturers, and retailers, each of whom may have differing priorities and capabilities. Another critical challenge is the efficient management of resources.

Chapter 8: Greening the Supply Chain: Ais Role in Reducing Carbon Emissions in Logistics

The modern logistics sector, crucial to worldwide trade, is also a major source of carbon emissions. The environmental impact is significant, from transporting goods over long distances to the energy-intensive activities in warehouses. With the pressing necessity to cut down greenhouse gas emissions and address climate change, the logistics industry is at a crucial point. Introducing artificial intelligence (AI), an impressive technology that has the ability to transform logistics and greatly reduce its environmental impact.

Chapter 9: Smart Warehousing: Automation And Real-Time Monitoring

Warehousing is an integral component of supply chain management, serving as a link between procurement and distribution. After witnessing a downfall during the COVID-19 pandemic, the 3PL market has experienced a rebound due to the adoption of digital solutions and technology. Gone are the days when organisations were using traditional warehouse management systems to manage and track the activities of a warehouse. Modern warehouse managing systems are software-based and are designed to be cost-effective, improve order fulfillment and delivery times, optimise inventory levels, improve lead times, reduce logistics costs and increase competitive advantage. This chapter focuses on introducing and analyzing the concept of smart warehousing. This chapter aims to enable the students to understand how the supply chain industry is embracing digital transformation by using Artificial Intelligence, IoT and analytics, advanced robotics and automation, RFID, automated guided vehicles (AGV) and Blockchain Technology.

Chapter 10: Enhancing Inventory Control and Reducing Holding Costs Using Generative AI

The role of generative AI in enhancing inventory control and reducing holding costs in supply chain management. Traditional inventory management methods often struggle to balance stock levels, leading to overstocking or understocking, which incurs additional costs. Generative AI, through advanced predictive models, allows businesses to simulate various demand scenarios, improving forecasting accuracy and optimizing stock levels. By incorporating real-time data and automating decision-making, generative AI helps reduce overstocking, minimizes holding costs, and improves operational efficiency. This paper discusses key applications of generative AI, including demand forecasting, stock optimization, and automated reordering processes, while also addressing challenges such as data quality and system integration. The successful implementation of AI in inventory control provides companies with a competitive edge in an increasingly complex and dynamic supply chain landscape

Chapter 11: Optimization of Inventory Management Using Artificial Neural Networks and K-Means Clustering for Cost Reduction and Improved Customer Service

This paper addresses the critical issue of optimizing inventory strategies using advanced machine learning techniques, particularly Artificial Neural Networks (ANN), to enhance customer service and reduce costs. Here's a brief summary and analysis of the concepts you mentioned: 1. Optimization of Stock Management: Inventory management plays a vital role in fulfilling customer demand while minimizing lead times and costs. Optimizing stock levels can balance supply with fluctuating market demand. 2. Artificial Neural Networks (ANN): ANN is employed to forecast the optimal stock levels. By analyzing past stock data, ANN models can predict the required inventory and reduce errors in forecasting, thus minimizing overstocking or under stocking. 3. K-Means Algorithm for Grouping Items: The k-means algorithm is used to classify raw materials and finished goods into homogeneous groups. These groups share similar characteristics, allowing for tailored inventory policies. For example, fast-moving items may have different inventory strategies than slow-moving ones.

Chapter 12: Generative AI in Innovative and Eco-Friendly Packaging Solutions

This chapter examines how generative AI is being used to solve innovative packaging solutions that minimize or reduce their impact on the environment. The concept here is to show how generative AI improves the packaging design by assessing sustainability, durability, and material utilization, which in turn enables the design and creation of smarter sustainable substitutes. AI algorithms allow the designers to run several hundreds of permutations across the design canvas, thus achieving packaging that is made from recycled or biodegradable material and a net total carbon footprint. The chapter also looks at how generative AI improves efficiency in package creation, from its design to its manufacture. Therefore, through the detailed chapter analysis, we showcase the appropriate use of packaging from the AI perspective and explore trends by considering generative AI as a critical tool that will help in achieving environmental sustainability goals as well as addressing global packaging challenges.

Chapter 13: Food Preservation and Utilization Strategies for the Indian Population: Food Preservation strategies for Indian Population

Food security remains a significant concern in India, where post-harvest losses, inefficient supply chains, and lack of awareness about preservation methods contribute to substantial food wastage. This research article explores various strategies for preserving and utilizing food effectively to ensure food security and sustainability. It highlights traditional and modern preservation techniques, government policies, community-driven initiatives, and the role of technology in minimizing food wastage.

CONCLUSION

This book contributes to the growing body of knowledge at the intersection of AI and sustainable food systems by being among the first to focus specifically on generative AI's role in FSCM. It offers a strategic and operational perspective, bridging academic theory and industry application. By highlighting real-world use cases, technological challenges, and future research directions, it provides a roadmap for adopting AI responsibly and effectively. The insights presented here aim to stimulate innovation, guide informed decision-making, and support the global transition to a smarter, more secure, and sustainable food supply chain.

Chapter 1
Industry 4.0 and AI Transformation in Food SCM

Madhu Arora
https://orcid.org/0000-0001-6015-6945
Symbiosis Institute of Computer Studies and Research, Symbiosis International University, Pune, India

Neeraj Anand
Chitkara Business School, Chitkara University, India

Parag R. Kaveri
Symbiosis Institute of Computer Studies and Research, Symbiosis International University, Pune, India

ABSTRACT

Artificial intelligence (AI) and Industry 4.0 are transforming food supply chain (FSC) by providing leading-edge solutions to enduring problems. This transformation uses latest technologies like blockchain, Internet of Things (IoT), AI-driven automation, and big data analytics to improve supply chain resilience, product and process quality along with significant improvement of operational efficiency. AI will help contribute towards waste reduction, improve inventory management while enabling predictive analytics for demand forecasting. The integration of AI and Industry 4.0 technologies in frozen FSC management is examined in this study, along with its advantages, disadvantages, and future prospects supporting the assumption that together AI and Industry 4.0 have the distinct ability to bring about significant transformation in the frozen FSC management through more effective decisions and collaboration. The study reflects on real benefits being waste reduction, cost reduction, and huge impact on the ecology.

DOI: 10.4018/979-8-3693-9856-2.ch001

INTRODUCTION

This chapter offers a logical flow to deliberate upon different aspects of employing Artificial Intelligence (AI) and Industry 4.0 to help bring about a digital transformation in the food supply chain. Since the present frozen FSC (food supply chain) is still at an evolving stage, dominant application of technology is needed for improved optimization of food supply chain management (SCM), especially considering the predicted substantial demand and growth in the sector. Further, each section and subsection are explored through relevant examples, case studies, and facts to ensure a full understanding of the subject matter being studied.

Overview of Artificial Intelligence

Artificial intelligence (AI) is a field that was established more than 65 years ago. The field began with high expectations and supercilious objectives, moved through several phases of popularity, and most recently saw a renaissance owing to the development of deep neural networks (Emmert-Streib et al., 2020). McCarthy first used the term artificial intelligence in 1956 (McCarthy et al., 2006) at the Dartmouth conference, which launched a focused effort in this field of study. As per a study by Crevier (1993), initially the key focus of artificial intelligence was reasoning and symbolic models before it shifted to expert systems and neural networks and expert systems. Deep neural networks, another resurgence of neural networks, were made possible by the availability of Big Data in the early 2000s. Today, a variety of sectors including speech recognition, robotics, facial recognition, finance, and healthcare have seen significant success with the help of AI.

The term "intelligence" in AI was defined as:

universal intelligence of agent $\pi = \Upsilon(\pi) = \sum \mu \in E2 - K(\mu) V \pi \mu$

where K - Kolmogorov complexity function, π - agent, E - environment set, μ - environment set, $V\pi\mu V\mu\pi$ - value function and $2-K(\mu)2-K(\mu)$ - algorithmic probability distribution over environment and $\Upsilon(\pi)$ - universal intelligence of agent π (Legg & Hutter, 2007).

The equation thus provides a way to quantify intelligence by measuring an agent's capacity to accomplish objectives in a variety of settings. However, the equation is highly complicated and unintuitive with exact computation of Kolmogorov complexity function K being a challenge therefore, a precise practical evaluation is not feasible. According to Performing such an intelligence test is also is a challenge as an agent may seem intelligent without truly possessing intelligence. That signifies the challenge in neither possessing an acceptable definition for "intelligence" nor having any test that can identify it in a reliable manner.

However, as per Russell (2016), there are two assumptions under Artificial Intelligence posits. According to one, it was possible for machines to behave intelligently while according to the other that strongly believes that such machines are actually capable of thinking and not only simulating it. Yampolskiy and Fox (2012) termed the second one as 'Artificial General Intelligence'. AI's key aim is to make machines smarter while other secondary objectives are comprehending intelligence and to make machines more beneficial. According to Kurzwell (1990), AI is "the art of creating machines that perform functions that require intelligence when performed by people".

The term artificial intelligence (AI) describes the process of creating computer systems that possess the ability and capability to carry out tasks including: problem-solving, language comprehension, judgement, analysis and selection, recognition of pictures, etc. that normally require human intelligence. The fundamental purpose of artificial intelligence (AI) is to enable machines to operate and respond intelligently by mimicking human cognitive functions using algorithms and data-driven procedures. Artificial intelligence (AI) can be broadly classified into two categories: general AI, which attempts to perform any intellectual work that a human can perform, and narrow AI, which is intended for specialised tasks (eg facial recognition). Presently, it is narrow AI that encompass the majority of applications.

Artificial intelligence (AI) is expected to dominate technology research in the future compared to other technology research areas, according to the IEEE Computer Society's technology prediction for 2023. The majority of AI research areas also have a strong correlation to human impact and a high chance of success (Chiroma, Hashem & Maray, 2024). AI to a great extent is based on the concept of machine learning which enables the machines to learn from experience thereby eliminating the need for explicit programming.

Overview of Industry 4.0

The digital revolution is radically changing how people live and work. Businesses today have been incorporating the Industry 4.0 technologies more and more since the last few years. Industry 4.0, also referred to as the fourth industrial revolution, is the result of industry adoption of ICT (information and communication technologies) advancements. I 4.0 is built on the integration of information technology and industry. H. Kagermann, a Professor of physics, coined the term 'Industry 4.0' during the 2011 Hanover Fair. It is a concept that can be comprehended as amalgamation of cyber-physical systems, production and logistics systems and application of IIoT (industrial internet of things) and IoS (internet of services) thereby leading to formation of new value chains, transformation in business models and process reorganization. (Kagermann et al., 2013)

IoT (internet-of-things), Big Data, Cloud computing, Augmented reality, autonomous systems, cybersecurity, additive manufacturing and cybersecurity are few of foundational pillars upon which industry 4.0 is established (Gajdzik et al., 2020). Intelligent machines, smart materials, and networked computers in an Industry 4.0 setting interact with the environment, communicate with one another, and ultimately make decisions with little intervention from humans (Ghobakhloo et al., 2023; Ching et al., 2022; Ghobakhloo, 2020). The digitisation of the manufacturing sector and value delivery channels may provide a number of advantages, such as increased manufacturing productivity and efficiency, business process innovation, enhanced knowledge sharing and collaborative working, and improved customer experience (Machado et al., 2020; Jena et al., 2020). Presently, the emphasis of Industry 4.0 is upon integration which will help significantly improve production efficiency, allow more responsive product adjustments. All such integrating components will result in a new sustainable business model. And through new technology, stakeholders can actively contribute to sustainable development. Figure 1 shows the adoption stages of Industry 4.0 technologies.

Figure 1. Stages of Adoption of Industry 4.0

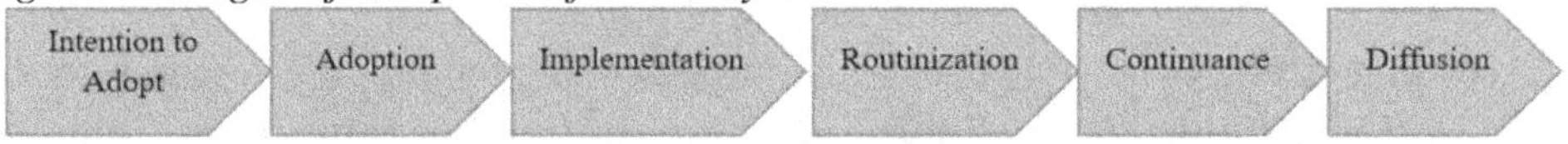

(Source: Queiroz et al., 2022)

Overview of Food Supply Chain

In addition to dealing with perishable items, the frozen food supply chain (SC) is complicated and involves many different parties, including farmers, wholesalers, food manufacturers, and retailers who operate independently (Arora, Kumar & Raju, 2023). By 2050, it is predicted that there shall be three billion people on our planet earth (Wunderlich & Martinez, 2018). This indicates the growing requirement and consumption of food products including frozen food. However, at the same time, climate change, global warming and other such critical issues are not only affecting agricultural production yields but are predicted to worsen during the next fifty years along with price increase. Hence, it becomes imperative to re-assess the current frozen food supply chain structure and operations so that it has the capability to meet the food demand.

The resources and processes used in the production, processing, marketing, consumption, and disposal of products from forestry and agriculture, as well as the required inputs and products produced at each stage, encompass the food systems

(FAO,2016). There are primarily key components including consumer behaviour, environment and consumer behaviour that constitute a food system. Infrastructure, innovation and technology act as environmental drivers of change.

The various processes and stakeholders responsible for moving goods from manufacturers to end-consumers while also including waste management together make a food supply chain (Hawkes & Ruel, 2006). Agriculture, processing, and packing together constitute food production; storage and transportation in distribution; and finally, trade includes food economics. Finally, disposal entails recycling, degradation, and nutritional loss.

However, there could be failure in the transition of a frozen food item from the stage of production to that of consumption leading to a disruption in the FSC. Such a disruption has the potential to impact not only the infrastructure of the food systems but can significantly impact its functioning too. The impact could be in terms of unfavourable operating outcomes, excessive transportation expenses, delays in orders, shortages in inventory and detrimental financial effects. The disruption could also lead to supply-demand imbalance, price volatility, operational capability variance, operational system volatility along with reduced availability of high-quality products (Rojas-Reyes et al., 2024).

SAFETY AND QUALITY IN FSC THROUGH USE OF AI AND INDUSTRY 4.0

An FSC (food supply chain) encompasses all activities including manufacturing, processing, packaging, distribution, and consumption activities. Since contamination or inefficiencies in frozen FSC can result in foodborne diseases, financial losses, and environmental impact, it is imperative that safety and quality be maintained along this chain. Adoption of Industry 4.0 and AI (Artificial Intelligence) technology has the potential to revolutionise how these issues are addressed.

FSC and AI Considering the dynamic and complex business global scenario, AI has become essential to the FSC's automation, optimisation, and improvement of safety and quality. It affects quality and safety in a number of ways (Refer figure 2).

Figure 2. FSC and AI impact on safety and quality

Predictive Analytics — Results in improved quality assurance

- AI - driven systems can examine historical data to anticipate potential quality problems
- Stakeholders in FSC can take preventative action by using ML algorithms to find trends in contamination or spoilage in specific environments

SCM monitoring & Traceability — Results in enhanced end-to-end traceability

- Application of technologies like blockchain and AI can help trace and trace ingredients right from point of production to point of consumption and that too real-time
- Ensures transparency and enables prompt detection of source of contamination in the event of

Quality control & Automated inspection — Results in significantly improved food safety

- Computer vision systems driven by AI are used to examine food products, detect impurities and contaminations while ensuring compliance of safety standards
- Such algorithms can quickly detect even the subtlest of the differences that identifies food

Predictive Maintenance — Results in early identification of malfunctions

- AI can also be used to forecast equipment failures in manufacturing, packaging and storage of food
- Through examination of operational data, systems can identify indications of malfunction even before a shutdown or product contamination occurs

(Source: Author)

FSC and Industry 4.0 The fourth industrial revolution, or Industry 4.0, is characterised by the integration of automation, the Internet of Things, and digitalisation of processes into conventional supply chain and manufacturing activities. Few of the chief applications in the frozen food supply chain are:

- IoT for Real-Time Monitoring: A key component of Industry 4.0, IoT sensors gather data on food goods in real-time, including temperature, humidity, and pressure, as they are being transported and stored. Such an information will help in preserving specific conditions to guarantee the quality and safety of food.
- Automation and Robotics: Industry 4.0 supports employing robots capable of robots that are capable of doing precise operations like sorting, packing, and portioning in the food processing and packaging industry. This ensures a constant product quality, lowers the possibility of contamination, and reduces mistakes made by humans.
- Smart warehousing and distribution: By integrating AI and IoT, it is possible to automate warehouses in a way that inventories can be managed in real-time thereby ensuring storage of food items in optimal conditions. Automated vehicles can be utilised for safe and effective distribution.

Food Safety through Industry 4.0 and AI As both AI and Industry 4.0 help automate monitoring, evaluation and response mechanisms, they now offer a new degree of control with respect to frozen food safety. Some of the ways are as listed below:

- Food safety Compliance: Through continuous monitoring of parameters and identification of non-compliances, AI empowered systems can ensure adherence of safety rules by the different production processes. This helps reducing possibility of human error or oversight.
- Improved inspection and food testing: Conventional food inspection techniques are tedious and susceptible to human error. Visual recognition software driven by AI can automatically identify contaminants, flaws, or foreign objects in food goods while they are being produced. With the help of AI-driven diagnostics, food safety testing can be expedited along with identification of harmful substances, allergies, etc. Not only can these systems handle enormous volumes of data more quickly but they can do so with high accuracy of detection. For example, computer vision technology, can keep a check on food products for deterioration, guaranteeing that only safe goods make it to market.
- Contamination prevention: Some repetitive operations including sorting, packaging, and processing can be performed by robotics with AI integration to achieve high-precision. By reducing human participation, these automated technologies will help lessen the possibility of contamination during the food production/processing process. AI-enabled predictive models can estimate the likelihood of contamination, forecast foodborne pathogen outbreaks, as well as improve food safety procedures. It can identify contamination hazards such as bacterial growth, and suggest preventive measures for preventing contamination. In case of safety violation, it can also help expedite recall by pinpointing the precise origin and distribution of products that are contaminated thereby enabling firms to reduce risks before they become more serious.
- Enhanced transparency and visibility: Real-time traceability of food items right from its source can be done with the help of blockchain technology and internet of things. Information regarding sourcing, processing and handling of food items is verified and available to consumers and other stakeholders resulting in increased trust. In the event of a foodborne illness outbreak, this facilitates prompt identification of the source of contamination, allowing for quicker recalls and hence limiting the spread of bacteria.
- Optimized food safety practices: Through application of Big data and data analytics, firms can find trends in food safety events, identify food chain vulnerabilities and improve risk forecasting. Owing to data-driven perceptiveness,

food safety procedures can be kept up-to-date with industry best practices thereby optimising quality control methods, safety procedures along with initiatives taken towards staff training for better handling of food products.

Although there are many prospects for improving food safety with Industry 4.0 and AI, there are also obstacles in the form of data privacy, high implementation costs, and the requirement for legislative frameworks that encourage the adoption of new technologies. But as AI becomes more widely available and regulatory organisations incorporate digital technology into food safety regulations, the frozen food sector is set to reach previously unheard-of levels of quality, safety, and consumer confidence.

In summary, the industry is changing as a result of the incorporation of AI and Industry 4.0 into food safety management. Food businesses may lower contamination risks, guarantee a safer and more transparent supply chain, and react to food safety issues more quickly by utilising cutting-edge technologies. This will eventually safeguard customers and enhance public health.

Responsiveness in FSC

The ability of a supply chain to promptly adjust to shifts in the market, client demands, and interruptions is referred to as responsiveness. Responsiveness in the frozen food supply chain is essential in today's globalised food sector to ensure quality, reduce risks and maintain a balance between supply and demand. As responsiveness has a direct impact on food safety, freshness, satisfaction of the consumer and overall efficiency of the food chain, it becomes even more crucial in the context of food supply chain. The present consumer is increasingly conscious of food safety regulations and therefore looks for safe, fresh and high-quality food items. In an FSC, responsiveness aids organisations in fulfilling these requirements by:

- Lead time reduction - Speedier delivery of fresh goods is ensured by capacity to respond promptly to consumer requirements, achieved through reduction in lead time.
- Address variability in demand - Often there is fluctuation in demand owing to promotional campaigns, seasonal shifts and health trends. A responsive chain can quickly adjust to such demand shifts.
- Enhance customer satisfaction - A responsive food chain can boost customer loyalty and increase confidence when quality or safety concerns in customer feedback or complaints are quickly addressed.

Businesses with responsive supply chains shall be in a better position to: introduce new goods through rapid adjustment of their food supply chains; adjust to shifting market conditions; have flexibility in offering wider range of products; and finally gain market share in the fiercely competitive food sector. However, responsiveness also requires intense collaboration between all food chain participants. Such strong collaboration will enable stakeholders to negotiate flexible arrangements that guarantee a consistent supply of materials even in the case of fluctuating demand. Manufacturers, distributors, and retailers can exchange real-time data throughout the supply chain to better synchronise activities and cut down on delays and bottlenecks.

The essence of responsiveness is the capacity to swiftly modify distribution plans, inventory levels, and manufacturing schedules. This is especially crucial in FSC because demand may change quickly while the shelf life is limited. Hence, agile inventory systems, flexible manufacturing/processing can be few key aspects considered to enhance responsiveness in the food supply chain.

Few strategies that can be formulated and implemented to enhance responsiveness include:

- Lean and Agile food chains - Lean food chains emphasise upon waste reduction and efficiency improvement while agile supply networks will help in enhanced adaptability to deal with abrupt changes in supply or demand.
- Demand forecasting - In order that a food chain remains responsive, accurate demand forecasting is highly crucial. Past data, current market trends and external factors such as social media, weather, etc. can be examined by employing AI-powered analytics and machine learning algorithms to forecast demand.
- Dynamic inventory management - Excess inventory can be avoided and waste can be reduced by using real-time data to modify inventory levels in response to changes in demand. This is especially crucial for perishable items.
- Decentralized food supply chains - By cutting down on transit times and improving the ability to react to local demand, decentralising the supply chain and utilising more local suppliers can enhance responsiveness. Through local sourcing, lead times can be shortened and demand changes may be adjusted more quickly. By having micro-fulfilment centres i.e. putting smaller distribution facilities near customers expedites delivery, particularly for goods that must be delivered quickly.

However, there are a few challenges in the frozen food supply chain such as regulatory compliances, cost considerations and supply chain disruptions. Unforeseen events, geopolitical conflicts, natural disasters, etc certainly can lead to challenges and disruptions in the food chain. Such disruption risks can be reduced through diversification of sources and creating strong backup plans. Investment in technology, infrastructure, and supplier relationships may not be easy for the small and medium-sized enterprises in the food sector. Also, it can be difficult to strike a balance between responsiveness and stringent regulatory requirements and compliances, especially when quick operational adjustments are needed.

Meeting customer expectations, guaranteeing food safety, and preserving a competitive edge in the frozen food industry all depend on responsiveness in the supply chain. By utilising technology, encouraging teamwork, and implementing adaptable supply chain tactics, businesses can enhance their capacity to react swiftly and efficiently to shifting market conditions. As the food industry develops further, responsiveness will continue to be a crucial component in guaranteeing a robust and effective supply chain.

Sustainability in FSC

The demand for sustainable products from consumers is rising with the increasing concern for the environment. Through enhanced consumer loyalty, and improved brand reputation, firms that use sustainable FSC practices frequently obtain a competitive advantage.

Environmental Impact Today everyone including governments, corporations and consumers are seriously concerned about sustainability related to frozen food supply chain. It is predicted that the demand for food will rise by 70% by 2050 considering the expanding global population thereby placing tremendous strain on the natural resources viz land, water and energy. There is pressing need for sustainable food supply chain practices to ensure that food production and distribution systems are able to meet present demands without endangering the potential of future generations. It will also help strike a balance between social equality, environmental preservation and economic prosperity.

The food supply chain has a significant impact on the environment as it can influence changes in climate, scarcity of water resource, biodiversity loss along with deforestation. Ecosystems can be deteriorated owing to unsustainable farming methods, food packaging and transportation. Approximately one-third of the world's greenhouse gas (GHG) emissions are a resultant of the FSC, with the largest contributions coming from the food processing, logistics, and agricultural sectors. Reducing emissions at every stage of the supply chain is the goal of sustainable practices. 70% of the world's freshwater use comes from agriculture, and ineffective irrigation systems frequently result in water waste. Sustainable FSC practices concentrate on enhancing water management and lowering consumption. Soil degradation and decreased agricultural output are the results of excessive use of chemical fertilisers, pesticides, and conventional agricultural practices. Sustainable techniques place a higher priority on soil health, encouraging organic farming and crop variety.

Social and Economic Impact Sustainability involves not just protecting the environment but also making sure that resources are distributed fairly. FSC sustainability tackles economic and social concerns. Low wages and unfavourable working conditions are commonplace for many employees in the food business, especially in agriculture. Human rights, social justice, and decent working conditions are given top priority in sustainable FSC practices. It is quite difficult for smallholder farmers to compete in international food markets. By encouraging fair trade, organic certification, and direct access to markets, sustainability initiatives aim to raise their standard of living.

However, there are quite a few challenges in attaining sustainability in food supply chain, as presented in Fig 3.

Figure 3. Key Challenges in Achieving Sustainability in FSC

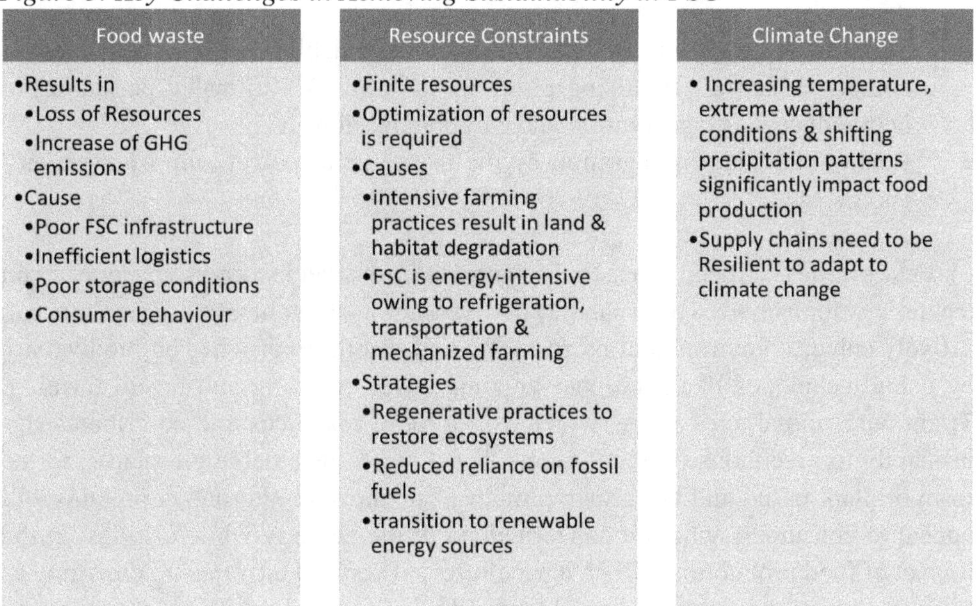

(Source: Author)

Some of the strategies that can be implemented to enhance sustainability in food supply chain are as given below:

- Sustainable agricultural practices including: (a) Organic farming that stresses upon use of natural fertilisers and pesticides like compost, biological pest control, crop rotation, etc; (b) Precision agriculture which utilizes technology such as drones, IoT sensors, and artificial intelligence to optimize usage of resources on the basis of real-time data and improved efficiency; (c) Agro-ecology as an approach promotes bio-diversity, soil fertility and carbon sequestration.

- Sustainable transportation & logistics including: (a) Shortened food supply chain through urban farming, local sourcing and decentralized food chains can help reduce carbon footprint while maintaining productivity; (b) Green logistics through the use of electric vehicles, AI optimized routes and use of low-carbon fuels.
- Reduced food wastage and losses through: (a) improved cold chain management by investing in cold chain infrastructure to preserve perishable products; (b) sustainable packaging solutions like biodegradable packaging, better food labelling practices and reduced plastic waste; (c) Circular economy approach that focus on minimising waste by recycling, reusing, and refurbishing food items.
- Sustainable packaging including; (a) biodegradable material so that it decomposes naturally reducing plastic pollution; (b) Minimalist packaging to help reduce waste generation and carbon emission.
- Renewable energy integration by the use of solar power, wind power, bioenergy, etc.

Future trends in frozen food supply chain could consider incorporating blockchain technology for increased transparency and traceability. Regenerative agriculture can actively enhance environment by restoring soil health, improving on biodiversity by using techniques like rotational grazing, cover cropping and no-till farming. Traditional animal agriculture, which uses a lot of resources and contributes significantly to greenhouse gas emissions, has a more sustainable alternative in the form of plant-based and lab-grown proteins. The move to alternative proteins will appeal to consumers who are concerned about the ecology while also lessen the impact of food production. Urban agriculture and vertical farming are emerging as ways to address the increasing need for food in urban settings. In addition to using less water and land, these methods lower transportation emissions by growing food near to the area of consumption.

Finally, to conclude, in order to solve the environmental, social, and economic issues related to farming and distribution, sustainability in the frozen food supply chain is essential. Businesses may create robust and environment-friendly supply chains by using sustainable agriculture methods, cutting down on food waste, enhancing resource management, and investing in green logistics and renewable energy. The implementation of sustainable FSC practices will become an essential part of the worldwide food chain, guaranteeing that it can meet the demands of both current and future generations, as consumer demand for sustainable products rises and environmental challenges increase.

Overall FSC resilience and optimization

In order to ensure that flow of food products right from the farmers to the consumers is steady and reliable, it is important that the food chain becomes resilient and optimised. A food supply chain is considered resilient if it can tolerate, adapt to, and bounce back from disruptions that occur in the chain while optimisation focusses on increasing supply network's sustainability, efficiency and waste reduction. Some of the strategies that can be implemented for building up resilience and optimisation include:

- Having a large network of domestic and international suppliers lowers reliance on any one source and the likelihood of supply shortages.
- Increase in shelf life of perishable items and a consistent supply even during disruptions by investing in improved storage methods and infrastructure including cold chains.
- Real-time monitoring through use of IoT, improved traceability through use of blockchain technology, and enhanced transparency through predictive analytics enables stakeholders to make well-informed decisions and reduce risks. This will also enable improved forecasting, cut down on delays, and improved inventory control.
- A substantial percentage of food is lost at different links on the chain. Such a wastage can be reduced by implementing better harvesting practices, enhancing logistics, and minimising deterioration during storage and transit.
- Long-term sustainability is facilitated by promoting the adoption of circular economy models, regenerative farming, and the implementation of eco-friendly techniques such lowering transportation-related carbon emissions.
- Throughout the supply chain, integrating robotics, automation, and AI-driven solutions improves resource allocation, lowers human error, and boosts efficiency.
- Develop legislation that promote sustainable farming methods, trade agreements, and prompt reaction to disturbances through collaboration amongst governments, corporations, and international organisations.

Additionally, the frozen food supply chain needs to adapt to changes in customer preferences, such as the rising demand for organic, plant-based, and locally produced components. For this are needed systems flexible enough to quickly adjust to shifts in customer demand without creating shortages or bottlenecks. The cold chain will need to be reinforced and its performance improved in order to increase consumer uptake and satisfaction (Arora, Kumar & Anand, 2022).

By offering insight into each stage of the supply chain, technologies may increase responsibility and build consumer trust. Finally, a global approach towards the supply network is an imperative today if one really wants to ensure a resilient and optimized food supply chain. Hence, cooperation and collaboration amongst the nations is also necessary to maintain the stability of the world's food commerce and place shared problems into practice.

FUTURE TRENDS AND CHALLENGES

It is essential to comprehend the future trends and challenges for all the stakeholders including governments so that they can ensure the resilience, sustainability, responsiveness and efficiency of the frozen food supply network. Figure 4 depicts the future trends and challenges.

Figure 4. Future trends

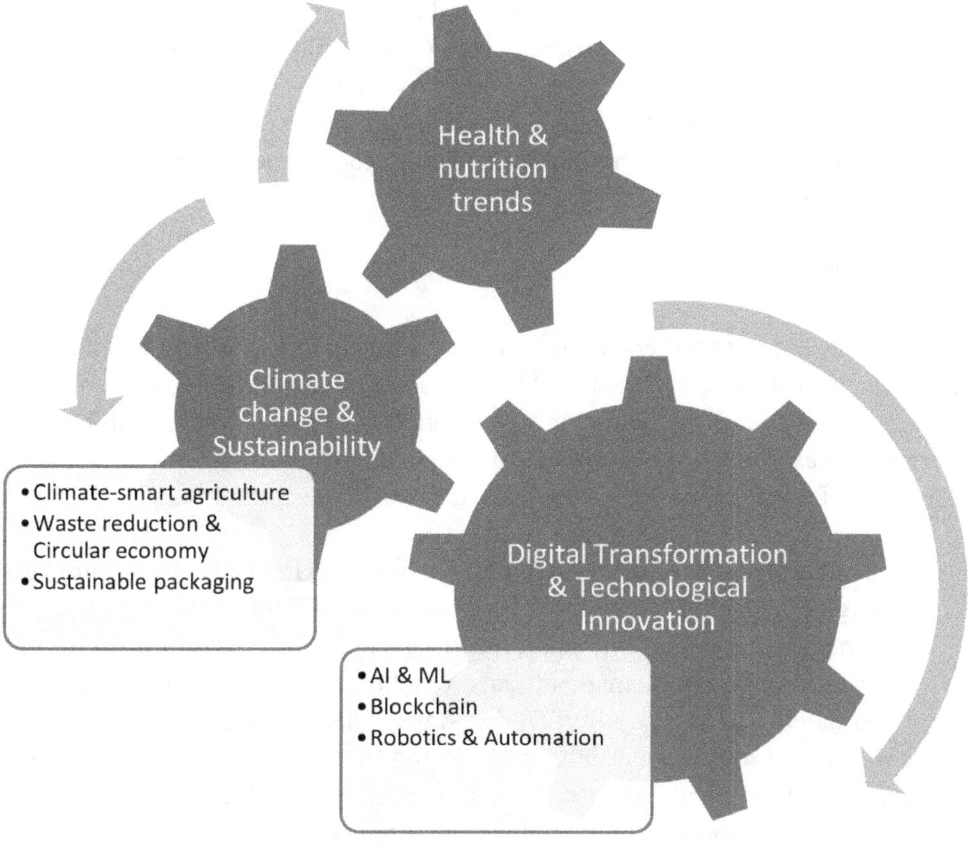

There could be numerous challenges including changing consumer preferences, geopolitical instability and risks, food access, safety and security, public health concerns, labour constraints, pandemic-led disruptions, etc. The COVID-19 pandemic exposed weaknesses in the world's food supply network, such as labour shortages and bottlenecks. In order to minimise interruptions during pandemics or other public health emergencies, supply chains in the future will need to be better equipped to handle global health crises. This includes having more robust contingency plans, better worker protections, and increased operational flexibility.

Labour shortages will be addressed by automating repetitive processes like harvesting, packing, and logistics, but doing so will also necessitate spending money on retraining employees to handle and use cutting-edge technology.

There will be an increased demand for qualified personnel with the ability to handle data analytics, run intricate systems, and supervise cutting-edge supply chain operations. Future trends suggest that talent development in data science, technological integration, and logistics will be given more importance.

GMO use is still a controversial topic, and discussions over food safety, labelling, and public acceptance are probably going to continue. Ensuring fair access to food is becoming more difficult as the world's population rises, especially in areas that are experiencing poverty, violence, and climate-related disturbances. Finding ways to protect vulnerable communities and solve food security will be crucial to the food supply chain's future.

Case Study Example

One of the biggest retailers in the world, Walmart has an extensive and intricate supply chain. Walmart implemented AI-powered solutions in order to stay competitive and tackle issues in the food supply chain, including waste reduction, improving demand forecasts, and guaranteeing food safety.

Walmart faced a number of challenges as given below:

- Demand Forecasting: In order to prevent overstocking or understocking and managing perishable food goods accurate demand forecasting is an absolute imperative. Using conventional forecasting techniques it was difficult to consider local preferences, market trends, weather, etc. into consideration.
- Food Safety and Traceability: During times when consumers want transparency in food source and safety, ensuring food safety is of utmost importance. It is not only difficult but also time-consuming to trace the origin of contamination or food-borne illnesses along the supply chain.
- Reduction of Food Waste: The vast waste generation by the retail food industry, expenses and environmental impact is a major concern.

Therefore, in order to maximise shelf life for perishable commodities and manage inventory, Walmart required an improved method.

Deployment of AI solutions

- Artificial Intelligence for Demand Forecasting: Walmart has adopted machine learning algorithms that examine enormous volumes of data, such as past sales, meteorological trends, consumer preferences, and area variables. Demand could be predicted more precisely by AI models than by conventional techniques. Walmart decreased predicting errors by more than 30% by utilising AI. A decrease in stockouts resulted in higher customer satisfaction while overstock conditions decreased, reducing perishable item waste.

- AI for Inventory and Waste Management: Walmart may use AI to better identify products that are about to expire and tailor promotions to sell them before they go bad. As an illustration, the AI system assists in dynamically modifying prices for perishable goods that are about to expire. AI-guided inventory tracking systems made sure that extra inventory was redistributed to other stores with higher demand.
- Blockchain and AI for Food Traceability: To improve food traceability, Walmart combined blockchain technology with artificial intelligence. Walmart used blockchain technology into their supply chain in collaboration with IBM to track the origin of food items:
    - AI-enabled tracking technologies can identify the origins of a pollution breakout in a matter of seconds rather than days.
    - By allowing consumers to scan products and identify the farm from which the food was derived, blockchain also promotes transparency, which increases accountability and confidence.
- AI for Warehouse and Logistics Optimisation: AI-powered robots are used at Walmart's warehouses to manage inventory, cutting down on the amount of time needed to track products and stock shelves. Additionally, AI streamlines delivery route planning, lowering delivery times and fuel expenses.

These strategies reduced food waste significantly, contributing to a more sustainable supply chain. The following outcomes were attained:

- Food Waste Reduction: During the first year of deployment, Walmart's AI technologies helped to reduce the waste of perishable food by 20%.
- Increased Forecasting Accuracy: Demand forecasting driven by AI has increased accuracy by 10–20%, resulting in fewer missed sales and higher customer satisfaction.
- Improved Traceability: Walmart was able to remove contaminated products more quickly thanks to blockchain-based AI technologies, which also greatly decreased the time it took to identify the source of contamination.
- Sustainability: Through waste reduction and logistical optimisation, Walmart's AI initiatives have improved the sustainability and environmental friendliness of its supply chain.

Thus, Walmart's food supply chain successfully incorporated AI, demonstrating the technology's revolutionary potential in the retail sector. Walmart has increased efficiency, boosted customer trust, and encouraged sustainability by tackling important issues including demand forecasting, waste reduction, and food safety. This case example shows how artificial intelligence (AI) may transform food industry supply chain management, offering advantages for operations and the environment.

CONCLUSION

The food supply chain's AI transformation offers a huge chance to improve system-wide efficiency, sustainability, and transparency. Stakeholders may optimise processes from agricultural production to consumer delivery by utilising AI technologies such as automation, machine learning, and predictive analytics. Enhancing crop output forecasts, cutting waste through accurate demand forecasting, guaranteeing food safety through real-time monitoring, and facilitating more individualised customer experiences are all possible with this shift.

But there are drawbacks to integrating AI as well, like worries about data privacy, the necessity of internet infrastructure in remote places, and the possible replacement of manual labour. The food sector, governments, and tech developers must work together to ensure that AI is used in a way that is sustainable, inclusive, and morally sound if the benefits are to be fully realised. In the end, AI could completely transform the world's food supply chain, assisting in supplying the world's expanding food demand in an increasingly resource-constrained environment.

REFERENCES

Arora, M., Kumar, R., & Anand, N. (2022). Analysis of frozen food adoption by the consumer in Uttarakhand, a state of India: An inferential statistics approach. *International Journal of Value Chain Management*, *13*(1), 88–111. DOI: 10.1504/IJVCM.2022.122164

Arora, M., Kumar, R., & Raju, T. B. (2023). Identification of issues in the cold chain of Indian frozen food. *International Journal of Logistics Economics and Globalisation*, *10*(1), 1–22. DOI: 10.1504/IJLEG.2023.130236

Ching, N. T., Ghobakhloo, M., Iranmanesh, M., Maroufkhani, P., & Asadi, S. (2022). Industry 4.0 applications for sustainable manufacturing: A systematic literature review and a roadmap to sustainable development. *Journal of Cleaner Production*, *334*, 130133. DOI: 10.1016/j.jclepro.2021.130133

Chiroma, H., Hashem, I. A. T., & Maray, M. (2024). Bibliometric analysis for artificial intelligence in the internet of medical things: mapping and performance analysis. *Frontiers in Artificial Intelligence, 7*, 1347815. Erboz, G. How to Define Industry 4.0: The Main Pillars of Industry 4.0. Available online: https://www.researchgate.net/publication/326557388_How_To_Define_Industry_40_Main_Pillars_Of_Industry_40 (accessed on 6 July 2020).

Emmert-Streib, F., Yli-Harja, O., & Dehmer, M. (2020). Artificial intelligence: A clarification of misconceptions, myths and desired status. *Frontiers in Artificial Intelligence*, *3*, 524339. DOI: 10.3389/frai.2020.524339 PMID: 33733197

FAO. Influencing Food Environments for Healthy Diets, FAO, Roma, (2016). https://www.fao.org/3/a-i6484e.pdf accessed on 15 Oct 2024.

Gajdzik, B., Grabowska, S., Saniuk, S., & Wieczorek, T. (2020). Sustainable development and industry 4.0: A bibliometric analysis identifying key scientific problems of the sustainable industry 4.0. *Energies*, *13*(16), 4254. DOI: 10.3390/en13164254

Ghobakhloo, M. (2020). Industry 4.0, digitization, and opportunities for sustainability. *Journal of Cleaner Production*, *252*, 119869. DOI: 10.1016/j.jclepro.2019.119869

Ghobakhloo, M., Iranmanesh, M., Foroughi, B., Tirkolaee, E. B., Asadi, S., & Amran, A. (2023). Industry 5.0 implications for inclusive sustainable manufacturing: An evidence-knowledge-based strategic roadmap. *Journal of Cleaner Production*, *417*, 138023. DOI: 10.1016/j.jclepro.2023.138023

Hawkes, C., & Ruel, M. (2006). Value chains for nutrition. Reshaping Agriculture for Nutrition and Health. 73-81.

Jena, M. C., Mishra, S. K., & Moharana, H. S. (2020). Application of Industry 4.0 to enhance sustainable manufacturing. *Environmental Progress & Sustainable Energy*, *39*(1), 13360. DOI: 10.1002/ep.13360

Kagermann, H., Helbig, J., Hellinger, A., & Wahlster, W. (2013). *Recommendations for implementing the strategic initiative INDUSTRIE 4.0: Securing the future of German manufacturing industry; final report of the Industrie 4.0 Working Group*. Forschungsunion.

Legg, S., & Hutter, M. (2007). Universal intelligence: A definition of machine intelligence. *Minds and Machines*, *17*(4), 391–444. DOI: 10.1007/s11023-007-9079-x

Machado, C. G., Winroth, M. P., & Ribeiro da Silva, E. H. D. (2020). Sustainable manufacturing in Industry 4.0: An emerging research agenda. *International Journal of Production Research*, *58*(5), 1462–1484. DOI: 10.1080/00207543.2019.1652777

Queiroz, M. M., Fosso Wamba, S., Chiappetta Jabbour, C. J., Lopes de Sousa Jabbour, A. B., & Machado, M. C. (2022). Adoption of Industry 4.0 technologies by organizations: A maturity levels perspective. *Annals of Operations Research*, 1–27. PMID: 36217321

Rojas-Reyes, J. J., Rivera-Cadavid, L., & Peña-Orozco, D. L. (2024). Disruptions in the food supply chain: A literature review. *Heliyon*, *10*(14), e34730. DOI: 10.1016/j.heliyon.2024.e34730 PMID: 39149054

Wunderlich, S. M., & Martinez, N. M. (2018). Conserving natural resources through food loss reduction: Production and consumption stages of the food supply chain. *International Soil and Water Conservation Research*, *6*(4), 331–339. DOI: 10.1016/j.iswcr.2018.06.002

Yampolskiy, R. V., & Fox, J. (2012). Artificial general intelligence and the human mental model. In *Singularity hypotheses* (pp. 129–145). Springer. DOI: 10.1007/978-3-642-32560-1_7

Chapter 2
Integrating AI-Enabled Food Safety, Compliance, and Future Innovations for Resilient Food Supply Chain

L. B. Muralidhar
https://orcid.org/0000-0003-3453-613X
Jain University, India

H. R. Swapna
https://orcid.org/0000-0003-0362-409X
Jain University, India

K. P. Sheeba
https://orcid.org/0009-0003-4707-5392
Jain University, India

Mohsina Hayat
https://orcid.org/0000-0003-1274-8713
Jain University, India

K. Nethravathi
https://orcid.org/0000-0003-4138-0276
Jain University, India

ABSTRACT

The chapter explores how AI technologies can be applied to overcome some of the

DOI: 10.4018/979-8-3693-9856-2.ch002

challenges facing quality assurance, compliance with regulation, and disruption of supply chains in the global food supply chain. This chapter further highlights that AI-powered systems, machine learning, and predictive analytics enhance supply chain resilience. The present chapter takes into consideration various other aspects of food safety, such as AI-based hazard detection, automated quality control, predictive maintenance in food processing equipment, AI-enhancing traceability of food items, compliance monitoring, real-time reporting, food safety-related decision support, incorporation of AI into standards and regulations on food safety, automation in compliance, real-time monitoring for regulatory compliance, AI-driven audit systems, risk management, smart contracts on food safety, and automated regulatory updates. The chapter emphasizes that AI plays a crucial role in shaping the future of food safety compliance, improving global food security and earning consumer trust.

INTRODUCTION

The modern food supply chain has been facing many challenges lately on how to assure quality, comply with regulations, and manage supply chain disruption. With the advent of artificial intelligence, this changed as novel solutions to these issues have been developed, revolutionizing how food safety and compliance are conducted. Active AI technologies, such as machine learning and predictive analytics, will help in improving resilience within the supply chain through real-time monitoring and putting in place mechanisms for rapid responses in case of eventualities(Smith, 2022).

AI-powered systems enhance traceability in food products even more with regulated full standards, hence their safety is warranted. For instance, temperature-sensitive shipments can be analyzed with the application of AI-driven analytics to spot those that may be exposed to spoilage and take measures so that food reaches consumers in good condition(Johnson & Lee, 2023). Also, artificial intelligence can help in routine quality control processes and reduce potential human errors in majority of the steps, increasing efficiency along the entire value chain(Doe, 2024).

Furthermore, with AI working its way into supply chain management, this could favor future food innovations that includes predictive modelling for demand forecasts coupled with logistics optimization. It is these kinds of innovations that will go the extra mile in reinforcing the resilience of supply chains and reducing risks associated with shortages of supplies in the supply chain and volatility in the markets(Miller et al., 2023).AI has much to say about the future of food safety, compliance, and supply chain resiliency, heralding new prospects toward enhancing global food security.

1. FOOD SAFETY

Food safety is vital in terms of public health; it involves handling, preparing, and storing food properly to avoid any kind of foodborne illness due to foods contaminated with pathogenic agents and chemicals(World Health Organization, 2020). The CDC reports that every year, the world records millions of cases due to food contamination(CDC, 2021). This necessitates the implementation of stringent safety measures, such as strict protocols, regular inspections, and adherence to standards like HACCP(Food and Agriculture Organization, 2019).Technological advances, including AI and blockchain, enhance risk monitoring and response (Chen et al., 2023).. Global trade has led to increasingly complex supply chains, necessitating collaboration from governments, industries, and consumers(Lee & Thompson, 2022). This is an area that calls for ongoing innovation to ward off contamination and protect consumers.

1.1 AI-based hazard detection in Food Safety

Food safety is a significant public health concern, prompting the need for more efficient detection of hazards. Despite their effectiveness, conventional food safety practices are time-consuming and prone to various types of errors. In this regard, AI technologies have emerged as promising tools that propose efficient mechanisms for the accurate detection of hazards.

AI implementation in Food Safety depends on Machine learning, deep learning, and computer vision . Machine learning algorithms recognize patterns in large datasets and predict possible hazards using supervised and unsupervised learning algorithms (Wang et al., 2021). Deep learning employs neural networks to identify intricate patterns in food safety data (Zhao et al., 2020). Computer vision, together with sensor technologies, also detects physical and chemical contamination of food.

Traditionally, critical control points are the cornerstone of food safety management using the Hazard Analysis and Critical Control Points model. In essence, the combination of HACCP and AI instantly automated the monitoring of control points, resulting in increased accuracy and speed of detection (Kumar & Bhatnagar, 2019). It is also crucial that AI analyses real-time data, which aligns with a modern risk-based approach to food safety.

Advocates argue that AI transforms hazard detection through the availability of insights in real time. The traditional methods, such as the microbiological testing method, have their limitations due to delayed results and constraints in the processing of data (Huang et al., 2022). AI is going to analyse sensor outputs, images, and genome sequences—hefty volumes of data—for quicker and more precise hazard detection. On the other hand, critics are concerned about data quality, model interpretability, and the requirement for domain-specific knowledge to appropriately train AI, according to Tian et al. (2021). Despite that, current research emphasizes that AI's advantages outweigh its limitations, especially in combination with people.

Supervised learning algorithms like Support Vector Machines and Random Forests classify food samples either as microbial contamination or about quality parameters (Patrignani et al., 2020). Unsupervised learning, for example, as in clustering, may help in identifying unknown hazards from production environments. The computer vision-based methods involving Convolutional Neural Network(CNNs) and Recurrent Neural Network(RNNs) have been successful in the detection of contaminants from image-based detection. CNNs detect foreign objects on production lines, while RNNs perform the monitoring of storage conditions through sensor data analysis (Zhao et al., 2020). Techniques such as computer vision combined with chemical sensors enhance hazard detection and thus enable control of the quality of food, microbial presence, and chemical composition all at once (Wang et al., 2021).

Predictive capabilities would involve analyzing historical data from diversified sources, including environmental factors to supply chain records. In fact, this proactive risk assessment is actually aligned with the new paradigm on food safety(-Kumar & Bhatnagar, 2019). AI implementation is extremely important regarding choosing sources and models for integration. As a matter of fact, high-quality data, ranging from sensors and lab results to supply chain records, is just vital for robust AI systems(Tian et al., 2021). Also, accuracy will always require regular validation with real-world samples(Patrignani et al., 2020). Human expertise in AI results with complex patterns is needed to interpret these results correctly and make judgments; hence, this will reduce the occurrence of false positives or negatives(Huang et al., 2022). Application of AI within food safety must be implemented in such a way to meet the regulatory requirements: the automated decisions meet the legal thresholds.

More recent developments are AI-enabled sensors for real-time quality monitoring. Spectroscopic sensors, together with the intervention of ML algorithms, detect chemical contaminants(Wang et al., 2021). Blockchain technology, in combination with AI, allows food products to be traced along the chain to pinpoint the source of contamination(Zhao et al., 2020). AI-powered microbial detection through DNA sequencing analytics yields better results than traditional culture-based methods in terms of speed and accuracy(Huang et al., 2022). AI in food safety has come to instil confidence among consumers, especially in high-risk foods that abound in the meat and dairy industries(Tian et al., 2021). AI-based hazard detection had been regarded as a game-changing approach in the face of food safety, contrasting inefficiency and inaccuracy or delayed solutions with its efficiency, accuracy, and real-time solution, notwithstanding problems related to quality and interpretability.

1.2 Automated quality control in food safety.

With the advancement in technology and increasing demand for high-quality food products, the rate at which Automated Quality Control is adopted in ensuring the safety of food products has increased tremendously. Various theoretical frameworks form the basis of AQC, including TQM and HACCP. In the case of TQM, there is the need to improve continuously through automation and reduced human error to enhance efficiency. For HACCP, automating these points provides real-time monitoring and corrective measures to reduce the risks associated with foodborne illnesses.

The big advantage of AQC is that it can provide consistent, objective evaluations without the subjectivity and variability often associated with human-led inspections. Additionally, the use of sensors, machine vision, and AI enables automated systems to deliver standardized measurements in order to enhance accuracy in assessment. AQC also allows for additional continuous monitoring of key control points 24/7, without human fatigue. It enhances traceability through supply chain management platforms by tracking food products from production to consumption and quickly identifying contamination sources.

Different models of AQC perform the function of detection and monitoring quality. Machine vision systems are increasingly being applied to defect inspection, such as colour variations and foreign objects, in real time. With AI algorithms, these work well in finding patterns that denote inferior quality. Global best practices in AQC implementation typically involve sensors, spectroscopy, ML, and IoT. Sensors detect chemical contaminants, while IoT allows for the monitoring at the center of all facilities.

Other recent developments also involve hyperspectral imaging combined with deep learning, which can detect spoilage and contamination by the chemical composition of foodstuffs. Robotic systems using food products on a processing line minimize cross-contamination in food processing. These robots are fitted with sensors and cameras that make sure only high-quality products proceed to processing. Predictive analytics, along with ML algorithms, study past data for predictive analysis of quality control issues to facilitate proactive management of food safety. It implements increased precision, homogeneity, and real-time analysis into food safety. With improvement in technology, AQC would therefore be an essential element of worldwide food safety practices.

1.3 Predictive maintenance in food processing equipment

Equipment failure in the food processing industry results in loss of production, compromises safety, and invites non-conformity from regulatory audits. Predictive maintenance has emerged as one of the proactive solutions in finding the potential malfunction of equipment before this actually translates to downtime.

PdM has its theoretical basis in Condition-Based Maintenance(CBM). Besides that, PdM also has its theoretical basis in Reliability-Centered Maintenance(RCM). CBM has a background in the idea that maintenance should be performed with respect to the actual condition of the equipment. For this purpose, continuous monitoring of the equipment must be carried out by sensors receiving information from temperature, vibration, and pressure conditions of the equipment (Jardine, Lin, & Banjevic, 2006). This information will later be analysed by machine learning algorithms in order to predict failures(Schwabacher, 2005). Contrarily, RCM focuses on system reliability in relation to only critical component failure modes that impact the specific food quality and safety standards(Moubray, 1997).

Predictive maintenance uses regression analysis, support vector machines, artificial neural networks, and deep learning methods as some of the ML techniques for analysing historical and real-time data to understand when a particular equipment breakdown could occur(Lee, Bagheri, & Kao, 2015). These models really function well in operating breakdowns related to food processing equipment, like regular mixers, boilers, and refrigeration units.

Integration of IoT and Industry 4.0 paradigms enhances PdM capability to a great extent. Sensor-equipped IoT devices can collect real-time data, thus allowing for remote monitoring and control (Xu, Xu, & Li, 2018). Critical parameters related to temperature, pressure, vibration, and flow rates are continuously monitored online and provide the necessary data for model training with high accuracy and subsequent condition-based monitoring (Ghosh, 2020). Integration of PdM with ERP systems further optimize the schedules of maintenance, parts availability, and inventory control. According to Al-Turki (2011), regularly updating the model with the entry of new data is required, especially as equipment behavior changes over time due to wear and tear. According to Schwabacher (2005), this is necessary because equipment behaviour changes over time due to wear and tear.

Advanced sensor systems, IoT-enabled devices, augmented reality, and digital twins are some of the latest few innovations in PdM. Machine learning algorithms analyze real-time sensor data to identify equipment anomalies, thereby helping maintenance teams to act noticeably if something needs to be fixed(Xu et al., 2018). AR on-site supports the on-site maintenance by visualizing internal components of equipment, thus providing real-time data during various repairs of the equipment(Kritzinger et al., 2018). Digital twins are virtual mimics of the equipment that simulate processes in order to forecast more precisely any needs for maintenance(Tao et al., 2018).

Studies prove that with food processing, PdM decreases downtime and the cost of maintenance. Ghosh (2020) identified 20-30% reduction in maintenance costs alone and up to a 50% decrease in the equipment's downtime. Integrating machine learning in the realm increases the accuracy of correct failure predictions, enhancing equipment effectiveness(Lee et al., 2015). The IoT and sensors also help in product quality improvement, enhancement, and other regulatory needs related to food safety when applied to PdM(Xu et al., 2018). Predictive maintenance represents a paradigm shift in food processing, defined by the utilization of condition-based maintenance, reliability-centred maintenance, machine learning, and the IoT for preventing equipment failures with an aim toward assuring continued production and ensuring the safety of food.

1.4 AI-enhanced food traceability

Food traceability has become quite important considering increasing concerns about food safety and quality, besides regulatory purposes. Data accuracy, timeliness, and scalability, which have always been issues in traditional systems, can be vastly improved using AI. AI has totally revolutionized food traceability by offering higher levels of data collection, analysis, and decision-making.

AI-based enhanced traceability systems depend on the working framework of FSCM, dealing with goods and information flow from farm to fork. FSCM hence ensures food quality and safety within the whole chain(Aung & Chang, 2014). AI optimizes data collection and communication across the supply chain within this framework. The other key-foundational theory and framework is DDDM that insists on the utilization of AI in studying the data of a supply chain in aid of determining some potential main hazards and food safety responses optimization. Kumar & Budhwar(Kumar & Budhwar, 2020)

Key technologies of AI in food traceability include machine learning, deep learning, natural language processing, and blockchain. ML algorithms like decision trees, support vectors, and random forests examine previous data sets from the supply chain and predict incidents using patterns derived(Manning et al., 2021). Deep learning, especially Convolutional Neural Networks, identifies visual defects in the product when it undergoes different processing and packaging on the way to the consumer(Zhong et al., 2021). The NLP processes labels and other regulatory documents that exist in textual data form for effective tracking in the supply chain(Wang et al., 2019)

Although not an AI model, blockchain is a complementary technology that facilitates a safe and decentralized ledger regarding recording supply chain transactions(Tian, 2017). Data, as stored on the Blockchain, gets analysed by AI for consistently raising the bar related to performance on aspects of transparency and traceability. IoT devices, sensors, and RFID tags record real time data across various nodes of the entire supply chain. The sensor-based information derived gets analysed for insight by the AI systems(Kumar & Budhwar, 2020). For AI-based traceability, the need for data is indeed very essential; it has to be accurate and timely, ensuring that the conditions of transport and storage are excellent(Verdouw et al., 2016).

Standardization of data from different stakeholders, in essence, facilitates ease of exchange and interoperations amongst diverse stakeholders(Aung & Chang, 2014). AI algorithms leverage both historical and real-time data to detect food contamination and spoilage risks, thus offering new possibilities for proactive risk management(Manning et al., 2021). AI-enhanced traceability systems ensure free-flowing information among supply chain players and help win consumer trust through regulatory compliance(Zhong et al., 2021).

Integrating AI with IoT sensors will support real-time monitoring of parameters such as temperature and humidity, which the AI analyzes to maintain food quality. AI integrated with blockchain provides a safe, transparent ledger for food product tracking, while AI goes deep into the blockchain data to identify discrepancies in the data without interference(Tian, 2017). Deep learning techniques in detecting defects through AI image recognition systems ensure that only quality products reach consumers(Zhong et al., 2021). AI-powered with NLP, therefore, reduces the time taken by manual verification to check compliance since it automates compliance checks by analyzing regulatory documents(Wang et al., 2019).

Certain studies have pinpointed the improvement in traceability brought about by AI in terms of precision and speed. AI-powered tracking and tracing better detects contamination sources, hence reducing food recalls, which improves public health(Aung & Chang, 2014. AI's predictive capability enhances risk management along the food supply chain(Kumar & Budhwar, 2020). Integrating AI into IoT allows for real-time monitoring; this reduces spoilage and waste(Manning et al., 2021). According to Manning et al., 2021, the integration of AI with blockchain ensures data integrity and boosts consumer confidence(Tian, 2017).

AI-based traceability takes the supply chain management of food safety, quality, and transparency to the next level. Integration within various AI models, IoT, and blockchain creates a very strong base for keeping track of the food products to ensure their safety and quality with continued innovation.

1.5 AI-powered compliance monitoring in food safety

Food safety compliance monitoring ensures that the food products meet regulatory and quality benchmark standards throughout the supply chain. While classically it was comprised of a series of manual inspections and pre-defined reactive approaches, today Artificial Intelligence has transformed this space by embedding automation, accuracy, and real-time decision-making into its integrity.

AI-driven compliance monitoring has hitherto been based on frameworks that are critical in nature, such as Hazard Analysis and Critical Control Points-building on the identification of critical points within the workflow, which logically represent locations where hazards are most likely to occur(Mortimore & Wallace, 2013). AI enhances HACCP through the automation of monitoring these points with data analytics and sensor technologies that enable continuing compliance and immediate responses to deviations. The AI approach to Risk-Based Food Safety Management assesses supply chain risks of this approach, prioritizes them, and can detect hazards in real time or predict them(Charlebois et al., 2016). AI models will even go to the extent of suggesting various preventive actions that could be taken to promote food safety practices.

The Data Driven Decision-Making (DDDM) framework, therefore, supports AI in compliance monitoring by the emphases it lays on data analysis and predictive modelling (Wang et al., 2019). In this respect, AI algorithms go over historical data and real-time data for detecting patterns of non-compliance, predicting potential safety issues, and recommending corrective actions.

Some key AI technologies incorporated into compliance monitoring include machine learning, natural language processing, and computer vision. Decision trees, neural networks, and deep learning-based ML models in SVMs scan data to find patterns in compliance and its violating areas(Jin et al., 2020). For instance, ML can monitor temperatures in cold storage facilities that could degrade food quality, thus generating alerts for them to take corrective action. NLP is used to analyze unstructured text data, such as inspection reports and regulatory documents, to ensure that they all meet the food safety standards(Chuang et al., 2021). AI-driven computer vision leverages image recognition systems like Convolutional Neural Networks to inspect food products and packaging for contaminants, defects, or labelling errors(Xu et al., 2018).

Predictive analytics empower the AI systems to estimate the future compliance violation and, therefore, early intervention that diminishes the risks of non-compliance(Charlebois et al., 2016).. Combination of diverse data from disparate sources such as IoT sensors, production logs, and supply chain databases provides an integrated insight into the food production cycle to monitor the compliance status more accurately. Food safety is improved through real-time monitoring with the use of AI and IoT sensors at critical points of control(Wang et al., 2019). This modern technological method provides automatic alerts when there are deviations in temperature, humidity, and pH levels(Jin et al., 2020).

Models are updated and constantly trained with new, fresh data so that accuracy is maintained and updated for ever-improving production processes and regulations(Chuang et al., 2021). AI can do automatic compliance reports, therefore simplifying regulatory audits and reducing charges imposed on food producers(Mortimore & Wallace, 2013). IoT sensors integrated with AI have transformed compliance monitoring wherein IoT devices constantly feed data into real-time analysis by AI algorithms, which ensure immediate follow-up corrective action upon the detection of non-compliance(Zhu et al., 2021).

The blockchain technology, with AI integration, offers enhanced traceability and compliance as all supply chain transactions are recorded in a secured ledger from which AI can then identify and trace non-compliance events(Tian, 2017). Augmented Reality (AR) is used in compliance inspections; AR provides the inspector with real-time data as well as AI-generated insights that point out the areas of concern(Xu et al., 2018). Production, storing, and distributing data-informed AI automatic risk assessments put a focus on checks for compliance to actively ensure safety(Charlebois et al., 2016)

Research has shown that AI increases the efficiency of compliance monitoring and makes it more effective. Jin et al. (2020) reported that with the use of AI-based monitoring, there was a 50% reduction in time for compliance checks, while at the same time, non-compliance instances were identified with greater accuracy. Chuang et al. (2021) proved that AI-powered NLP systems automate the compliance process with lesser human errors. Zhu et al. (2021) mentioned the AI-IoT subject that will provide good scope for real-time monitoring of compliance matters to ensure prompt responses.

AI-powered monitoring of compliance is transforming food safety in aspects such as efficiency, increased accuracy, and real-time insights. Integrating AI models with IoT and blockchain technologies can support efficient and proactive monitoring that introduces more safety and transparency in the food supply chain.

Figure 1. Integrating AI-Enabled Food Compliance

1.6 Real-time reporting in AI-Enabled Food Safety

The food industry is full of strict standards regarding safety; the risk of foodborne illness and contamination is very high. AI-enhanced real-time reporting has become an important means to monitor, detect, and solve food-safety problems, enabling companies to take proactive action against critical regulatory and reputation issues involving consumer trust.

HACCP, or Hazard Analysis and Critical Control Points, is a preventive approach where the identification and control of hazards within food production are its major focuses(Mortimore & Wallace, 2013). AI-powered real-time reporting tends to complement HACCP by extending their important aspects: monitoring critical control points on a continuous basis, with instantaneous detection of deviation from safe operating parameters. Risk-Based Food Safety Management emphasizes dynamic assessments at all stages of production, storage, and distribution, analyzed through streams of data by AI for real-time insight and early risk identification(Charlebois et al., 2016).

Real-time reporting can moreover help in developing supply chain transparency, ensuring timely information flow throughout the stages in time(Aung & Chang, 2014). The AI systems integrate various data from IoT sensors and cameras into comprehensive reports on contamination or temperature breaches as they occur. ML models look for food safety risks; for example, spoilage may take place in storage facilities by applying such models as SVM, decision trees, and neural networks on large datasets Jin et al., 2020). Computer vision makes use of CNNs to scan products for contaminants or defects and correct labels while automatically raising reports on the detection of anomalies(Xu et al., 2018).

NLP models examine unstructured textual data, such as regulatory documents and inspection reports, for possible non-compliances(Chuang et al., 2021). IoT sensors with AI could perform on-site data processing, which means tracking environmental conditions and issuing alerts immediately in cases of discrepancy(Zhu et al., 2021). Indeed, automation of warnings is crucial to notify personnel in real time about any deviation(Jin et al., 2020). These systems will only be effective if they are designed according to HACCP, FDA, and ISO standards, by which real-time compliance checks and corrective measures will become possible(Mortimore & Wallace, 2013).

Standardization of the data formats across the food supply chain will guarantee precise real-time analysis by artificial intelligence; this will improve the interoperability and accuracy of the captured data(Charlebois et al., 2016). IoT sensors integrated with Artificial Intelligence have revolutionized food safety reporting. Some of the parameters monitored by these sensors in a continuous basis include temperature, humidity, and pH. These sensors examine the hazards detected and send appropriate alerts immediately(Verdouw et al., 2016). Having AI combined with blockchain provides secure, immutable of supply chain transactions to ensure better compliance and traceability(Tian, 2017). In this respect, edge AI processes data locally on devices for instant detection of contaminants and triggering corrective actions(Zhu et al., 2021).

AI and AR further support the process by providing insights right on the site(Xu et al., 2018). Several literatures have corroborated the same, locating that real-time monitoring through AI reduces the time to respond to safety risks by 40% and reduces wastage of food (Jin et al., 2020). AI-IoT integratedly provides instant alerts when timely corrective actions are being taken against unwanted variations (Zhu et al., 2021).

Tian (2017) also pointed out the added value provided by blockchain in terms of traceability systems: a more accurate recording, coupled with transparency, would be an immutable record useful for audits. As stated by Charlebois et al. (2016), the value of AI for supply chain management at present is its capability to provide real-time reporting, thus enabling dynamic risk assessments. Integration of AI models-machine learning, computer vision, and natural language processing-with IoT and blockchain maintains immediate decision-making for reduced risks about food safety. Best practices in continuous monitoring, automated alerts, and data standardization-solution sets also enhance newer innovations such as edge AI-whenever possible.

1.7 AI-assisted decision support for food safety

Food safety remains one of the paramount public health concerns around the world, with various estimates indicating around 600 million cases of foodborne illness annually(WHO, 2020). The rising complexity in the production and distribution of food imposes a greater need for more sophisticated detection, monitoring, and response approaches. AI has increasingly played an important role in enhancing the current state of food safety management through early detection, predictive modelling, and real-time decision support.

The key frameworks include: Risk Analysis-such as the triad of risk assessment, management, and communication-in which AI provides accuracy in data analytics towards hazard identification(Roesel et al., 2019); and DSS that allow the processing of data in real time and automates decisions(Min, 2019). Big Data Analytics Paradigm allows AI to utilize large volumes of data from the industry in identifying patterns that depict contamination(Galvez et al., 2018). Beside that, IoT devices integrate AI mechanisms while processing the data to achieve insight into monitoring through real-time data(Zhao et al., 2019).

Other AI models in foodborne pathogen detection and classification include several machine learning procedures using support vectors, random forests, and neural networks(Li et al., 2020). Predictive models forecast incidents of the named foodborne illness, while predictive microbiology simulates pathogen growth under various conditions(Ross & McMeekin, 2003). Computer vision, with the use of AI, identifies spoilage or contamination through visual appearance(Feng et al., 2020). NLP models extract useful information from unstructured data, such as social media, to identify the next potentially emerging risks(Huang et al., 2021).

Proponents say that AI improves food safety by making the identification of impurities and their sources very fast. Thus, the processes related to quality control can also be automated, which is considered much more efficient than traditional methods in relating data to increasing the precision of risk assessment(Min, 2019; Galvez et al., 2018)

However, most of the advantages of AI in ensuring food safety also have several disadvantages(Roesel et al., 2019). Some of the disadvantages include quality and privacy issues about data that most AI models must be fed with large and high-quality datasets if their performances are to be realized(Huang et al., 2021). On the other hand, most food industries have inconsistent and biased data. Besides, it is hard to interpret some AI models due to their complicating nature, for instance in deep learning, which acts like a "black box" (Li et al., 2020). These include the integration of IoT devices for real-time monitoring(Zhao et al.,2019), or a mix of AI and blockchain to make tracing more feasible and transparent, as mentioned by (Tian, 2017).

AI-powered computer vision enhances quality inspection along the production lines by improving speed and accuracy in contaminant detection(Feng et al., 2020). AI-based surveillance for foodborne illness from public health data and social media can provide outbreak detection, early, and timely interventions(Huang et al., 2021).

The AI influence on food safety is quite significant since research has unraveled many hidden aspects. For example, Galvez et al. (2018) discovered that AI analytics improve pathogen outbreak predictions through the identification of critical control points. Similarly, Li et al. (2020) found out that machine learning models ensure better performance reputedly than traditional microbiological methods related to pathogen detection. Moreover, AI integrated IoT systems provide real-time monitoring and thus can reduce the risks of food spoilage (Zhao et al., 2019). Standardization in data collection and sharing nevertheless holds the key to AI's success(Roesel et al., 2019).Though challenges do persist-especially regarding data quality and regarding model interpretability-the overall benefit of AI to food safety is undeniably inesti; what is needed now is a marshaling of efforts toward protocol standardization and technology integration.

1.8 Integration of AI with food safety standards

Food safety standards could not be more crucial to ensure that foodstuffs are free from contamination for the sake of public health. AI has much more advanced capabilities to implement and monitor such standards, especially in long and complex global supply chains. The Risk Analysis Framework involves risk assessment, management, and communication, and the application of AI will elevate effective decision-making on risk management by predictive analytics. (FAO/WHO 2006; Galvez et al. 2018). Improved HACCP systems with machine learning from sensor data can identify deviations in real-time food processing(Lu et al., 2019).

The Compliance Management Paradigm uses AI to ensure the realization of standards such as ISO 22000 and GFSI through automation of monitoring and reporting (Zhao et al., 2020). AI also exploits the power of Big Data Analytics for traceability. Supervised models in the form of SVMs and random forests classify contaminants to keep things compliant(Li et al., 2020). An unsupervised approach identifies anomalies that have been incorporated into food quality. IoT sensors, together with AI analytics, ensure that cold-chain compliance is observed through real-time monitoring of the conditions and performing alerts in case of deviation conditions(Zhao et al., 2020).

NLP-based approaches assess inspection reports, audits, and regulations for their compliance status regarding food safety standards, thereby making documentation assessment automatic(Huang et al., 2021). AI-based predictive models simulate the growth of pathogens to support microbiological safety practices(Ross & McMeekin,

2003). As argued, AI improves efficiency and accuracy in standards monitoring, the identification of risks, and proactive measures that could be taken accordingly(Lu et al., 2019; Li et al., 2020). Yet, challenges in data quality, standardization, and interpretability of AI models still exist(Zhao et al., 2020). As complex AI models need transparency-after all, food safety professionals need to understand the insight into the AI decisions to make a call on compliance(Galvez et al., 2018).

AI-powered systems detect safety deviations at the critical control points in food items production with sensor data and machine learning algorithms that warn managers in time(Lu et al., 2019). Combined with blockchain, AI enables better traceability by analyzing immutable supply chain records for compliance according to standards such as ISO 22005(Tian, 2017). Innovations include everything from the automated analysis of audit reports through natural language processing for areas of non-compliance(Huang et al., 2021). AI models also employ real-time risk assessments through various data to suggest certain control measures(Ross & McMeekin, 2003).

Where compliance and better safety outcomes are concerned, research has proven that the incorporation of AI technologies into food safety standards does improve them significantly. Galvez et al. (2018) showed how AI analytics improve traceability and pathogen detection that underpin microbiology standards. On the other hand, Lu et al. (2019) pointed out that the IoT systems enabled by AI in real-time monitoring to confirm HACCP compliance are effective. AI permits the risk to be communicated through reports automatically. According to Zhao et al. (2020), this is possible. For effective integration of AI, however, there has to be data standardization and a way in which the same regulatory processes collaborate across borders(Huang et al., 2021). It is also of utmost importance that some forthcoming challenges regarding the model interpretability, among other concerns of regulatory harmonization, be considered if the food safety system's AI-driven enhancement is in any respect to be optimized.

2. COMPLIANCE

Food safety is a very important matter when it comes to the public health issue. Economic compliance with standards guarantees that products are safe, nutritional, and of excellent quality to consume. Compliance in food safety involves a broad array of regulations, practices, and procedures laid down to lessen the risk associated with the production, processing, distribution, and consumption of food. Food safety essentially has basic standards laid down-both at the national and international levels. Agencies such as the WHO, FAO, and various national regulatory bodies-for example, the FDA in the United States, or the FSSAI in India-have laid down basic standards and other regulations to safeguard consumers.

Non-compliance can lead to severe consequences such as product recall, legal action, financial losses, and loss of brand reputation. Hence, compliance with the standards for food safety is not only a legal duty but also a commitment to ensuring the consumers get healthy and safe food and is a way of gaining confidence among consumers in the brand. Compliance in food safety standards is defined as an act of proactive and organized approach towards ensuring the safety of quality food products. Basically, the adoption of best practices, enforcement of regulations, and its continuous monitoring will protect the consumers and ensure public health.

2.1 Compliance Automation in Food Safety

Despite this, international food supply chains are complex and have continuously driven the demand for automation in food safety compliance. Instead, this encompasses advanced technologies such as AI, machine learning, blockchain, and IoT in stringent assurance of compliance with regulations regarding food safety. Several theoretical frameworks found a basis for compliance automation. Among these key models is the Risk Management Framework that focuses on the identification, assessment, and mitigation of food safety risks(Gupta, 2018). RMF will support the introduction of automated systems for ensuring compliance, focusing on continuous monitoring and proactive detection of risk, crucial for food safety throughout the value chain.

It is also complimented by Systems Theory, where it considers food safety compliance to be an integrated system comprising stakeholders such as producers, manufacturers, regulators, and consumers. Automation of processes in compliance, according to systems theory, improves better communication and coordination along the value chain to assure better food safety(Kumar & Singh, 2019). These frameworks help guide the integration of such technologies, which may include monitoring, detecting potential hazards, and being in compliance with the standards that regulate food safety, thereby enhancing supply chain management.

The heart of most compliance automation systems is the HACCP model, which uses digital methods to monitor CCPs in real time and permits instant corrective actions if there were any deviations(Manning, 2020). Automation cuts down on human error, hence making compliance more efficient. Advocates say compliance automation adheres to standards with higher exactitude compared to manual methods(Lemke & Petersen, 2020). This would facilitate real-time data collection to help when any incident of food safety occurs, hence trying to speed up decision-making.

Yet, critics point to such limitations as the high implementation cost for SMEs, potential data security risks, and the integration challenges of automation into existing systems(Smith et al., 2021).. Despite all these issues, automation improves traceability, data accuracy, and adherence to regulation. IoT devices installed at production facilities provide the ability to monitor, in real-time, critical conditions such as temperature and contamination(Baldwin, 2022). Blockchain technology traces food products throughout their journey in the supply chain and provides an unalterable record for compliance(Tian, 2017).

Digital platforms for automated data logging decrease manual errors and document data more accurately(Chen et al., 2019). AI and ML algorithms allow for identifying possible risks before actual contamination occurs(Singh et al., 2020). The innovations of smart sensors, AI-driven inspection tools, and blockchain offer holistic solutions to monitor and control the risks associated with food safety(Lemke & Petersen, 2020; Tian, 2017).

Recent studies have indicated that automation of compliance improves the standards of food safety. Singh et al. (2020) determined how AI and ML improve the accuracy in the identification of risks, hence helping formulate robust policies of food safety. Baldwin(2022) has identified the contribution of IoT in the improvement of real-time monitoring for compliance management effectively. Smith et al. (2021), on the other hand, have identified organizational preparedness, technological, and training issues that may impede overall success. Whereas automation requires large investments, the real-time monitoring, enhanced traceability, and predictive analytics enabled by automation support a proactive approach to managing food safety using frameworks such as Risk Management and Systems Theory.

2.2 Real-Time Monitoring for Regulatory Standards in Food Safety

Real-time monitoring in food safety involves constant observation of the critical parameters in the supply chain to create real-time compliance with regulations. New technologies, such as IoT, sensors, AI, and blockchain, enable rapid detection of hazards for quicker responses to minimize health risks. Most of the models on food safety have been based on the Hazard Analysis and Critical Control Points system, in which real-time monitoring automates tracking CCPs, lowering the contamination risks of food products(Manning & Soon, 2019).

According to Systems Theory, compliance is an integrated process, and real-time monitoring integrates information at every stage in the supply chain for complete decision-making coordination(Kumar & Singh, 2019). Real-time risk assessment enhances the Risk Management Framework whereby hazards could be mitigated before it happens(Gupta, 2018). Real-time monitoring has been said to enable compliance with strict regulations following instantaneous feedback pertaining to temperature, humidity, and contamination; therefore, it reduces illness and recall(-Tian et al., 2020).

The challenges are high implementation costs for SMEs and requirements for technical expertise(Smith & Johnson, 2021). However, advantages such as improved safety, better compliance, and increased consumer trust outweigh costs(Manning & Soon, 2019). IoT sensors are state-of-the-art for continuous measurement of key parameters as they enable automated data logging for regulatory audits(Baldwin, 2022; Chen et al., 2019).

AI and ML models analyze real-time data to predict the occurrence of any potential hazard in order to take proactive risk management measures for further improvement in regulatory compliance(Singh et al., 2020). Blockchain adds that extra bit of secure and immutable record to the food supply chain, hence improving traceability and transparency about regulatory standards and product recalls(Tian, 2017).

Other technological innovations in real-time monitoring include smart sensors that can detect microbial contaminants for immediate identification of pathogens(-Kumar & Singh, 2019). Edge computing processes data right at the source of origin to reduce latency and increase responsiveness in managing food safety(Lemke & Petersen, 2020). Predictive models powered by AI provide actionable insights through data trend analysis to predict equipment failures or possible contamination for proactive measures towards compliance(Singh et al., 2020)

Research proves real-time monitoring works. Tian et al. (2020) reported that IoT sensors can enhance the detection of hazards, which leads to sustained compliance with standards. Indeed, Baldwin (2022) noted that the efficiencies derived from advanced sensors and AI provide instant data elements for making decisions that enhance the levels of food safety. However, Smith and Johnson (2021) said that it will be successful only when the organization is ready, supported by technology readiness and the training and preparation of manpower. Although there is considerable drawback, real-time monitoring has been imperative for food safety, as aided by HACCP, Systems Theory, and Risk Management Framework. Innovations like IoT, AI, and blockchain make it stronger in modern food safety management.

2.3 AI-Driven Audit Systems in Food Safety

AI-driven audit systems in food safety represent the so-called transformative approach to compliance with regulatory standards. The system deploys AI and ML algorithms to automate audits, which is very much necessary for accuracy, speed, and consistency. One possible framework can be within the RMF, considering identification, assessment, and mitigation of risks within or across the food supply chain(Gupta, 2018). AI-driven systems meet the criteria for RMF with the use of on-premise constant monitoring and predictive analysis through food-safety data to further enhance the efficiency of risk-based audits.

Secondly, Systems Theory will also apply because it looks at food safety as a system in itself-populated by various interdependent stakeholders. Artificial Intelligence-driven audit systems take a holistic view of compliance across the value chain. This, in turn, improves coordination and communication(Kumar & Singh, 2019). Machine learning models, such as decision trees and neural networks, look through large volumes of data to spot cases of non-compliance. It predicts possible hazards and further refines its accuracy on detecting both over time(Smith et al., 2021).

Proponents argue that AI-driven audits allow for real-time analysis, immediate hazard detection, hence improving response times and reducing risks(Lemke & Petersen, 2020). AI also standardizes audit processes, reducing human error and inconsistency(Chen et al., 2019).However critics point out a number of challenges, including high implementation costs on SMEs and raising concerns about data privacy(Williams & Taylor, 2022). Integration with any pre-existing FSMS, such as HACCP or ISO 22000, has been mentioned as essential to ensure ease and smoothness in data collection and its analysis(Manning & Soon, 2019). Centralized repositories are necessary so that the AI models get access to real and timely information(Tian et al., 2020).

With evolving food safety risks and regulations, AI models have to be constantly fed with new data to retain their fitness for emergent risk detection and to keep pace with changing regulatory environments (Singh & Sharma, 2021). While such AI-driven audit systems confer tremendous benefits, human expertise cannot be completely foregone. Current best practices include the use of AI systems as support through decision-making, wherein reports generated by the AI system are reviewed by experts themselves (Lemke & Petersen, 2020).

Other recent developments involve predictive AI models. These may analyze historical and real-time data to identify potential risks(Smith et al., 2021). The technologies also involve computer vision for the automatic inspection of products and environments, adding more objectivity to the processes(Chen et al., 2019). The analysis of documentation has also been performed with NLP. Relevant information is extracted from that documentation, in addition to identifying compliance issues(Williams & Taylor, 2022).

Recent studies confirm the fact that AI-powered auditing schemes enhance the compliance of food safety. Singh and Sharma (2021) indicate that AI-based auditing enhances detection capability related to violations in a complex supply chain. Lemke and Petersen (2020) established that such systems reduce audit times and costs associated with audits, with results being more harmonious and accurate. On the other hand, Smith et al. (2021) consider that the effectiveness of the audit using

AI depends on data quality and accessibility, which needs extensive practice in the field of managing data. Thus, though there are some challenges, like implementation costs and data security, AI-driven audits provide the following advantages: real-time analysis and predictive capabilities, along with increased compliance-all these ideas underpin both Risk Management and Systems Theory.

2.4 Risk Management Through AI in Food Safety

Food safety encompasses difficult challenges on a global scale concerning the food industry because of the intricacies of the supply chain, the regulatory demands, and the consumer expectations. Traditional risk management, often consisting of inspectors and manual checks, is hardly satisfactory in such cases(Kumar et al., 2020). AI has emerged as a transformational solution offering tools for betterment in food safety regarding improved risk management

Theoretically, AI applications to food safety are based on the risk analysis framework of the Codex Alimentarius Commission, which presents risk assessment, management, and communication as interlinked components in risk analysis(FAO/WHO, 2009). AI accelerates such risk assessments and allows for real-time communication. Moreover, Decision Theory underpinning AI modeling is rooted in the works of Kahneman & Tversky (1979) for predictive analytics.

These ML models predict outbreaks of foodborne illnesses and classify contaminants, as in the case of supervised algorithms: SVM and Decision Trees(Zhao et al., 2019). Unsupervised learning helps to identify new risk patterns in large datasets. Deep Learning, specifically CNNs, detects contaminants by image analysis(Liu et al., 2021). NLP techniques analyze textual data from safety reports and feedback for the identification of emerging risks(Gong et al., 2020)..

Accordingly, best practices involve the use of various data sources there, such as supply chain records and IoT sensor data for integrated risk assessment (Wang and Liu, 2020). Real-time monitoring in upper streams-such as temperature control in food storage-allows predictive detection of hazards and, as an end result, reduces the risk of diseases and losses due to goods recall(Mahmoudi et al., 2022).

AI-powered platforms improve food safety through transparent communication among various stakeholders-regulators, producers, and consumers-to promote compliance and build confidence(Aung & Chang, 2014). Therefore, blockchain technology is now being introduced for traceability, where the AI-powered system can trace a product right from its farm origin to the fork, with tamper-proof records(Kamilaris et al., 2019). Integration of RPA with AI maintains hygiene in food processing and reduces the potential risk of contamination.

Studies have been able to establish how much effect AI has on managing the risk of food safety. Kumar et al. (2020) established that AI reduces the detection time of foodborne pathogens by as much as an estimate of 50%, compared to conventional methods applied. AI image processing has more than 90% correctness for foreign substance detection within the stream of food products(Liu et al., 2021). Fair enough, data privacy, the need for large datasets, and problems related to algorithmic biases are created hence the need for collaboration and ethics in AI deployment(Gong et al., 2020). It also calls for continued research in supportive models of AI for adaptability in meeting evolving needs in food safety.

2.5 Smart Contracts for Food Safety Compliance

Modern food supply chains are complex, and a variety of regulations make them out of compliance. Classic track record methods are not only inefficient but also highly prone to non-compliance. Using self-executing contracts-which have the rules of compliance programme-codified into the code-can automate and enforce food safety regulations(Christidis & Devetsikiotis, 2016).

Agency Theory can be postulated by using analogies between the relationships of stakeholders and their need for enforcement of compliance(Eisenhardt, 1989). Smart contracts lower information asymmetry and agency costs through trustless interactions. Moreover, transaction cost theory supports smart contracts because they reduce some of the reasons that attract costs to manage compliance, such as record-keeping and reporting(Williamson, 1981).

Running on blockchain networks, smart contracts provide a compliant, tamper-proof log of events for enhanced transparency and traceability(Kamilaris et al., 2019). Smart contracts automate any and all compliance-related actions. Such examples include automated releases of payment in the event of successful quality checks of the produce recorded on the blockchain(Saberi et al., 2019). Self-executable mechanisms from the contract trigger punishments in case of non-compliance, such as freezing any transaction if the shipment temperatures, recorded through IoT, fall out of accepted range(Tian, 2017).

IoT sensors and RFID tags need integration with blockchain for accurate recording of data on compliance monitoring. Supply chains benefit from interoperability between different data systems, which enable fast information exchange, thus ensuring increased efficiency in smart contracts(Kouhizadeh et al., 2021). Also, embedding the rule of food safety into the rule of smart contracts will help determine the quality requirements of the product as well as storage and traceability by referring to the global benchmark standard such as provided by FSMA and Codex Alimentarius(Tsang et al., 2019). Compliance is ensured by the automated rules that give a transparent view to any stakeholder of the compliance records(Galvez et al., 2018). Techniques of cryptography include zero-knowledge proofs that achieve data privacy while maintaining transparency.

Smart Contract-oriented Architecture allows modular management for different compliance aspects, including certification and recall. The integration of AI with smart contracts will provide the potential to predict real-time risks and make dynamic adjustments in compliance(Kshetri, 2018). Blockchain-based smart contracts have been found to enhance traceability by reducing fraud cases by as much as 30% (Tian, 2017) and simplifying compliance reporting(Kamilaris et al., 2019).. It also presents standardization problems with regards to data format and high computational costs, which must be underlined so that more diffusion happens in practice(Saberi et al., 2019).

2.6 Automated Regulatory Updates in Food Safety

Food safety is an important public health concern globally since the complex supply chains have eased regulatory compliance. Considerably, automated regulatory updates ensure that standards within the food industry are efficiently updated. Among the significant frameworks is the Risk-Based Preventive Control that focuses on hazard identification and prevention. This is supported by automation, which has completely enhanced compliance(Unnevehr & Jensen, 1999). FSMA's paradigm moves to a more preventive management approach where automated updates on timely compliance tracking shall take place and contamination risk(FDA, 2011).

Proponents of automated updates of regulatory matters in food safety argue that compliance checking by hand is time-consuming, error-prone, and cannot keep pace with continuously updated regulations(Marucheck et al., 2011). Automation addresses these issues by providing for continuous monitoring of regulatory changes, reducing human errors, and offering instant access to updated standards. A key model in the compliance market for this will be Compliance-as-a-Service, wherein cloud-based platforms track regulations, interpret implications, and pass on the translation of what needs to be done to businesses(Almeida et al., 2020). With CaaS, in that regard, new regulations are responded to at a much faster pace than before, so very minimal compliance gaps are exposed.

Decision Support Systems also support food safety regulation using AI and machine learning to analyze data, predict risks, and suggest actions (Sun et al., 2018). This model processes big data for the purpose of making sure that protocols on food safety answer to the current regulations. Best practices in order to implement automated updates through integrating RIS into food management include real-time alerts and compliance checklists and analytics(Taylor, 2018). Blockchain increases traceability and accountability by recording such food safety data on decentralized ledgers, creating immutable compliance records(Tian, 2017).

Innovations like NLP translate regulatory texts and integrate compliance needs into the management system automatically to reduce human interpretation(Rostami et al., 2021). Intelligent predictive analytics powered by AI forecasts regulation changes, and simultaneously it assesses impacts to let tune compliance in a proactive manner(Li et al., 2022). According to research, it has been found that automated compliance systems reduce food safety violations by 30%, increase real-time monitoring, and responsiveness(Wang & Chen, 2023). They also earn consumer confidence with their traceable compliance records, important in increasing their market competitiveness(Hutter, 2011).

Fully automated updates of regulatory knowledge in food safety have been an iconic development toward continuous evolution in risk management and compliance with regulations. Add to this models-like Risk-based Preventive Controls, Compliance-as-a-Service and innovative technologies like NLP, artificial intelligence-powered predictive analytics, all merge to seal a new reality in the fields of food safety regulation. Other best practices, like Regulatory Intelligence Systems and Blockchain-enabled technologies, even smoothen the processes of compliance. Studies have shown that such automated systems have huge impacts on improving food safety results and gaining consumer confidence. Therefore, these automated updates are not only a strategic imperative for the food business but also for the protection of public health.

2.7 AI-Driven Risk Assessment in Food Safety

The application of AI in risk assessment is an enabling force for the food safety paradigm to move toward more dynamic and accurate methods of identifying and mitigating risks that could not be reached by traditional assessment methodologies(FAO/WHO, 2006). It's in tune with the Risk Analysis Framework, which uses advanced data analysis to enhance the ability of risk assessment, management, and communication. It analyzes HACCP's critical control points with real-time data analysis and predictive modelling(Unnevehr & Jensen, 1999). The Precautionary Principle also benefits from the use of AI in processing large datasets that predict any potential issues in food safety before their occurrence, hence pro-active measures can be taken in advance(Klinke & Renn, 2002).

The AI-driven flexible approach overcomes some of the flaws of conventional risk assessment, which rely upon past data and delay the detection of perils(Rostami et al., 2021). The ML models are focused on supervised and unsupervised learning to develop patterns from diverse data showing the signals of forthcoming hazards(Kang et al., 2020). Predictive Risk Assessment Models take in data on weather conditions and supply chain behaviour so that contamination incidents can be predicted and pre-emptive risk mitigation is allowed(Tian, 2017). The DSS enhance this by supporting a risk management system that provides real-time warnings and suggested actions based on AI-enabled risk assessments(Sun et al., 2018).

The best practices for AI-driven risk assessment emphasize data quality, algorithm transparency, and collaboration across sectors. High-quality data are required for any reliable assessment, along with cleaning protocols and the integration of various sources of data(Rostami et al., 2021). These best practices, together with the periodic updating of databases, may form the very foundation on which AI-driven food safety risk assessments could thrive.

However, algorithmic transparency is highly important in AI-driven risk assessment because it engenders stakeholder trust and, most importantly, underpins adherence to regulating standards through a valued, auditable decision-making process(Veiga et al., 2019). Collaboration by food businesses, regulators, and tech providers across the food sector is also basic to robust AI models that develop holistic and much-improved food safety responses(Huang et al., 2022). Innovations like NLP extract the relevant information from regulatory texts, scientific literature, and even social media. In this respect, AI models are kept up to date with the development of new risks(Zhou et al., 2021). Furthermore, visual inspections to detect spoilage or contamination are automated with a computer vision system(Kang et al., 2020).

AI-driven risk assessment enhances food safety outcomes. In this respect, Wang et al. (2023) report that AI predictive models detect pathogens with higher than 95% accuracy, whereas assessment time is largely reduced. Li et al. (2022) reported that AI is appropriate to deal with some of the tangled supply chain risks while providing a full-scale assessment by analyzing various pieces of information such as temperature and transportation records. Moreover, AI can optimize the strategic distribution of resources: companies that apply AI can focus their interventions on high-risk areas, optimizing interventions concerning food safety. Veiga et al. (2019).

In sum, risk assessment powered by AI will be performed in concert with models such as HACCP. This will raise the accuracy levels, rates of performance, and efficiencies to unprecedented levels. Success will be underpinned by leading practices for assuring data quality, transparency, and collaboration. Research has confirmed that AI is indeed bound to be transformative in delivering more precise risk assessments and proactive management strategies.

2.8 Reducing Human Errors in Compliance Checks in Food Safety

Human errors in compliance checking pose a serious risk to food safety as it results in foodborne illness, recalls of food items, and economic loss. The complex nature of food safety regulations, most compliance checking is too often performed manually makes the occurrence of mistakes more frequent. Based on Human Error

Theory, with Reason's model, 1990, for example, errors fall into slips, lapses, and violations, all arising from fatigue and poor training. Risk management theory such as the HACCP system emphasizes identifying and controlling potential hazards in food safety(Unnevehr & Jensen, 1999). This integration of automated technologies into the HACCP will minimize the errors that may occur, ensuring consistency in monitoring and decision support.

Manual compliance processes usually are labour-intensive, error-prone, and often inconsistent, raising the risk of contamination(Marucheck et al., 2011). Automation will contribute to increasing environmental accuracy, streamlining such processes, and keeping them in compliance with complex and ever-changing regulations(Holleran et al., 1999). One of the main models is the Compliance Automation Model, which keeps track of food safety through automated tools(Rostami et al., 2021). It increases less reliance on manual checks and tracking status of compliance through alerts for corrective actions. This model reduces human error by automating repetitive work.

The AI-driven Risk Assessment Model applies AI and ML to a wide array of data emanating from supply chains and production environments to identify hazards, predict risks, and arrive at suggestions regarding preventive measures(Li et al., 2022). This model further enhances decision-making with its real-time insights, minimizing human error in risk assessment. Implementation of best practices, such as SOPs, ensures that the execution of compliance tasks is uniform and clear to minimize variability in checks(Taylor, 2018). When implemented in combination with automation, SOPs increase cohesion and accuracy in compliance checks.

Meanwhile, training and education are conspicuous by their equal importance. Routine and simulation-based training using AI-driven virtual environments contribute to enhancing situational awareness(Kang et al., 2020). Automation in data collection and monitoring with the help of IoT sensors and digital record-keeping further minimizes errors by ruling out manual entry of data, enabling real-time monitoring(Tian, 2017). This makes sure that compliance records are accurate and documentation-related mistakes are at a minimum

Recent advances in food safety compliance have both AI and ML for enhanced precision with minimal human interference. AI-powered tools encompass NLP algorithms that go through regulatory documents and extract the requirements of compliance, thereafter integrating them into digital systems, enabling upkeep and actual reduction of the risk of non-compliance by businesses(Zhou et al., 2021). Further, traceability and accountability are improved through the blockchain facility of compliance record storage and verification in the decentralized ledger. By this approach, data is tamper-proof, making it difficult for records to be manipulated with errors(Tian, 2017)

Research has identified such influences of technology on reducing errors in compliance. Wang et al. (2023) established that automated systems of compliance reduced 40% of documentation errors as compared to manual checks, hence improving the accuracy of record keeping. AI-driven compliance checks identified hazards more effectively in comparison to human inspectors through the analysis of real-time data, thus enabling immediate corrective actions. AI-based simulation training significantly improved personnel performance(Rostami et al., 2021). For instance, Kang et al. (2020), in their study, have considered that perspectives on hazard recognition and record maintenance were high among the staff members trained using AI-based simulations than conventionally trained staff members.

Application of conceptual frameworks, such as Human Error Theory and HACCP, lends support to the development of such automated models. AI, blockchain, and automated monitoring are some of the critical elements to achieve no-error compliance and protect public health by virtue of regulatory standards.

3. FUTURE INNOVATIONS

Future developments in the resilient food supply chain lie at the intersection of technology, sustainability, and adaptability. It is in such developments the food system shall be developed to resist disruptions, reduce waste, and secure a food future for all. Building a resilient food supply chain is critical to answer such global challenges as climate change, population growth, and urbanization, all of which have implications for food security, nutrition, and sustainability. The focus of future innovations shall be to develop adaptive and robust systems that can match the changing demands of the world.

Therefore, a technological approach is integral to such a system driven by technologies in artificial intelligence, machine learning, and big data analytics. These enable correct demand forecasting, real-time tracking of products through the chain, and data-driven decision-making for cost optimization, efficient production, minimization of waste, and correct inventory management. For instance, predictive models within AI can forecast changes in demand that are used to adjust production. Blockchain technology offers transparency and traceability from farm to fork, besides mitigating issues of food fraud and safety due to its practicality in building trust. Robotics and automation in agriculture involve planting, harvesting, and food packaging, reducing human error and increasing workforce efficiency. Fully automated systems, involving self-driving tractors or drones, have advantages like resource optimization with minimal waste and a lower carbon footprint in food production.

Innovations in vertical farming and controlled-environment agriculture reduce land and water usage, thus ensuring less dependency on climatic conditions. On the other hand, alternative proteins-from plant-based to cultured meats-address food security with reduced environmental pressures from traditional livestock farming. Also, it requires sustainable packaging, adoption of circular economy practices, and renewable energy toward a resilient supply chain. This innovation in biodegradable packaging and renewable energy use reduces the environmental footprint and enhances long-term viability. It is important that these are adopted as a measure of security towards a sustainable food future.

3.1 AI for Future-Proof Compliance Systems in Food Safety

The food industry is, therefore, stretched by the demand to ensure that safety and compliance are met within a continuously changing set of regulations, besides complex global supply chains. Traditional compliance systems encompass laborious tasks, which are highly susceptible to human error, resulting in too many delays and inaccuracies. Artificial Intelligence now brings automated, data-driven solutions to advance the efficiency, accuracy, and adaptability of the compliance system.

AI implementation is also in line with the RMF in food safety, which identifies hazard identification, risk assessment, and mitigation(Schneider et al., 2021). AI technologies apply big data analytics to identify risks, predict hazards, and suggest corrective actions with real-time suggestions. The SCRM model furthers controlling risks along the value chain of supplies(Tse & Tan, 2020). AI-powered systems offer end-to-end visibility, enabling organizations to identify and fix instances of non-compliance with much ease.

Key AI developments, such as ML algorithms, analyze production, supply chain data, and quality control checks for patterns and anomalies(Kankanamge et al., 2022). Predictive analytics strengthen HACCP by predicting non-compliances that may happen along with equipment failures; food safety management is thus no longer reactive but proactive(Manning & Soon, 2016). NLP or Natural Language Processing automatically conducts analysis on the regulatory documents to keep the companies updated with global regulations through extracting relevant compliance information embedded therein(Smith et al., 2021). Automation for critical regulatory intelligence, thereby, becomes an imperative for multinational companies in treading diverse landscapes of regulation. In fact, AI radically transforms food safety compliance through its unprecedented analysis and predictive capabilities.

Blockchain integrated with AI in Food Safety Compliance: Integrating blockchain with AI in considering food safety compliance, both traceability and transparency are enhanced. Blockchain provides an irreversible record of supply chain transactions, while AI analyzes big data derived from this to track any compliance breach or fraud attempts in line with standards such as the Global Food Safety Initiative, GFSI(Caro et al., 2018). Studies have depicted that AI-based compliance systems guarantee the speed and accuracy of the inspections. For instance, Zhang et al. (2020) demonstrated how AI image recognition can rapidly detect food defects with greater accuracy than human inspectors. Similarly, other scholars, Kankanamage et al. (2022), have determined that applying AI algorithms to social media and public health data could predict seizures of foodborne illness in good time and thus allow warnings.

Data collection and management are key for any AI because through trend analysis, informed decisions may be made based on them(Tse & Tan, 2020). AI systems need redevelopment and updating with new inflows and changing regulations(Schneider et al., 2021). Collaboration with regulators can result in the development of industry-wide AI standards for food safety(Manning & Soon, 2016). This could enable companies to move toward proactive compliance management from reactive compliance management, entrenching resilience into food safety systems using AI in concert with blockchain.

Integration With IoT for Real-Time Data Capture in Food Safety

IoT integration into the food supply chain revolutionized food safety due to enhanced monitoring, traceability, and compliance(Khan et al., 2021). IoT devices are basically interconnected devices that continuously share data in real time on the production, processing, transportation, and storage of food. It also meets the HACCP framework-structure because such technology allows for continuous monitoring at points of critical control with sensors tracking among other contamination risks, temperature, humidity, and pH values(Manning & Soon, 2016).

IoT has also contributed to the Risk Management Framework through the automation of data capture for dynamic risk assessment and mitigation(Schneider et al., 2021).

Thus, IoT in chain supply management provides end-to-end visibility and traceability related to the standards of food safety from farms through consumer channels(Tse & Tan, 2020). Also, during this process, some critical parameters are tracked by IoT sensors in production and storage that may trigger corrective measures if any deviations are detected. Like, for instance, the temperature sensors in cold chain logistics can monitor and automatically regulate cooling systems before any incidence of spoilage occurs at all (Bai et al., 2020).

Combined with blockchain, IoT provides increased transparency through tamper-proof records by way of RFID tags, GPS trackers, and QR codes, which allow for supporting regulatory compliance with efficient recall processes(Lin et al., 2022). Besides, data from IoT sensors analyzed by machine learning enables predictive maintenance that reduces equipment failures and offers reduced risks for food safety(Jin et al., 2019). The proactive approach reduces downtime and maintains high product quality.

IoT integration in the food supply chain promotes IoF, whereby through mobile applications or smart labels, a consumer can view in real-time all information concerning its origin, handling, and compliance status(Gheorghe et al., 2021). It is such transparency that will build trust among consumers about the safety of their food. However, IoT effectiveness can only be realized when there is usage of standardized data format and protocols for smooth data exchange(Khan et al., 2021). Interoperability of IoT systems, blockchain, and analytics tools ensures integrated monitoring for traceability. Cybersecurity has been ensured by robust security measures such as encryption through secure access controls in order to avoid data breaches of sensitive information in food safety(Lin et al., 2022).

The integration of IoT with AI-driven analytics had provided the ability for automatic decisions to be made based on real-time data, with less human interference and at better outcomes(Bai et al., 2020). Implementing continuous system updates and staff training in IoT technologies are just a few basic reasons why it is indispensable to remain updated with evolving regulations(Schneider et al., 2021). IoT works effectively in food safety according to empirical studies. Bai et al. (2020) found that IoT-based monitoring reduces spoilage during transport, while Jin et al. (2019) demonstrated to reduce 30% of the incidents related to food safety due to the implementation of predictive maintenance.

Gheorghe et al. (2021) showed that due to IoT-enabled traceability, there was increased consumer trust, and Lin et al. (2022) described how times for recalls were reduced by 70% upon integrating IoT with blockchain.

Predictive Compliance Through Machine Learning in Food Safety

The food industry faces growing demands to meet high standards of food safety and compliance because the food supply chain is increasingly becoming more complex and the regulatory needs evolve. Predictive compliance is an active strategy that will promote better food safety through machine-learning-driven data analysis and trendspotting, effectively meeting a likely occurrence of a compliance problem(Schneider et al., 2021).

ML-based predictive compliance supports the structure of the RMF to further improve risk identification, assessment, and mitigation. ML, through analyzing of historical and real-time data, predicts non-compliance events for early risk mitigation. The model of HACCP also reaps some benefit from ML since it continuously monitors a CCP for predictions of safety standard deviations, moving food safety management from being reactive to proactive(Manning & Soon, 2016).

In supply chain management, ML optimizes the process through analysis of data from the supply chain structure, identifies patterns that point to potential compliance issues and maintains standards for safety(Tse & Tan, 2020). The most common supervised ML algorithms are the decision tree and SVM that use historical food-safety data to facilitate the model in predicting the future risks, for instance, microbial contamination(Zhu et al., 2020). Salient advantages will be provided by unsupervised learning, based on some algorithms such as k-means clustering to identify from the data of food safety those hidden patterns and anomalies that help in finding previously unknown risks such as unusual variations in temperature while storage or transportation (Kankanamge et al., (2022).

Reinforcement learning algorithms would help in optimizing the food safety processes through real-time adaptation to changes in production conditions, thus maintaining compliance in automated quality control systems(Chen et al., 2021). Machine learning models based on sensor data of temperature, pH, and moisture levels predict product quality and detect deviations from compliance standards to enable early corrective actions(Liu et al., 2019).

ML processes data from environmental monitoring systems, like microbial testing and air quality sensors, to predict risks related to contamination for the support of preventive measures against the incidence of pathogens such as Salmonella and Listeria(Zhu et al., 2020). The use of machine learning in supply chain management will analyse data on supplier performance, transportation conditions, and storage environments to identify high-risk segments where intervention is required (Tse & Tan, 2020). Natural Language Processing is a branch of ML that helps keep track of texts for regulatory matters by performing automation of required compliance checks, and producing alerts when new regulations are on the cards(Smith et al., 2021).

The success of ML in predictive compliance depends very strongly on the quality and completeness of the data. It pulls together data from environmental monitoring, production, supply chain operations, and regulatory reports, all of which are indispensable for good models(Kankanamge et al., 2022). Models must be trained and updated continuously with the latest information to enjoy better predictive accuracy(Chen et al., 2021).

While ML provides the insights, human judgement is still required to interpret those results and make context-sensitive decisions, in other words, a collaborative approach to compliance management(Schneider et al., 2021). Also, explainable AI techniques support the acceptance of regulatory needs by interpreting the rationale for the predictions(Liu et al., 2019). Studies also show that predictive compliance, based on ML, significantly improves food safety. For example, Zhu et al. (2020) reported that microbial contamination prediction was over 90%, and Liu et al. (2019) presented a 40% reduction in non-compliance events after the introduction of the ML-based quality control system.

According to research, Kankanamge et al. (2022) state that ML algorithms in the sector of supply chain risk analysis enable high-risk suppliers and predict disruptions, for which the practitioner can take pre-emptive action. Smith et al. (2021) explain how compliance systems powered by NLP automatically process updates in regulations, thereby making compliance management efficient. Predictive compliance run using ML will make the approach toward food safety management proactive, ensuring quality in data, training models, collaboration between humans and AI, transparency, and more for better regulatory compliance and public health.

Automating Compliance Workflow with AI in Food Safety

Food industries operate in a strict regulatory environment that protects public health and instills confidence in consumers. Smith et al. (2021) highlighted that traditional compliance workflows, involves a lot of human effort, audits, and heavy documentation, which is time-consuming, prone to errors, and resource-intensive(Smith et al., 2021). With the advancement in AI, there is an increased possibility of automating these workflows for better accuracy, efficiency, and regulatory compliance.

In food safety management, both RMF and HACCP need to be identified. The former was for risk identification and mitigation by Schneider et al. ; while the latter monitors critical control points that assist in preventing hazards (Manning & Soon., 2016) AI enhances these frameworks by moving real-time monitoring, data analysis, and decision support forward; thus, enabling timely compliance actions. Another important model is continuous compliance, which continual monitoring and verification are involved. AI-driven automation for this approach means 'continuously analysing operation data to detect compliance violations and derive corrective actions' (Chen & Zhang, 2020).

AI applications, including NLP, read the regulatory documents and guidelines to let an organization automatically align workflows with the current regulations(Smith et al., 2021). Machine learning models tap into both historical and real-time data to forecast compliance issues related to microbial contamination risk(Kankanamge et al., 2022). Finally, Robotic Process Automation smoothes the procedures and makes them less prone to human error by guaranteeing homogeneous execution processes of an organization(Brown et al., 2021).

AI-powered IoT sensors permit real-time surveillance over production conditions, notice deviation, and call for corrective action(Bai et al., 2020). AI-aided risk assessment tools classify risks with regard to priority based on historical and real-time input(Kankanamge et al., 2022). Automated internal audits using AI identify gaps that need to be covered by proactive improvement(Chen & Zhang, 2020). Integration of diverse data sources is required for the development of correct AI analytics and reporting(Bai et al., 2020).

AI models, in particular, have to be refreshed regularly with new input data and changes in regulatory provisions for accuracy. Retraining machine learning algorithms together with compliance professional feedback enables them to adapt to changing requirements(Kankanamge et al., 2022). Similarly, AI-driven workflows should be made transparent to build trust; explainable AI techniques justify automated decisions, at least during audits (Smith et al., 2021). Similarly, while AI automates compliance, human judgment remains an important dimension necessary for interpreting complex situations. Therefore, a good approach would be to let the AI provide the insights while the professionals apply the contextual knowledge; this will ensure that compliance is effective(Chen & Zhang, 2020).

AI-driven automation leads to enhanced efficiency. Smith et al. indicate that the NLP cuts interpretation time for regulatory requirements by 60% (Smith et al., 2021). In this respect, Bai et al. comment that the incorporation of AI-IoT increased compliance with hygiene and temperature requirements-reducing the level of non-compliance by 35% (Bai et al., 2020). Other research indicated that through the machine learning model, prediction of compliance issues had an accuracy of 85% (Kankanamge et al., 2022) and RPA reduces manual errors and workforce by 40% (Brown et al., 2021).

AI-Driven Predictive Analytics for Food Demand

The food industry uses AI-driven prediction analytics to estimate consumer demand, which optimizes supply chain operations and reduces food waste, improving profitability. According to Li & Zhang (2021), several studies on the very issue seem to have established that traditional methods of forecasting usually lack capturing the complication of modern consumers' behavior(Kamble et al., 2020). AI uses machine learning and deep learning to scan large volumes of data to identify patterns in demand much more efficiently.

A fine example of AI-driven analytics is DFF, integrating the factors of seasonality, market trends, and promotions with socio-economic factors(Wang & Li, 2019). In this respect, AI-enhanced forecasts used within an integrated SCM Model will contribute to inventory management, production planning, and logistics(Tse & Tan, 2020). AI refines the JIT Inventory Management system by reducing overstocking and stockouts(Kamble et al., 2020).

Popular supervised learning algorithms for demand forecasting include linear regression analysis, support vector machines, and neural networks. These models learn from past sales data and external variables like seasonality, weather, and promotions to predict future demand(Li & Zhang, 2021). Neural networks will learn nonlinear relationships between inputs and demand to provide more accurate forecasts in highly complex markets

Deep learning methods, such as LSTM networks, perform best while operating with time-series data and capturing daily, weekly, and seasonal fluctuations(Wang & Li, 2019). Their capabilities for learning from sequential patterns make them suitable for nuanced demand forecasting. Reinforcement learning further improves the forecast by correcting its strategy in real time to optimize pricing, promotion, and inventory toward demand(Kamble et al., 2020).

Artificial intelligence-driven predictive analytics draws data from multiple sources: POs, e-commerce websites, and even social media and weather forecasts(Tse & Tan, 2020). This multi-sourced data provides holistic forecasts, which, in turn, track and record market shifts in real time. The Demand Sensing AI helps enhance near-term forecasts by analyzing recent sales data, promos, and social sentiment carried out through social media; this enables quick adjustments in inventory and supply chain strategies(Wang & Li, 2019).

Using AI-powered forecasting allows them to engage in personalized marketing and dynamic pricing strategies by estimating the demand for products, optimizing promotions, and reducing excess inventories(Li & Zhang, 2021). Complex versions of AI even do automatic inventory management; it suggests reorder points, thereby minimizing stocking costs and potential spoilage(Kamble et al., 2020). The accuracy of AI-based forecasting depends upon data quality. It means that structured and unstructured data-integrating companies should be employed with data preprocessing to enhance model robustness(Wang & Li, 2019). Models may be tested against their best performance by using regression, neural networks, and ensemble methods(Li & Zhang, 2021).

Although AI can make analysis on volumes of data and come up with predictions, human judgments on strategic decisions are called for in interpreting results. In short, demand planners realize better efficiency if they apply AI insights in the development of strategies on inventory, marketing, and supply chain issues(Kamble et al., 2020). Scenario analysis, together with AI in forecasting, will enable companies to simulate market scenarios in the event of, for example, an economic crisis or a change in weather that may allow flexibility in decisions(Tse & Tan, 2020). Indeed, AI-driven analytics has promoted precision in the demand forecasting of food. Wang and Li (2019) found that deep learning models such as LSTM networks reduce the forecast error by as high as 25% in contrast to traditional statistical models that show weaker performances while handling seasonal and nonlinear trends.

Li and Zhang (2021) proved that the inclusion of real-time data sources like weather reports and social media increased the prediction accuracy by 20% against the model using only historical sales data. Kamble et al. (2020) reported a 30% reduction in stockouts and a 15% decrease in excess inventory through AI-driven demand sensing. While Tse and Tan (2020) found that AI forecasting resulted in a 20% improvement in the periodic inventory turnover ratio. Altogether, AI-driven predictive analytics enables corporations to achieve three broad objectives: optimization of inventory levels, improvement of supply chain efficiency, and adaptation to changes in the market. Quality of data, model selection, collaboration between humans and AI are part of the bouquet of key practices at the core heart of the journey to success, or a more sustainable food system.

Machine Learning Models for Food Supply Chain Optimization

Supply chains for food should be optimized to fulfill food security, minimize the loss of food, and cater to the demand of a market which dynamically changes. Traditional SCM faces issues on three grounds: variability in demand due to seasonality, perishability of products, and global influences(Tsolakis et al., 2020). Machine learning can analyze vast volumes of information by running complex algorithms to identify underlying patterns, thus offering advanced insights into food supply chain optimization.

Some of the key areas where the SCM Framework is believed to be useful in the development and optimization of food supply chains are demand forecasting, production planning, inventory, and logistic management(Wang et al., 2020). Integration of ML into SCM provides data-driven decision-making in input analysis by enhancing general efficiency and resilience of supply chains. ML will make the JIT Inventory Management Model enable accurate demand prediction, optimize the production schedule, and handle perishable goods accordingly(Fan et al., 2018).

As put by Tsolakis et al. (2020), the SSCM framework extends the traditional SCM to address environmental and social considerations. ML models support the SSCM by analyzing resource usage, carbon footprint, and waste for promoting sustainable operations. Supervised learning models comprise linear regression, decision trees, random forests, and SVM, extending their applications in demand forecasting, inventory management, and route optimization by Kamble et al. (2021). Clustering algorithms, as an unsupervised learning approach, assist in supplier segmentation and demand classification to allow for customized supply chain strategies(Tsolakis et al., 2020)..

Reinforcement learning models make supply chain operations optimized in terms of decision-making while learning policies, including warehouse management, routing, and inventory replenishment(Wang et al., 2020). RL adjusts inventories dynamically to real-time data in order to reach a balanced stock. The deep learning models include CNNs and RNNs, which are used for quality control by images or time-series forecasting, respectively. CNNs assess the quality of food by images, while RNNs, such as LSTM, depict demand patterns(Fan et al., 2018). ML models analyze sales data, market trends, and weather to predict demand, optimizing inventory and waste reduction. They also consider all options for transportation routes to minimize cost and environmental effects.

These unsupervised learning models, such as clustering and principal component analysis, identify optimum suppliers based on rating criteria related to cost, quality, reliability, and sustainability of products. Tsolakis et al. (2020) also established that ML models evaluate the risks of a supplier through data sets every year about its financial health, regulatory compliance, and market reputation. Again, deep learning models, especially CNN, facilitate automated quality inspection in food production through detection of defects and contaminants in finished and partial products(Fan et al., 2018). This automation guarantees quality and safety consistency, further entrenching consumer confidence.

Effective ML requires good-quality and standardized data coming from sales, production, inventory, logistics, or environmental monitoring(Wang et al., 2020). Integration of data coming from the supply chain forms just one single dataset to which complete analytics can be applied. Different optimization tasks require various ML models; therefore, companies need to try a variety of supervised, unsupervised, and reinforcement learning algorithms in order to derive the best-fitting models for their needs(Kamble et al., 2021). It is at this point that customization of the models to business processes and market dynamics specific to it should be done, to maximize effectiveness.

Dynamic factors affecting the food supply chain include seasonal demand, changes in consumer preference, and disruptions like natural calamities. Continuous learning using real-time data allows ML models to adapt immediately, hence optimizing operations(Fan et al., 2018). Although ML can provide insights through its findings, human expertise is still crucial in interpreting results and making strategic decisions to come up with better optimization strategies(Tsolakis et al., 2020).

Proof from research has been able to highlight how ML has optimized the supply chain. Kamble et al. (2021) were able to identify that ML-based demand forecasting increased accuracy by 30%, which led to efficient inventory management and further resulted in a reduction in related costs. Wang et al. (2020) proved how reinforcement learning reduced transportation costs by 15% and further enhanced the delivery time by an additional 20% with the integration of real-time data. Tsolakis et al. (2020) identified that unsupervised learning models are enhancing the selection of suppliers for building supply chain robustness. Fan et al. (2018) indicated that deep learning in automatic quality inspection reduces the level of defects and contamination by a quarter.

Overall, ML models significantly enhance supply chain efficiency, flexibility, and sustainability by improving forecasting, inventory management, logistics, supplier selection, and quality control.

Blockchain for Secure Traceability in Food Safety

Food safety and security in global supply chains are at risk, given the existence of contamination fraud and mislabeling (Tian, 2017). In this respect, blockchain presents a potential solution in the form of a decentralized, tamper-evident ledger for recording every transaction and movement, therefore enhancing transparency and trust. Blockchain serves the Supply Chain Management (SCM) Framework, focusing on the agenda of visibility and traceability(Kamble et al., 2020). It builds up an immutable record of a product's history from farm to fork. Coupling blockchain with the Food Safety Management System allows timely monitoring and verification of critical control points, thereby giving a quick response to food safety hazards(Li et al., 2021).

Blockchain technologies enhance the RMF of food safety through its function of providing secured real-time data concerning the origin, condition, and handling practices of the food product, thus helping the identification, assessment, and mitigation of risks by the food industry(Tian, 2017). The decentralization paradigm in blockchain allows access to a shared ledger by all amounts in the supply chain: producers, processors, distributors, and retailers. This reduces information asymmetry among supply chain partners and enhances trust(Wang et al., 2021). Such transparency encourages the validation of the authenticity and safety of the food products by the stakeholders.

Immutability in blockchain ensures that once recorded data is not changed, hence avoiding tampering and fraud. This is highly important in ensuring that information touching on the origin, processing, storage conditions, and transportation history of products is accurate(Tian, 2017). Smart contracts are basically a set of agreements programmed into the blockchain, which automate actions based on pre-set rules when an event occurs. Smart contracts carry out quality inspections, beliefs in certifications, and send notifications in case the standards are not met for food safety, hence simplifying compliance issues(Casino et al., 2019). It helps in end-to-end traceability by recording every transaction in a supply chain. Products tagged with unique identifiers, such as QR codes linked to blockchain, enable the stakeholders to track their journey in real time. This enables fast action during recalls(Wang et al., 2021). Blockchain also eliminates food counterfeit because it creates a verifiable record; hence, it is challenging to tamper with information related to products or fraudulently present their superiority(Kamble et al., 2020).

Integration of blockchain with IoT sensors enables real-time accessing of conditions such as temperature and humidity to meet the set standards of safety(Tian, 2017).. It automates the verification of certifications and thereby increases the confidence of consumers in food labels(Casino et al., 2019. In ensuring maximum effectiveness in food safety, companies are supposed to standardize and then integrate all the data that has been captured along the supply chain. This requires the use of common data formats, protocols, and interoperable systems so that data can flow smoothly(Wang et al., 2021). The system should be jointly developed by all participants in the supply chain, and it needs to be a single platform where records are accurate and transparent(Li et al., 2021).

Such systems need to process volumes of lightweight transactions efficiently, using consensus mechanisms for scalability, such as proof-of-stake(Casino et al., 2019). Techniques for privacy preservation will keep encrypted information private, using permissioned blockchains, while still allowing openness in transparency(Kamble et al., 2020). Research indicates blockchain improves traceability and increases trust. Tian (2017) did find the time taken for tracking down a product's origin reduced from days to seconds, which bettered food safety and minimized any economic losses that might have happened during recalls.

Kamble et al.,(2020) indicated that immutability and transparency of blockchain increased the authenticity of food, decreased fraud in food by up to 30%, while there was also an increased consumer trust where traceability brought about by blockchain was made public. Wang et al.(2021) reported that the integration of IoT sensors with blockchain in this case ensures compliance with set standards of temperature and humidity by 25%, decreased rate of spoilage, thereby enhancing the safety of food consumed. Casino et al. (2019) stated that blockchain-driven smart contracts had automated the processes of compliance checks, thereby reducing administrative work by 40%.

With Blockchain, food safety gains traceability through the decentralized tamper-proof ledge. The features of decentralization, immutability in the ledger, and smart contracts also support risk management and compliance. Above all, data needs to be standardized, collaboration fully developed, scalability assured, and publicity/privacy protected for successful complete implementation, which continues to ensure that blockchain supports a sea-change in food safety management.

AI-Powered Food Supply Chain Resilience

The structure of the food supply is complex and vulnerable to a shock from numerous factors, including climate change, geopolitical conflict, and public health crises like the COVID-19 pandemic(Sharma et al., 2020). Supply chain resilience-or the ability to predict, respond to, and recover from disruption-is integral to both food security and business sustainability. AI enhances SCR by examining data in real-time and predictive modelling and through decision-making on an automated basis.

These opportunities allow business organizations to explore risks more accurately and adapt to them in time(Ivanov & Dolgui, 2021).

The Dynamic Capabilities Framework instigates the basis of organizational adaptation to shifting environments. AI supports that by analyzing large datasets for market shift detection, consumer behavior, and newly emerging risks, hence proactive adjustments may be enabled(Tukamuhabwa et al., 2015). AI integration also goes in line with the Digital Twin Paradigm, where the twin of the supply chains is digitally simulated and thus allows for advanced response strategies(Ivanov & Dolgui, 2021). Then, machine learning algorithms, including regression-based and decision trees, allow prediction of supply chain disruptions. Correspondingly, managers are capable of taking several actions to prevent such disruption from occurring(Sarker et al., 2022).

For instance, deep learning models use LSTM networks and CNN, which process volumes of data to provide an accurate forecast of demand. This enables an enterprise to optimize their inventories and production schedules accordingly(Ruan et al., 2019). This reduces waste and helps companies avoid stockouts, besides making them truly responsive in case of disruptions. Reinforcement learning optimizes inventory management, routing, and production in real-time by adapting to dynamically changing conditions while striving to realize recovery in supply chains(Ivanov & Dolgui, 2021). Unstructured data from sources like news and social media are analysed for early warnings of disruptions, such as natural disasters or political unrest, by NLP Sharma et al., 2020). With AI-powered sensors and IoT devices, monitoring becomes real-time through the detection of food parameter anomalies and the initiation of automated responses in case of deviations(Sarker et al., 2022).

AI-powered digital twins simulate supply chain processes, enabling firms to run disruption scenarios-such as natural disasters or supplier collapse-and devise robust contingency plans(Ivanov & Dolgui, 2021). By testing contingent strategies in a virtual environment, firms seek to optimize responses to real-life disruptions. Accordingly, reinforcement learning models presented in this study have the ability to optimally adjust the supply chain process dynamically based on real-time conditions through inventory replenishment, production scheduling, and distribution(Ruan et al., 2019). The main intent of the adaptive approach is the potential to minimize losses while sustaining service levels in case of disruptions.

These AI models assess supply chain risks around reliability in supplies and market volatility based on both historical and real-time data(Sarker et al., 2022). Such automation of identified risks creates actionable insights to implement mitigations in the forms of sourcing other supplies or adjusting production schedules. In this regard, integrated, fine-quality data from different sources is the backbone of AI functionality, integrated through data governance practices for accuracy, consistency, and timeliness(Sarker et al., 2022).

Supply chains operate in dynamic environments driven by market trends, consumer behaviour, and geopolitical temperatures. AI models must be retrained regularly to retain predictive accuracy(Ruan et al., 2019). Incorporating feedback loops can enable a model to learn from previous disruptions for better responses. Most importantly, human judgment plays a crucial role in distilling insights that are generated by AI and in an ethical and effective application thereof(Ivanov & Dolgui, 2021).

Research supports the considerable impact of AI on strengthening supply chain resilience. Sharma et al. (2020) found that AI-driven risk monitoring tools, using NLP and ML models, ensured a disruption prediction rate that was 30 percent higher than before and hence proactive mitigation is possible. Ivanov and Dolgui (2021) showed that AI-powered digital twins had reduced the supply chain recovery time up to 20% in some cases. Sarker et al. (2022) reported that the implementation of RL models even reduced stockouts and excess inventory by 20% in dynamic inventory management. Ruan et al. (2019) argued that deep learning models improved accuracy in demand forecasts by 18%, further improving production planning and reducing waste. It is observed that the inclusion of AI into supply chain strategies will enhance the visibility of the supply chain, risks, and operation efficiency, thus being substantively important in the future for the resilience of food supply chains.

The Meta Model

Source: Authors Generated

The model has AI Safety at the center, with artificial intelligence right at the core of insights into food safety and compliance. Further out are capabilities around Predictive Analytics, which protect the core and have meaning in decision-making processes related to the supply chain. "Real-Time" nodes are connected to nodes on food safety, showing that through AI, real-time monitoring of compliance issues is possible:. This theme knits predictive analytics with real-time actionable insight in ensuring compliance across the supply chain. It brings in the concept of Seal-Time Data Traceability and Sear-Time for Security Data Tracing, emphasizing secured and time-sensitive data management to trace the stages of food products on different levels of the supply chain. Smart Contracts for Compliance Chain incorporates blockchain to compliance transparency in ensuring predicament-free trust, reflecting yet another point of intersection between AI, blockchain, and data security. Real-time enhancement of supply safety evidences the use of AI in strengthening protocols relative to the supplies' safety in motion. Predictive-Based Compliance suggests that, through the use of back-history data and predictive models, AI may anticipate events that could lead to safety and compliance issues and take the necessary steps to prevent them. The term Smart Compliance Chain would suggest a footing of AI and IoT devices-that is, the Internet of Things-for automation in compliance. It is believed from the network that AI integrates with other technologies to manage compliance in real time in order to make the smooth going of the supply chain process possible. Big Data integrated into this model would mean the footings of AI in large data sets to increase their precision in food compliance, tracing, and prediction.

Meta-model presented explains that the modern backbone of a resilient food supply chain is based on AI-enabled systems interlinked in real time through data, predictive analytics, and secured traceability. Integration of various technologies, such as blockchain, smart contracts, and IoT into one system, is foreseen to enhance compliance, safety, and efficiency. Core would remain the use of AI for predicting, monitoring, and safeguarding the whole food supply chain to ensure further developments will also be toward resilience.

CONCLUSION

Developing a resilient food supply chain-one that can respond to troubled global market conditions-requires the integration of AI-powered food safety and compliance and innovative technologies. Powered by predictive analytics, real-time monitoring of automated workflows of compliance activities, and data-driven decision-making, AI truly empowers food supply chains for proactive risk management, product quality, and regulatory adjustment. Such improvements have gone a long way in increasing food safety and authenticity, besides taking care of sustainability through reduced food waste and optimized resource utilization. Supply chain transparency, efficiency, and resilience will be further enhanced with AI-driven blockchain in traceability, machine learning models for demand forecasting, and smart IoT integration. In embracing AI and associated technologies, the food industry can guarantee a safer, more sustainable, and agile supply chain that is more effectively positioned to respond to changing consumer needs and regulatory requirements.

REFERENCES

Al-Turki, U. (2011). A framework for strategic planning in maintenance. *Journal of Quality in Maintenance Engineering, 17*(2), 150–162. DOI: 10.1108/13552511111134583

Almeida, F., Santos, J. D., & Monteiro, J. A. (2020). Compliance-as-a-Service (CaaS): A model to streamline regulatory compliance in cloud computing. *Journal of Cloud Computing (Heidelberg, Germany), 9*(1), 1–14.

Aung, M. M., & Chang, Y. S. (2014). Traceability in a food supply chain: Safety and quality perspectives. *Food Control, 39*, 172–184. DOI: 10.1016/j.foodcont.2013.11.007

Bai, C., Dallasega, P., Orzes, G., & Sarkis, J. (2020). Industry 4.0 technologies assessment: A sustainability perspective. *International Journal of Production Economics, 229*, 107776. DOI: 10.1016/j.ijpe.2020.107776

Baldwin, C. (2022). The Role of IoT in Enhancing Food Safety Compliance. *Journal of Food Quality, 35*(2), 121–135.

Caro, M. P., Ali, M. S., Vecchio, M., & Giaffreda, R. (2018). Blockchain-based traceability in Agri-Food supply chain management: A practical implementation. International Conference on Internet of Things (iThings), 1-8. DOI: 10.1109/IOT-TUSCANY.2018.8373021

Casino, F., Dasaklis, T. K., & Patsakis, C. (2019). A systematic literature review of blockchain-based applications: Current status, classification, and open issues. *Telematics and Informatics, 36*, 55–81. DOI: 10.1016/j.tele.2018.11.006

Chen, J., Wang, Y., & Li, X. (2021). Reinforcement learning in food safety management: A dynamic decision-making approach. *Journal of Food Engineering, 293*, 110409.

Chen, Y., et al. (2023). Technology Advancements in Food Safety Monitoring. Journal of Food Safety Technology.

Chen, Y., Luo, L., & Zhang, H. (2019). Automated Data Logging in Food Safety Management: A Practical Guide. *Food Control, 98*, 285–290.

Chuang, C., Li, C., & Shen, X. (2021). Natural language processing (NLP) for compliance monitoring in food safety: A systematic review. *Food Control, 130*, 108324.

Doe, J. (2024). Automation and Quality Control in Food Supply Chains. International Supply Chain Review.

Fan, Z., Yu, W., & Sun, L. (2018). Deep learning for food supply chain: Opportunities and challenges. *Computers and Electronics in Agriculture, 157*, 268–277.

Food and Agriculture Organization. (2019). Hazard Analysis and Critical Control Points (HACCP). *System.*

Food and Drug Administration (FDA). (2011). Food Safety Modernization Act (FSMA). Retrieved from Galvez, J. F., Mejuto, J. C., & Simal-Gandara, J. (2018). Future challenges on the use of blockchain for food traceability analysis. *Trends in Analytical Chemistry, 107*, 222–232.

Ghosh, S. (2020). IoT-enabled predictive maintenance: Application in food processing industry. *Journal of Food Engineering, 268*, 109737.

Gong, L., Xiao, D., Huang, H., & Wang, R. (2020). Application of artificial intelligence in food safety. *Trends in Food Science & Technology, 102*, 160–172.

Gupta, R. (2018). Risk Management in Food Safety: A Framework for Compliance. *Food Protection Trends, 38*(4), 297–305.

Holleran, E., Bredahl, M. E., & Zaibet, L. (1999). Private incentives for adopting food safety and quality assurance. *Food Policy, 24*(6), 669–683. DOI: 10.1016/S0306-9192(99)00071-8

Huang, H., Lu, J., & Wang, H. (2021). Using AI and NLP to analyze social media for emerging food safety risks. *Food Control, 123*, 107893.

Ivanov, D., & Dolgui, A. (2021). A digital supply chain twin for managing the disruption risks and resilience in the era of Industry 4.0. *Production Planning and Control, 32*(9), 775–788. DOI: 10.1080/09537287.2020.1768450

Jardine, A. K., Lin, D., & Banjevic, D. (2006). A review on machinery diagnostics and prognostics implementing condition-based maintenance. *Mechanical Systems and Signal Processing, 20*(7), 1483–1510. DOI: 10.1016/j.ymssp.2005.09.012

Jin, X., Shi, X., Wang, J., & Zhu, L. (2020). Application of artificial intelligence in food safety: A review. *Trends in Food Science & Technology, 99*, 94–103.

Johnson, M., & Lee, S. (2023). AI-Driven Analytics for Food Safety Compliance. *Journal of Food Safety.*

Kamble, S. S., Gunasekaran, A., & Arha, H. (2020). Understanding the Blockchain technology adoption in supply chains-Indian context. *International Journal of Production Research, 58*(7), 2100–2123.

Kamble, S. S., Gunasekaran, A., & Gawankar, S. A. (2021). Achieving sustainable performance in a data-driven agriculture supply chain: A review for research and applications. *Sustainable Production and Consumption*, *24*, 83–98.

Kankanamge, N. D., Lee, E., & Zhang, L. (2022). Predictive food safety risk analysis using machine learning: A systematic review. *Food Control*, *132*, 108543.

Kritzinger, W., Karner, M., Traar, G., Henjes, J., & Sihn, W. (2018). Digital twin in manufacturing: A categorical literature review and classification. *IFAC-PapersOnLine*, *51*(11), 1016–1022. DOI: 10.1016/j.ifacol.2018.08.474

Kumar, A., & Budhwar, N. (2020). Artificial Intelligence and Traceability in Food Supply Chain: A Framework for Future Research. *The Journal of Supply Chain Management*, *15*(3), 245–262.

Kumar, P., & Singh, A. (2019). Systems Theory and Food Safety: An Interdisciplinary Approach. *International Journal of Food Sciences*, *54*(1), 55–62.

Lee, H., & Thompson, D. (2022). Global Food Safety and Supply Chain Management. International Journal of Food Safety.

Lee, J., Bagheri, B., & Kao, H. A. (2015). A cyber-physical systems architecture for Industry 4.0-based manufacturing systems. *Manufacturing Letters*, *3*, 18–23. DOI: 10.1016/j.mfglet.2014.12.001

Li, J., Huang, Q., & Lin, W. (2022). AI-powered predictive analytics for food safety regulation. *Food Control*, *130*, 108302.

Li, J., Wang, S., & Song, S. (2021). An exploration into the mechanism of blockchain application in food supply chain traceability. *Journal of Cleaner Production*, *290*, 125921.

Li, Q., Zhang, X., & Liu, Y. (2020). Machine learning-based detection of foodborne pathogens: A review. *Foodborne Pathogens and Disease*, *17*(4), 216–227.

Liu, S., Song, X., & Hu, Z. (2019). Machine learning applications in predictive quality control of food products. *Journal of Food Quality*, *2019*, 1–11.

Mahmoudi, A., Adl, M., & Barbalho, S. M. (2022). The role of IoT and AI in food safety. *Food Control*, *134*, 108691.

Manning, L., & Soon, J. M. (2016). Food safety, food fraud, and food defense: A fast-evolving literature. *Current Opinion in Food Science*, *10*, 114–120. PMID: 26934423

Miller, R., et al. (2023). Future Innovations in AI for Supply Chain Resilience. Global Logistics Magazine.

Min, H. (2019). Artificial intelligence in supply chain management: Theory and applications. *International Journal of Logistics Research and Applications*, *22*(3), 267–275.

Mortimore, S., & Wallace, C. (2013). *HACCP: A Practical Approach*. Springer. DOI: 10.1007/978-1-4614-5028-3

Moubray, J. (1997). *Reliability-centered maintenance*. Industrial Press.

Ruan, J., Shi, Y., Han, D., & Wen, Q. (2019). Deep learning for supply chain demand forecasting: Methods, challenges, and directions. *Computers & Industrial Engineering*, *127*, 225–234.

Sarker, M. N. I., Wu, M., & Liu, X. (2022). Resilient supply chain management through AI-powered risk analysis in the post-COVID era. *International Journal of Logistics Research and Applications*, *25*(3), 475–492.

Schneider, K., van der Vorst, J. G., & Garcia-Martinez, M. (2021). The impact of digitalization on food supply chains: Insights from the European Union. *Sustainability*, *13*(16), 9150.

Schwabacher, M. (2005). A survey of data-driven prognostics. In AIAA Infotech@ Aerospace Conference (p. 7002). DOI: 10.2514/6.2005-7002

Sharma, M., Luthra, S., Joshi, S., & Kumar, A. (2020). Developing a framework for enhancing survivability of sustainable supply chains during and post-COVID-19 pandemic. *International Journal of Logistics Management*, *31*(4), 869–885.

Smith, A. (2022). AI in Food Supply Chains: Enhancing Resilience. Food Tech Journal.

Smith, J., Johnson, R. D., & Williams, C. (2021). Navigating food safety regulations with artificial intelligence: An application of natural language processing in compliance management. *Journal of Food Protection*, *84*(3), 451–460.

Tao, F., Zhang, H., Liu, A., & Nee, A. Y. C. (2018). Digital twin in industry: State-of-the-art. *IEEE Transactions on Industrial Informatics*, *15*(4), 2405–2415. DOI: 10.1109/TII.2018.2873186

Tian, F. (2017). A supply chain traceability system for food safety based on HACCP, blockchain & Internet of Things. *IEEE International Conference on Service Systems and Service Management*, 1-6.

Tse, Y. K., & Tan, K. H. (2020). Managing food supply chain risks: The interplay of demand and supply management practices. *Transportation Research Part E, Logistics and Transportation Review*, *136*, 101887.

Tukamuhabwa, B. R., Stevenson, M., Busby, J., & Zorzini, M. (2015). Supply chain resilience: Definition, review and theoretical foundations for future research. *International Journal of Production Research*, *53*(18), 5592–5623. DOI: 10.1080/00207543.2015.1037934

Wang, X., Lin, X., & Zhang, Q. (2021). Integrating blockchain and Internet of Things for food traceability: Research trends, challenges, and future prospects. *Journal of Food Engineering*, *292*, 110369.

Wang, Y., Zhang, Q., & Li, Z. (2020). A machine learning-based approach for optimizing food supply chain operations. *Journal of Cleaner Production*, *275*, 123482.

World Health Organization. (2020).. . *Food Safety (Tokyo, Japan)*.

Xu, L. D., Xu, E. L., & Li, L. (2018). Industry 4.0: State of the art and future trends. *International Journal of Production Research*, *56*(8), 2941–2962. DOI: 10.1080/00207543.2018.1444806

Zhao, Y., Zhan, Y., & Liu, G. (2019). IoT and AI technologies for food safety surveillance: A review. *Sensors (Basel)*, *19*(19), 4185. PMID: 31561597

Chapter 3
Generative AI Effect on Circular Economy Sustainability by Mediation of Food Supply Chain Agility

Shashi Kant
https://orcid.org/0000-0003-4722-5736
Bule Hora University, Ethiopia

Tamire Ashuro
Bule Hora University, Ethiopia

Metasebia Adula
https://orcid.org/0000-0001-5732-2850
Bule Hora University, Ethiopia

Zerihun Kinde Alemu
https://orcid.org/0000-0002-8546-2882
Ethiopian Civil Service University, Ethiopia

ABSTRACT

In order to comprehend how these elements support sustainable supply chain practices, this study examines the interactions between Food Supply Chain Agility (FSCA), Generative AI (GAI), and Circular Economy Sustainability (CES). Agile supply chains are better positioned to adopt sustainable practices, according to the research, which uses a mixed-methods approach and finds a substantial positive as-

DOI: 10.4018/979-8-3693-9856-2.ch003

sociation between FSCA and CES. The results also demonstrate how important GAI is in promoting supply chain agility, which helps businesses become more responsive and efficient. In order to accomplish sustainable goals, the study emphasizes the significance of using cutting-edge technology and encouraging agility, providing insightful information for both managers and researchers. In the conclusion, the study advances the theoretical knowledge of sustainable supply chains and offers useful suggestions for businesses.

INTRODUCTION

In particular, the food supply chain offers a revolutionary potential to improve sustainability through the convergence of generative AI and the circular economy (Irfan et al., 2024). There is an urgent need for creative solutions that support sustainability as the world's food systems deal with hitherto unseen problems (Li et al., 2023), such as population increase and rising demand, resource depletion, and climate change (Madanaguli et al., 2024). A workable framework to deal with these problems is the circular economy idea, which places an emphasis on resource efficiency, waste minimization, and the continuing use of resources (Suali et al., 2024). The agility of food supply chains may be greatly increased by using generative AI, which can improve decision-making and optimize operations. This will promote a more sustainable method of producing and distributing food (Wu et al., 2024). Although the circular economy has its roots in early ecological economics, it really took off in the twenty-first century when sustainability became a pressing worldwide concern (Yaroson et al., 2024). The significance of technology in promoting circular practices is being more and more highlighted in empirical research, with generative AI playing a key role in enabling innovations that increase the resilience and responsiveness of supply chains (Fonsec et al., 2024). Studies have shown, for example, that AI-driven models may eliminate food waste, streamline logistics, and forecast changes in demand—all of which are consistent with the ideas of a circular economy (Akhtar et al., 2024).

The successful use of generative AI in circular economy activities within the food supply chain is hampered by a number of real-world issues, despite the technology's encouraging promise (Shafique et al., 2024). These include stakeholder opposition to change, a lack of data consistency, technical obstacles, and inadequate governmental frameworks to foster innovation (Zhang et al., 2024). Also, there are gaps in the literature that need to be filled since current research frequently ignores the precise interactions between generative AI capabilities and the circular economy (Panigrahi et al., 2023). By methodically investigating how generative AI may manage food supply chain agility and support a circular economy, this work seeks to close these gaps. This study will give future academics and policymakers a thorough framework by combining the body of existing knowledge and identifying real-world issues. It will outline practical methods for successfully utilizing generative AI, guaranteeing that food systems are both sustainable and effective (Zheng et al., 2024).

"How can the integration of generative AI enhance the agility of food supply chains to promote sustainability within the circular economy?" is the main research question that drives this investigation. This inquiry is important because it tackles the dual problems of circular economy principles and food supply chain optimization. The study intends to help academics and policymakers navigate the complexity of sustainability in food systems by examining this interaction and offering important insights and suggestions.

STATEMENT OF PROBLEM

Inefficiencies, waste, and a substantial environmental impact define the current status of food supply chains, which runs counter to the idealized notion of a sustainable circular economy (Suali et al., 2024). Food systems would function with little waste in this ideal situation, making use of resources in a way that encourages sustainability and regeneration (Irfan et al., 2024). The reality, however, shows a wealth of conflicting data emphasizing the difficulties in adopting this strategy. For example, even though generative AI provides optimization capabilities, many food supply chains are still reluctant to implement these technologies, which lead to ongoing inefficiencies and resource waste (Yaroson et al., 2024). Significant gaps still exist despite an increasing amount of empirical research demonstrating generative AI's promise to improve supply chain sustainability and agility (Zhang et al., 2024). Regarding the true effect of AI on circular economy practices, the findings of several researches are not entirely clear (Shafique et al., 2024). The absence of coherent frameworks that combine the concepts of the circular economy with generative AI might be the cause of these discrepancies (Fonsec et al., 2024). Additionally, academics and practitioners lack clear insights or models to follow due to information gaps surrounding the precise processes via which generative AI might impact the agility of food supply chains (Wu et al., 2024).

Contextual gaps are especially important since most of the research that has been done so far has tended to concentrate on certain areas or industries, ignoring the variety of problems that food supply chains confront across the world. This inability to comprehend context might result in solutions that are not generally applicable, which reduces their efficacy. The operationalization of generative AI in real-world contexts is further hampered by pragmatic issues such as insufficient infrastructure, data silos, and stakeholder opposition, which further widens the gap between theory and practice (Akhtar et al., 2024). By offering a thorough examination of how generative AI might moderate the agility of food supply chains to improve the circular economy, this study seeks to close these complex gaps. This study will shed light on the ways that generative AI might promote sustainability by creating an integrative framework that harmonizes theoretical understanding with real-world implementations. Additionally, it will feature case studies and best practices that show how these technologies may be successfully used in a variety of circumstances, providing future academics and policymakers with practical advice (Zheng et al., 2024).

By doing this, our study will not only close gaps in the literature but also give parties attempting to negotiate the intricacies of sustainable food systems a clear path forward. The results will enable academics to pursue new lines of investigation and provide policymakers with useful tactics to improve agricultural supply chain sustainability by clarifying the potential of generative AI inside the circular economy.

RELATED THEORIES

Several theories underpin the interplay between generative AI, food supply chain agility, and the circular economy, each contributing unique perspectives that enhance our understanding of this complex landscape.

Systems Theory

According to systems theory, businesses and their surroundings are interdependent webs of connections. This idea highlights the significance of seeing food supply chains as a whole system rather than as separate components (Suali et al., 2024). It emphasizes how generative AI may help stakeholders communicate and share data in real time, improving the responsiveness and robustness of the system as a whole. Researchers can investigate how modifications to one link in the supply chain affect other links, ultimately resulting in more sustainable behaviors, by using Systems Theory (Wu et al., 2024).

Resource-Based View (RBV)

The Resource-Based View (RBV) focuses on how businesses may use their resources effectively to gain a competitive edge (Irfan et al., 2024). Regarding the circular economy, RBV emphasizes how important it is to use technology, such generative AI, to maximize resource use and reduce waste (Adula et al., 2025). According to this hypothesis, businesses that include AI into their operations may improve resource efficiency and conform to the tenets of the circular economy. The RBV offers a prism through which to examine how food supply networks might be made more sustainable by utilizing technology (Madanaguli et al., 2024).

Innovation Diffusion Theory

The adoption and dissemination of innovative technologies within a community or organization are examined under innovation diffusion theory. This theory emphasizes the obstacles and enablers of technology adoption, making it especially pertinent when thinking about the use of generative AI in food supply chains. Knowing the elements that affect the spread of AI technology might help explain why some businesses effectively use generative AI while others don't. This theory also sheds light on how peers pressure and social networks might hasten the adoption of sustainable behaviors (Akhtar et al., 2024).

Circular Economy Theory

The theory of the circular economy itself provides a fundamental framework for comprehending the goals and tenets of sustainability. It promotes a paradigm change away from the conventional linear economy and toward one that prioritizes closed-loop systems, waste reduction, and resource regeneration. This theory offers a paradigm for evaluating how AI technologies might assist circular initiatives in food supply chains, providing the platform for examining the integration of generative AI (Panigrahi et al., 2023).

EMPIRICAL LITERATURE REVIEW

Generative AI and Circular Economy Sustainability

By radically changing how resources are used and managed across several processes, generative AI has become a key player in advancing sustainability within the circular economy (Zheng et al., 2024). Generative AI can forecast changes in demand, improve resource allocation, and reduce waste in food supply chains by utilizing sophisticated algorithms and machine learning skills (Shafique et al., 2024). In a circular economy, which prioritizes the constant use of resources and waste minimization, this optimization is essential (Madanaguli et al., 2024). Generative AI, for example, can evaluate enormous volumes of data from supply chain activities, helping businesses better predict demand and modify production plans accordingly. This improves overall resource efficiency by bringing output into line with real consumption patterns and lowering overproduction and the waste that goes along with it (Panigrahi et al., 2023). Also, generative AI makes it easier to create creative product ideas that put sustainability first. Businesses may produce goods that use resources more effectively and are simpler to recycle or repurpose by utilizing approaches like generative design, which employs algorithms to investigate a wide range of design choices (Suali et al., 2024). In addition to reducing the negative effects of industrial processes on the environment, this strategy promotes a change in perspective toward long-term and circular design from the beginning. Waste potential is greatly decreased when items are developed with their full lifespan in mind, which is in complete harmony with the principles of the circular economy (Wu et al., 2024).

Also, by encouraging openness and information exchange among stakeholders, generative AI improves cooperation across the supply chain (Kant et al., 2025). Real-time resource and waste monitoring is made possible by this technology, which facilitates improved decision-making and more flexible reactions to shifting circumstances. For instance, generative AI can recommend different applications or redistribution tactics if an excess of a specific ingredient is found, minimizing waste and encouraging a more circular approach to resource management (Madanaguli et al., 2024). Nevertheless, there are certain difficulties in incorporating generative AI into circular economy procedures. The successful use of AI solutions may be hampered by problems including data privacy, obstacles to technological adoption, and the requirement for a strong infrastructure. The potential advantages of generative AI in promoting sustainability within the circular economy are significant, notwithstanding these obstacles (Irfan et al., 2024). Generative AI acts as a catalyst for revolutionary change in resource management, ultimately propelling the shift to a more sustainable and circular economy by increasing productivity, encouraging creative designs, and strengthening teamwork. Generative AI has the potential to be extremely important in helping businesses meet their environmental objectives and improve operational efficiency as they become more aware of the significance of sustainability (Akhtar et al., 2024).

H1: Generative AI has a significant effect on circular economy sustainability

Food Supply Chain Agility and Circular Economy Sustainability

Agility in the food supply chain is essential for improving sustainability in the context of the circular economy. The capacity of a supply chain to react quickly and efficiently to shifts in market circumstances, supply interruptions, and demand is known as agility (Suali et al., 2024). An agile food supply chain may greatly support sustainable practices in the framework of the circular economy, which emphasizes increasing resource efficiency and decreasing waste (Nayal et al., 2024). Agile supply chains may minimize overproduction and needless waste by swiftly adjusting to shifting consumer preferences and market conditions, which is very compatible with the fundamental ideas of circularity (Zheng et al., 2024). The effect of supply chain agility on waste reduction is one of its main advantages. Inefficient conventional linear supply chains can result in surplus inventory (Yaroson et al., 2024). Agile supply chains, on the other hand, use real-time data and sophisticated forecasting methods to maximize inventory levels and simplify processes (Shafique et al., 2024). Agile suppliers, for example, can modify their production plans in reaction to demand variations, reducing surplus and guaranteeing that food items are delivered fresh and eaten before they go bad (Kant & Adula, 2024). Because less funds are invested in unsold products, this responsiveness not only improves economic viability but also lessens environmental waste (Wu et al., 2024).

Additionally, agility makes it possible for food supply chains to more successfully adopt circular methods. Agile systems, for instance, can make it easier to recover and reuse waste materials and byproducts. Supply chains are better able to spot chances for upcycling trash into new goods or dispersing excess food to people in need when they can quickly adjust to changes (Fonsec et al., 2024). By tackling food poverty, this not only promotes social sustainability but also supports the goals of the circular economy by extending the useful life of resources (Suali et al., 2024). Additionally, an agile food supply chain encourages stakeholder engagement, which is critical to the adoption of circular processes. Agile supply chains may foster a common knowledge of sustainability objectives and difficulties by improving communication and collaboration among manufacturers, distributors, retailers, and consumers (Madanaguli et al., 2024). In addition to facilitating the creation of closed-loop systems, where waste from one process becomes input for another, this collaboration can result in creative solutions, such as cooperative recycling and composting projects. Because they allow for a more comprehensive approach to resource management, such cooperative activities are essential for accelerating the shift towards a circular economy (Akhtar et al., 2024).

H2: Food Supply Chain agility has a significant effect on Circular Economy sustainability

Generative AI, Circular Economy Sustainability, Mediation of Food Supply Chain Agility

The connection between generative AI and the sustainability of the circular economy, which is mediated by the agility of the food supply chain, demonstrates how technology may revolutionize sustainable practices (Wang et al., 2024). Advanced algorithms that can evaluate enormous information, create predictive models, and streamline procedures are referred to as "generative AI." This improves decision-making in food supply chains (Suali et al., 2024). When generative AI is included into these supply chains, it gives businesses the capabilities they need to react faster to shifting consumer preferences, market situations, and resource availability. Supply chain agility, which is crucial for reducing waste and maximizing resource use—two fundamental principles of the circular economy—is characterized by this enhanced reactivity (Panigrahi et al., 2023). Food supply chains may better match their operations with the ideas of the circular economy as they become more flexible via the use of generative AI. For example, better forecasting and inventory management are made possible by real-time data analytics, which lowers waste and overproduction (Irfan et al., 2024). This flexibility makes it easier to spot chances for material recycling or repurposing, extending the time that resources are in use. Agile supply chains can also swiftly adjust to interruptions like changes in customer demand or supplier shortages, preserving operational effectiveness while reducing environmental effects (Akhtar et al., 2024).

The relationship's mediation component implies that the advantages of generative AI for the sustainability of the circular economy come via improved supply chain agility rather than directly. To put it another way, generative AI offers the means for optimization, but the agility that results is what turns these tools into sustainable, successful practices. This mediation effect emphasizes how crucial it is to have adaptable and responsive supply chains in order to achieve the circular economy's sustainability objectives (Wu et al., 2024). In the end, this interaction between supply chain agility, generative AI, and circular economy sustainability shows a dynamic framework in which technology acts as a catalyst for long-term change (Fonsec et al., 2024). Organizations may significantly increase resource efficiency and waste reduction by utilizing generative AI to boost agility (Li et al., 2023). This will help create a more sustainable food system that adheres to circularity principles. In addition to highlighting the significance of using cutting-edge technology, this link demonstrates how crucial it is for businesses to prioritize agility as a strategic approach in their quest for sustainability (Madanaguli et al., 2024).

> H3: The relationship between generative AI and circular economy sustainability is mediated by food supply chain agility.

CONCEPTUAL FRAMEWORK

Figure 1. Conceptual framework

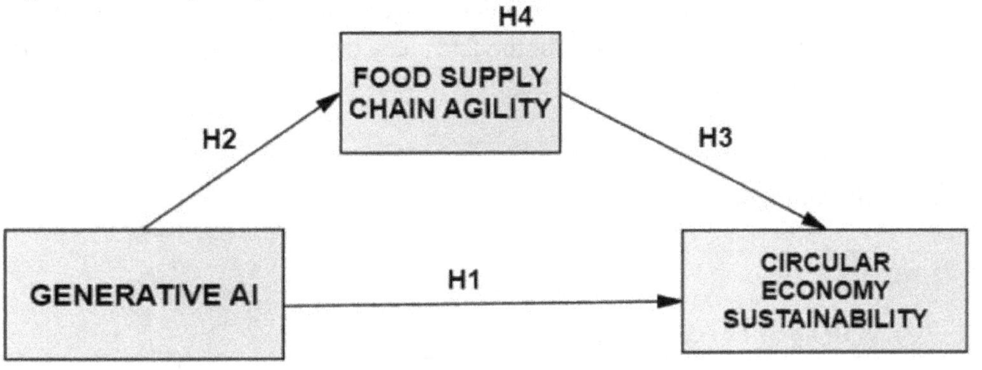

METHODOLOGY

This study employed a quantitative research approach to investigate the relationship between generative AI, food supply chain agility, and sustainability within the framework of the circular economy. The research utilized a structured survey instrument, which was developed based on a comprehensive review of the literature. The survey instrument underwent a rigorous process of scale development to ensure its validity and reliability. Initial items were generated from existing theories and empirical studies related to generative AI, food supply chain dynamics, and circular economy principles. A sample size of 390 participants was targeted for data collection, encompassing stakeholders from various segments of the food supply chain, including producers, distributors, retailers, and policymakers. This diverse sample was essential to capture a holistic view of the dynamics at play and ensure that the findings were representative of the broader industry context.

To analyze the data, the study first conducted a Kaiser-Meyer-Olkin (KMO) test to assess the adequacy of the sample for factor analysis. The KMO measure indicated a sufficient level of sampling adequacy, confirming that the data were appropriate for further analysis. Following this, Exploratory Factor Analysis (EFA) was performed to identify underlying structures among the survey items and to validate the dimensionality of the constructs related to generative AI, supply chain agility, and circular economy practices. The EFA results provided insights into the grouping of items and helped refine the measurement scales. Subsequently, Structural Equation Modeling (SEM) was applied using AMOS software to test the hypothesized relationships within the proposed model. SEM allowed for the assessment of complex relationships among multiple variables simultaneously, providing a robust framework for understanding how generative AI mediates the agility of food supply chains and contributes to sustainability outcomes. The model fit indices were evaluated to ascertain the adequacy of the model, and adjustments were made as necessary to improve the fit.

DATA ANALYSIS

Table 1. KMO and Bartlett's Test

Kaiser-Meyer-Olkin Measure of Sampling Adequacy.		**.850**
Bartlett's Test of Sphericity	Approx. Chi-Square	1419.738
	df	91
	Sig.	.000

Source: Authors

The data analysis in table 1 commenced with the application of the Kaiser-Meyer-Olkin (KMO) Measure of Sampling Adequacy and Bartlett's Test of Sphericity to evaluate the suitability of the dataset for factor analysis. The KMO value was recorded at 0.850, indicating a good level of sampling adequacy. This suggests that the variables in the dataset share a significant amount of variance, making them appropriate for further analysis. Additionally, Bartlett's Test of Sphericity yielded an approximate Chi-Square value of 1419.738 with 91 degrees of freedom (df), resulting in a significance level (Sig.) of 0.000. This significance value, being less than 0.05, allows for the rejection of the null hypothesis that the correlation matrix is an identity matrix. Consequently, these results confirm that there are significant correlations among the variables, reinforcing the appropriateness of conducting factor analysis on the dataset. Overall, both tests strongly support the conclusion that the data is suitable for exploring the underlying structures and relationships among the variables.

Table 2. Total variance explained

Component	Initial Eigenvalues			Extraction Sums of Squared Loadings			Rotation Sums of Squared Loadings		
	Total	% of Variance	Cumulative %	Total	% of Variance	Cumulative %	Total	% of Variance	Cumulative %
1	4.51	32.235	32.235	4.51	32.235	32.235	3.00	21.451	21.451
2	1.37	9.787	42.022	1.37	9.787	42.022	2.16	15.491	36.942
3	1.27	9.087	51.110	1.27	9.087	51.110	1.98	14.168	51.110
4	.949	6.782	57.892						
5	.871	6.223	64.115						
6	.773	5.519	69.634						

Extraction Method: Principal Component Analysis.
Source: Authors

Table 2 presents the results of the Total Variance Explained analysis, which was conducted using Principal Component Analysis (PCA) to identify the underlying structure of the dataset. The analysis reveals that the first component accounts for an eigenvalue of 4.51, representing 32.235% of the total variance. This initial component significantly contributes to the overall understanding of the data. The second component has an eigenvalue of 1.37, accounting for 9.787% of the variance, while the third component explains 9.087% of the variance with an eigenvalue of 1.27. Cumulatively, these three components together account for 51.110% of the total variance, indicating that they capture a substantial portion of the information within the dataset. The fourth component has an eigenvalue of 0.949, contributing 6.782% to the variance, while the fifth and sixth components account for 6.223% and 5.519%, respectively. Although these components explain additional variance, the cumulative percentage remains below the threshold typically considered sufficient for practical interpretation. The extraction method employed was Principal Component Analysis, which effectively elucidates the dimensionality of the data. Overall, these findings suggest that a relatively small number of components can sufficiently represent the variability in the dataset, allowing for a more focused analysis moving forward.

Confirmatory Principle Component Examination

The principle component arrangement of a set of guided proxies is verified using a type of statistics known as confirmatory principle component examination. Based on the underlying frame job and correlatives among observed proxies, investigations can evaluate hypotheses utilizing CFA. To evaluate the hypothesis that there is a association among the proxies that are being steered and the latent conceptions that underpin them, the inquiry employed CFA under table 3.

Table 3. Covariances

Covariance			Approximation	S.E.	C.R.	P	Hy.
Food Supply Chain agility	**<-->**	**Circular economy sustainability**	**.242**	**.021**	**8.231**	***	**H2**
Food Supply Chain agility	<-->	**Generative AI**	.303	.032	7.692	***	**H3**
Generative AI	<-->	**Circular economy sustainability**	.254	.24	7.831	***	**H1**

Source: Authors

Table 3 presents the covariance analysis among the variables: Food Supply Chain Agility, Generative AI, and Circular Economy Sustainability. The results indicate a significant positive covariance between Food Supply Chain Agility and Circular Economy Sustainability, with a covariance value of 0.242 and a critical ratio (C.R.) of 8.231, which is statistically significant at the *** level ($p < 0.001$). This finding supports Hypothesis 2 (H2), suggesting that greater agility in the food supply chain is associated with enhanced sustainability outcomes within the circular economy. Also, there is a strong positive covariance of 0.303 between Food Supply Chain Agility and Generative AI, with a C.R. of 7.692, also statistically significant at the *** level ($p < 0.001$). This result corroborates Hypothesis 3 (H3), indicating that increased agility in the food supply chain is positively influenced by the implementation of generative AI technologies. Lastly, the covariance between Generative AI and Circular Economy Sustainability is measured at 0.254, with a C.R. of 7.831, which is again statistically significant at the *** level ($p < 0.001$). This finding supports Hypothesis 1 (H1), suggesting that generative AI contributes positively to sustainability outcomes in the circular economy. Overall, the significant covariances among these variables highlight the interrelated nature of food supply chain agility, generative AI, and circular economy sustainability, emphasizing the importance of leveraging technology to enhance sustainable practices within supply chains.

Table 4. Validity concern

	CR	AVE	MSV	MaxR(H)	FSCA	GAI	CES
FSCA	0.732	0.643	0.219	0.852	**0.654**		
GAI	0.758	0.657	0.132	0.763	0.171	**0.665**	
CES	0.793	0.603	0.274	0.797	0.383	0.193	**0.684**

Note: FSCA= Food Supply Chain Agility; GAI=Generative Artificial Intelgence; CES= Circular Economy Sustainability

Source: Authors

Table 4 outlines the validity metrics for the constructs involved in the study: Food Supply Chain Agility (FSCA), Generative Artificial Intelligence (GAI), and Circular Economy Sustainability (CES). The Composite Reliability (CR) values indicate good internal consistency among the constructs, with FSCA at **0.732**, GAI at **0.758**, and CES at **0.793**. Generally, CR values above **0.70** are considered acceptable, suggesting that all constructs exhibit strong reliability. The Average Variance Extracted (AVE) values for FSCA, GAI, and CES are **0.643**, **0.657**, and **0.603**, respectively. AVE values above **0.50** indicate that the constructs explain a sufficient amount of variance in their indicators, which further supports the validity of the measurement model. The Maximum Shared Variance (MSV) values, which measure the highest correlation between any two constructs, are **0.219** for FSCA, **0.132** for GAI, and **0.274** for CES. These values suggest that the constructs are sufficiently distinct from one another, addressing potential concerns regarding discriminant validity. Finally, the Maximum Reliability (MaxR(H)) values indicate the overall reliability of the constructs, with FSCA at **0.852**, GAI at **0.763**, and CES at **0.797**. These values reinforce the robustness of the measures used in the study. Collectively, the metrics presented in Table 4 suggest that the constructs are valid and reliable, providing a strong foundation for further analysis and interpretation of the relationships among food supply chain agility, generative AI, and circular economy sustainability.

Mediating Role Examination

Figure 2. Structural Equation Model

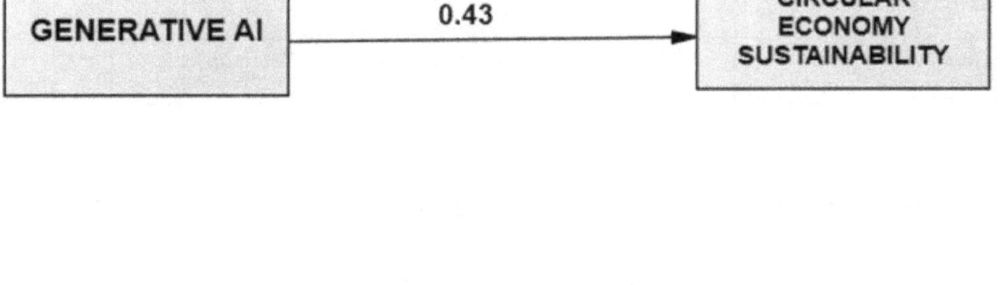

Model Fit Indices

Table 5. Indices for Model Fit

Sig.	Chi-Sq	RMR	Fitness Goodness	Fitness Confirmatory	TLI	RMSEA
0.004	1.954	.036	0.930	0.912	0.911	.028

Source: Authors

Table 5 presents the indices for model fit, which are crucial for assessing the overall adequacy of the proposed model. The significance level (Sig.) is reported at **0.004**, indicating that the model fit is statistically significant. The Chi-Square value of **1.954** suggests a good fit, as lower values typically reflect a better fit between the observed and expected data. The Root Mean Square Residual (RMR) is recorded at **0.036**, which is well below the commonly accepted threshold of 0.08, indicating that the differences between the observed and predicted values are minimal. The goodness-of-fit indices show a Fitness Goodness (GFI) of **0.930** and a Fitness Confirmatory (CFI) of **0.912**, both of which exceed the acceptable cutoff of 0.90, suggesting that the model fits the data well. Additionally, the Tucker-Lewis Index (TLI) is reported at **0.911**, again indicating a strong model fit, as values closer to 1 are preferred. The Root Mean Square Error of Approximation (RMSEA) stands at **0.028**, which is significantly below the threshold of 0.06, suggesting a close fit of the model in relation to the degrees of freedom. Overall, the indices presented in Table 5 indicate that the model demonstrates a high level of fit, supporting the validity of the relationships posited among the constructs in the study.

Table 6. Regression Examination

Relative			Approx.	S.E.	C.R.	P	Ass.
Food Supply Chain agility	**<-->**	**Circular economy sustainability**	**.864**	**.143**	**7.583**	***	**H2**
Food Supply Chain agility	<-->	**Generative AI**	.328	.131	2.853	***	H3
Generative AI	<-->	**Circular economy sustainability**	.957	.192	7.958	***	H1

Source: Authors

Table 6 summarizes the results of the regression examination, highlighting the relationships among Food Supply Chain Agility (FSCA), Generative AI (GAI), and Circular Economy Sustainability (CES). The analysis reveals a strong positive regression coefficient of **0.864** between FSCA and CES, with a critical ratio (C.R.)

of **7.583**, which is statistically significant at the *** level ($p < 0.001$). This finding supports Hypothesis 2 (H2), indicating that increased agility in the food supply chain is associated with substantial improvements in sustainability practices within the circular economy. Additionally, the regression coefficient between FSCA and GAI is **0.328**, with a C.R. of **2.853**, also significant at the *** level ($p < 0.001$). This result confirms Hypothesis 3 (H3), suggesting that enhanced agility in the food supply chain positively influences the adoption and effectiveness of generative AI technologies. Also, the relationship between GAI and CES is highlighted by a regression coefficient of **0.957** and a C.R. of **7.958**, which is statistically significant at the *** level ($p < 0.001$). This finding supports Hypothesis 1 (H1), indicating that generative AI plays a crucial role in fostering sustainability outcomes in the circular economy. Overall, the results presented in Table 6 illustrate the significant and positive interconnections among the constructs, emphasizing the importance of food supply chain agility and generative AI in promoting circular economy sustainability.

Table 7. Mediating role Effect

	Influence	**worth**	**Path Influence**
Generative AI → Circular economy sustainability	Direct Influence	.43	Direct influence stated
Generative AI → Food Supply Chain agility → Circular economy sustainability	Indirect Influence	.74*.67=.48	Indirect Influence Ensued
	Whole influence	.91	Partial mediation

Source: Authors

Table 7 outlines the mediating role of Generative AI (GAI) in the relationship between Food Supply Chain Agility (FSCA) and Circular Economy Sustainability (CES). The analysis reveals a direct influence of GAI on CES, quantified at 0.43, indicating a substantial direct effect of generative AI on sustainability outcomes. This direct influence underscores the critical role of generative AI in enhancing circular economy practices. Moreover, the table presents the indirect influence of GAI through FSCA, calculated as 0.74 (the influence of GAI on FSCA) multiplied by 0.67 (the influence of FSCA on CES), resulting in an indirect influence of 0.48. This finding emphasizes the mediating role that FSCA plays in the relationship between GAI and CES, suggesting that generative AI not only directly contributes to sustainability but also enhances it by improving supply chain agility. The total influence, combining both direct and indirect effects, amounts to 0.91, indicating that the overall impact of GAI on CES is significant. The analysis concludes with the identification of a partial mediation effect, suggesting that while generative AI contributes directly to sustainability, its influence is also significantly mediated by the agility of the food supply chain. This highlights the interconnectedness of these constructs and the importance of both agility and technology in achieving sustainability goals within the circular economy.

DISCUSSION

The study's conclusions offer important new information about how Food Supply Chain Agility (FSCA), Generative AI (GAI), and Circular Economy Sustainability (CES) interact. The large direct effect of GAI on CES, as well as the significant positive correlations and regression analyses, highlight the crucial role that FSCA plays in improving sustainability results. This implies that supply chains' adaptability greatly supports sustainable practices within the circular economy in addition to making it easier for new technologies to be adopted. Additionally, FSCA's mediating function emphasizes how crucial it is to create flexible supply chains that can swiftly adjust to changes and integrate cutting-edge technology like generative AI. The results show that although generative AI directly supports sustainability, its benefits are increased when combined with agile supply chain tactics. In order to optimize their sustainability efforts, firms must invest in both technology and agility, as shown by this dual influence. The durability of these interactions is further supported by the statistical significance of the model fit indices, which show that the suggested model successfully reflects the dynamics between the components. This research provides useful insights for companies looking to improve their supply chain procedures by illustrating the relationship between agility, technology, and sustainability. In order to position themselves to attain better sustainability in their operations, organizations should concentrate on creating agile frameworks that make use of generative AI. All things considered, this study adds to the expanding corpus of research on sustainable supply chains and emphasizes how important creativity and flexibility are to the advancement of sustainability programs.

CONCLUSION

To sum up, this study clarifies the important linkages between Circular Economy Sustainability (CES), Generative AI (GAI), and Food Supply Chain Agility (FSCA). According to the results, FSCA is an essential enabler for the successful use of generative AI, which enhances sustainability results. The idea that organizational agility is crucial for maximizing the potential of cutting-edge technology to promote sustainability in supply chains is supported by the evidence of both direct and indirect impacts. Also, the statistical analyses confirm the reliability and validity of the constructs used, guaranteeing the strength and significance of the associations found. The FSCA's partial mediation role emphasizes how important it is for companies to develop flexible supply chain procedures that can quickly adapt to market demands and environmental issues, in addition to using cutting-edge technology like generative AI. By offering a sophisticated understanding of how agility and technology interact to support sustainable behaviors within the circular economy, this study adds to the body of current work. It provides useful advice for businesses looking to improve their sustainability initiatives and promotes a strategic strategy that combines technical innovation and agility. In the end, the study highlights how crucial these components are to attaining sustainable supply chain practices, opening the door for more research in this crucial field.

MANAGERIAL IMPLICATIONS

The study's conclusions have important management ramifications for businesses looking to improve supply chain sustainability. First and foremost, managers should give priority to creating agile supply chain procedures, as indicated by the positive correlation between Food Supply Chain Agility (FSCA) and Circular Economy Sustainability (CES). Organizations may better adjust to shifting customer expectations and market situations by cultivating flexibility and responsiveness, which will result in a more sustainable operational framework.

Additionally, the significance of investing in cutting-edge technology is shown by the significant impact that Generative AI (GAI) has on sustainable results. To increase resource efficiency, make better decisions, and streamline operations, managers should think about incorporating generative AI into their supply chain procedures. In addition to advancing sustainable objectives, this integration puts businesses in a competitive position in a market that is changing quickly. Also, the recognized mediating function of FSCA suggests that firms must make sure their supply chains are flexible enough to take advantage of these advancements. Adopting new technology alone is not enough. This necessitates training and development programs that provide staff members the abilities they need to function in a changing workplace.

Lastly, managers may approach strategy formation holistically by comprehending how agility, technology, and sustainability are intertwined. Organizations may achieve long-term success in a market that is becoming more environmentally sensitive by coordinating their operational goals with sustainability objectives thanks to this integrated viewpoint. All things considered, accepting these consequences will improve organizational performance and favorably impact industry-wide sustainability initiatives.

PRACTICAL IMPLICATIONS

The study's conclusions have a number of useful ramifications for businesses looking to improve the sustainability of their supply chains. First off, companies should adopt strategies that encourage adaptability and responsiveness in their supply chains, since Food Supply Chain Agility (FSCA) plays a crucial role in promoting Circular Economy Sustainability (CES). Adopting agile approaches, like lean management practices, which enable faster response to shifting customer preferences and market dynamics, may be one way to achieve this. Also, the results highlight how important it is to include Generative AI (GAI) into supply chain processes. AI technology implementation should be investigated by organizations in order to increase demand forecasting, inventory management, and overall operational efficiency. Businesses may decrease waste and make better decisions that support sustainability objectives by utilizing generative AI. The survey also emphasizes the value of training and development initiatives that provide staff members the know-how to handle cutting-edge technologies and agile methodologies. This human capital investment is crucial for creating an innovative and flexible culture inside the company and making sure that employees are equipped to use new tools efficiently. Finally, companies have to think about creating interdisciplinary groups that can work together on sustainability projects. Businesses may develop integrated strategies that match supply chain agility with technology improvements and environmental aims by dismantling organizational silos and promoting cooperation. By doing these doable actions, businesses will be able to improve their sustainability initiatives and get a competitive edge in a market that is becoming more environmentally sensitive.

THEORETICAL IMPLICATIONS

By clarifying the intricate connections between Food Supply Chain Agility (FSCA), Generative AI (GAI), and Circular Economy Sustainability (CES), this work adds to the theoretical discussion on supply chain management. The study adds to current ideas on agility and sustainability by providing empirical validation of these relationships and demonstrating how technological advancements might work as stimulants for sustainable supply chain practices. The discovery that FSCA acts as a mediator between GAI and CES provides a more sophisticated understanding of the ways in which agility affects sustainability results. According to these results, future studies should examine how agility and technology interact in many settings. This might result in the creation of new theoretical frameworks that more thoroughly include these concepts. The study also emphasizes how crucial a systems viewpoint is to supply chain studies. It invites academics to think about the larger ecosystem in which supply chains function by demonstrating how several components combine to drive sustainability. This viewpoint might encourage further research into additional variables that can affect the connections between agility, technology, and sustainability, expanding the corpus of knowledge in this area.

RECOMMENDATIONS

The study's conclusions allow for the formulation of a number of suggestions for businesses looking to improve the sustainability of their supply chains through technological innovation and agility. First and foremost, businesses must make an investment in creating agile supply chain procedures. Adopting adaptable procedures and frameworks that enable quick adjustments to shifting customer preferences and market situations is part of this. By putting techniques like Lean or Agile supply chain frameworks into practice, businesses may improve their ability to adapt to disturbances, which promotes sustainability and resilience. Second, companies have to give integrating generative AI into their supply chain processes top priority. This may entail carrying out pilot studies to assess how well AI technologies work in fields like waste reduction, inventory control, and demand forecasting. Businesses may improve operational efficiency and promote sustainable practices without overburdening their current procedures by progressively integrating AI technology. Additionally, it is advised that businesses set up continuous training and development initiatives to give staff members the know-how to successfully use new technology. In order to ensure that the workforce is ready to adopt agile processes and cutting-edge technology, which will eventually result in better performance and sustainability results, this investment in human capital is crucial. Lastly, companies should establish cross-functional teams devoted to sustainability projects in order to promote a collaborative culture. Businesses may use a variety of viewpoints and areas of expertise to create creative solutions that balance supply chain agility with sustainability goals by promoting multidisciplinary cooperation. In addition to improving organizational performance, this all-encompassing strategy will support the industry's larger environmental objectives.

FUTURE DIRECTIONS

The study's conclusions pave the way for a number of future investigations into sustainability and supply chain management. Examining the long-term effects of incorporating generative AI into industries other than food supply chains is one exciting avenue. Examining how various businesses use and modify these technologies can reveal important information about best practices and difficulties that arise in various settings. Also, future research might look at how corporate culture affects the adoption of technical advancements and agile techniques. Organizations may be able to modify their tactics to create an atmosphere that is supportive of sustainability and agility by having a better understanding of how cultural variables affect these dynamics. Examining how external variables, such market forces or regulatory frameworks, affect the connection between Food Supply Chain Agility (FSCA) and Circular Economy Sustainability (CES) is another crucial subject. Examining the ways in which these outside factors impact internal procedures may offer a more thorough comprehension of the sustainable environment. Additionally, longitudinal research might be used to evaluate how the link between FSCA, GAI, and CES changes over time. As businesses advance in their sustainability journeys, this would enable academics to grasp how these structures interact and capture their dynamic character. Lastly, multidisciplinary study that integrates knowledge from information technology, environmental science, and supply chain management may be possible. These partnerships may result in novel models and frameworks that tackle the challenges of sustainable supply chains in a world that is changing quickly. All things considered, these potential paths can greatly advance our understanding of and proficiency with sustainable supply chain management.

REFERENCES

Adula, M., Kant, S., & Geda, H. D. (2025). Innovation Management Landscape for Digital Economy Transformation Through Innovation Resilience in Ethiopia. In *Innovation Management for a Resilient Digital Economy* (pp. 177–204). IGI Global Scientific Publishing.

Akhtar, P., Ghouri, A. M., Ashraf, A., Lim, J. J., Khan, N. R., & Ma, S. (2024). Smart product platforming powered by AI and Generative AI: Personalization for the circular economy. *International Journal of Production Economics*, *273*, 109283. DOI: 10.1016/j.ijpe.2024.109283

Irfan, M., Sulehri, N. A., & Manickiam, N. (2024). Digital threads in turbulent times: Unraveling technostress and cleaner production in the food industry. *Frontiers in Robotics and AI*, *10*, 1293904. DOI: 10.3389/frobt.2023.1293904 PMID: 38273894

Kant, S., & Adula, M. (2024). AI Learning and Work Attitude Mediation Between Reward and Organizational Support in Ethiopia. In *Generative AI for Transformational Management* (pp. 109–136). IGI Global. DOI: 10.4018/979-8-3693-5578-7.ch005

Kant, S., & Adula, M. (2024). Mediated by AI-Based Generative Re-Enforcement Learning and Work Attitude: Are Intrinsic Rewards Transforming Employee Perceived Organizational Support? In *Generative AI for Transformational Management* (pp. 83-108). IGI Global.

Kant, S., Niguse, T., & Adula, M. (2025). Family Business Entrepreneurial Orientation Effect on Improving Entrepreneurial Processes by Mediation of Open Innovation in the Horn of Africa. In *AI, Corporate Social Responsibility, and Marketing in Modern Organizations* (pp. 1-30). IGI Global Scientific Publishing.

Li, N., Liu, D., & Boadu, F. (2023). The impact of digital supply chain capabilities on enterprise sustainable competitive performance: An ambidextrous view. *Industrial Management & Data Systems*, *123*(6), 1670–1689. DOI: 10.1108/IMDS-11-2022-0699

Madanaguli, A., Sjödin, D., Parida, V., & Mikalef, P. (2024). Artificial intelligence capabilities for circular business models: Research synthesis and future agenda. *Technological Forecasting and Social Change*, *200*, 123189. DOI: 10.1016/j.techfore.2023.123189

Nayal, K., Raut, R. D., Kumar, M., Paul, S. K., & Narkhede, B. E. (2024). Role of Artificial Intelligence Capability in the Interrelation Between Manufacturing Strategies and Operational Resilience. *Global Journal of Flexible Systems Managment*, *25*(1), 137–162. DOI: 10.1007/s40171-023-00367-8

Panigrahi, R. R., Shrivastava, A. K., Qureshi, K. M., Mewada, B. G., Alghamdi, S. Y., Almakayeel, N., Almuflih, A. S., & Qureshi, M. R. N. (2023). AI Chatbot adoption in SMEs for sustainable manufacturing supply chain performance: A mediational research in an emerging country. *Sustainability (Basel), 15*(18), 13743. DOI: 10.3390/su151813743

Shafique, M. N., Adeel, U., & Rashid, A. (2024). The Synergy Between Industry 5.0 and Circular Economy for Sustainable Performance in the Chinese Manufacturing Industry. *Sustainability (Basel), 16*(22), 9952. DOI: 10.3390/su16229952

Suali, A. S., Srai, J. S., & Tsolakis, N. (2024). The role of digital platforms in e-commerce food supply chain resilience under exogenous disruptions. *Supply Chain Management, 29*(3), 573–601. DOI: 10.1108/SCM-02-2023-0064

Wang, S., Ghadge, A., & Aktas, E. (2024). Digital transformation in food supply chains: An implementation framework. *Supply Chain Management, 29*(2), 328–350. DOI: 10.1108/SCM-09-2023-0463

Wu, H., Liu, J., & Liang, B. (2024). AI-Driven Supply Chain Transformation in Industry 5.0: Enhancing Resilience and Sustainability. *Journal of the Knowledge Economy, 16*(1), 1–43. DOI: 10.1007/s13132-024-01999-6

Yaroson, E. V., Chowdhury, S., Mangla, S. K., & Dey, P. K. (2024). Unearthing the interplay between organisational resources, knowledge and industry 4.0 analytical decision support tools to achieve sustainability and supply chain wellbeing. *Annals of Operations Research, 342*(2), 1–48. DOI: 10.1007/s10479-024-05845-5

Zhang, Q., Ullah, A., Ashraf, S., & Abdullah, M. (2024). Synergistic Impact of Internet of Things and Big-Data-Driven Supply Chain on Sustainable Firm Performance. *Sustainability (Basel), 16*(13), 5717. DOI: 10.3390/su16135717

Zheng, G., Haq, M. Z. U., Huo, B., Zhang, Y., & Yue, X. (2024). Leveraging intellectual capital for building a supply chain circular economy system: A knowledge-based view. *International Journal of Production Economics, 272*, 109225. DOI: 10.1016/j.ijpe.2024.109225

KEY TERMS AND DEFINITIONS

Circular Economy Sustainability: An economic model aimed at minimizing waste and making the most of resources.

Food Supply Chain Agility: The ability of a food supply chain to rapidly adapt to changes in demand, supply, and market conditions while maintaining efficiency and effectiveness.

Generative AI: A subset of artificial intelligence that uses algorithms to generate new content, solutions, or predictions based on existing data.

Chapter 4
Enhancing Traceability and Transparency in Food Supply Chains Through Generative AI

Vishwajit K. Barbudhe
https://orcid.org/0000-0002-2200-4980
Sandip Intitute of Technology and Research Center, Nashik, India

Shraddha N. Zanjat
https://orcid.org/0009-0009-6379-3996
SOET, Sandip University, India

Bhavana S. Karmore
G.H. Raisoni University, Amaravati, India

ABSTRACT

This raises the need for traceability while ensuring its responsible use, a daunting task. Emerging technologies like blockchain and AI, including GPT-2, offer potential solutions. GPT-2, with its capabilities in generation synthesis, machine learning, and document automation, addresses practical challenges such as synthetic data generation, product simulation, and document processing. By applying these technologies, traceability and visibility in food supply chains can be significantly improved. Virtual product models can replicate physical characteristics and provenance, enhancing food safety, sustainability, and consumer confidence. GPT-2's ability to automate documents and generate realistic simulations strengthens supply chain transparency, ensuring secure and sustainable operations while building trust among consumers.

DOI: 10.4018/979-8-3693-9856-2.ch004

INTRODUCTION

The modern food supply chain deals with expanding issues relating to traceability and transparency. It is fueled by swelling consumer expectations for comprehensive information about their meals' origin, negotiating, handling, and ensuring sustenance safety and quality. Blockchain engineering has emerged as a promising solution, enabling protected and tamper-evident record-keeping across the supply chain. However, implementing blockchain-centered traceability structures has encountered numerous barriers, such as necessitating seamless integration with prevailing infrastructure and the trouble of engaging all stakeholders in embracing this groundbreaking technology. Combining machine learning modeling can boost traceability and transparency by automatizing data aggregation, review, and reporting across the complete food supply chain to address these hurdles. Deep generative models can be leveraged to fabricate in-depth digital recreations of goods, chronicling their journey from farm to dining room table and furnishing real-time insights to all engaged stakeholders. These digital duplicates can seize comprehensive specifics regarding the product's source, processing steps, handling procedures, and any quality-related happenings throughout the supply chain. By generating these digital replicas, deep learning models can offer a detailed and transparent record of the product's life cycle, empowering stakeholders to make informed choices and ensure greater accountability. Additionally, these AI systems can generate exhaustive reports on supply chain operations, distinguish potential issues, and propose improvements, ultimately enhancing overall traceability and transparency for customers, producers, and regulators.

Integrating generative AI and blockchain tracking frameworks can significantly amplify transparency frameworks within food networks. By automating the selection and review of data, these AI designs can help overcome challenges associated with manual data entry work and ensure the precision and dependability of tracking information. Moreover, having the capability to fabricate comprehensive digital representations of goods and generate exhaustive reports can furnish stakeholders with unprecedented visibility and insights into the entire supply chain operation, allowing them to make more informed decisions and take preemptive steps to address potential issues.

While blockchain and AI each offer benefits to food supply chain oversight independently, combining their strengths may generate a robust solution far greater than the sum of its parts. By collecting data across the distributed network and analyzing it through automated and creative means, faults could be predicted and nipped in the bud. At the same time, a complete and verifiable record ensures accountability. Each participant, whether producer, shipper, grocer, or consumer, benefits from access to real-time updates and a digital replica that tells the whole story from seed to table. Through modeling and simulation, the generative aspects of AI help pinpoint areas for improvement before issues emerge, empowering proactive problem-solving. With a more transparent flow chronicled in code, trust, and responsibility are built into the framework. Ultimately, this confluence of technologies may engineer integrity into the process, yielding a resilient, responsive system that surpasses current and future safety, quality, and reassurance needs.

Figure 1. Food supply chain

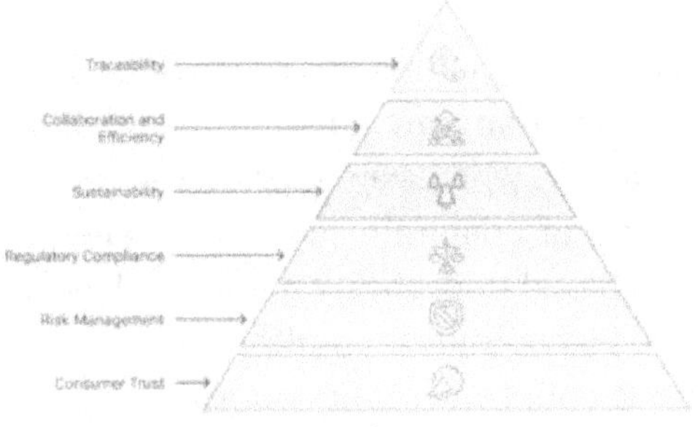

Transparency throughout supply chains is crucial in cultivating consumer assurance, mitigating hazards, ensuring adherence to guidelines, advancing sustainability goals, enabling collaboration between parties, and providing traceability. Today's customers are increasingly interested in where goods originate and want to comprehend if products are ethically sourced. Transparency between brands and patrons establishes trust, particularly for corporations openly imparting intricate supply chain details. With visibility into each phase, enterprises can proactively handle potential issues to safeguard against vulnerabilities and protect all stakeholders. Many industries face regulations concerning sourcing and safety that transparency allows to demonstrate compliance with, circumvent penalties, and boost reputation. This holds significant relevance for food supply chain openness. As ecological issues rise in awareness, all interested parties and customers demand eco-friendly practices. Transparency empowers illustrating such efforts and monitoring the impact on a long-term basis. When all parties share the same data through transparent and open communication, it facilitates teamwork, potentially delivering ameliorated effectiveness, decreased expenses, and optimized decision-making using supply chain transparency software solutions. Regarding product recalls or safety problems, openness permits quickly pinpointing impacted products to protect public health. This traceability is crucial to protect clients and maintain brand integrity, as seen through initiatives like the SBA programmatic supply chain transparency study, which aims to foster cooperation between all entities for more sustainable outcomes.

The potential use of blockchain technologies in food supply chains has been extensively analyzed in recent years, as it offers several advantages over conventional traceability systems when fully implemented. Blockchain's immutable ledger, distributed architecture, and smart contract functionality can enhance data transparency, reduce the risk of tampering, and facilitate seamless information sharing among all supply chain participants when all parties adopt new technologies. Existing studies have indicated that blockchain-based traceability may help strengthen food safety, accountability, and consumer trust by creating an incorruptible record detailing the journey of raw materials from their origin to the final delivery of products to end users, preventing misleading information and protecting public health. Specifically, blockchain's decentralized and transparent nature enables tracing the complete path of a product in an open manner, which can improve traceability by permitting more efficient tracking and recall of contaminated or substandard goods. Nonetheless, implementing blockchain-based traceability platforms has faced several obstacles, including integrating new technologies seamlessly into present infrastructures and requiring all stakeholders to adopt emerging solutions. To address such hurdles, integrating generative artificial intelligence models can notably enhance the capabilities of blockchain-based traceability systems when expertly applied. Generative AI can automate data compilation and examination processes through machine learning, ensuring comprehensive and accurate record-keeping across the continually evolving supply chain. These AI systems can acquire input from various sources, such as IoT sensors, enterprise platforms, and manual entries, and they can be assimilated smoothly and securely into the blockchain-based traceability platform through testing and oversight. Moreover, generative AI can analyze accumulated data to pinpoint patterns, detect anomalies, and create thorough reports, supplying stakeholders with valuable insights to make informed decisions and undertake corrective actions when necessary to better serve all supply chain members and customers.

INTEGRATING GENERATIVE AI WITH BLOCKCHAIN-BASED TRACEABILITY

Integrating generative AI with blockchain traceability could offer substantial advantages for complex food supply networks. First, AI may automatically accumulate and examine data, reducing manual entry errors and ensuring the accuracy and reliability of traceability records. This AI can produce comprehensive digital representations of products, capturing granular details about origins, handling, manufacturing, and any anomalies encountered along the way. These virtual models provide a transparent and credible document of a product's journey, empowering stakeholders to make informed decisions and guarantee accountability.

Additionally, AI has the potential to analyze the accumulated traceability records on the blockchain and generate in-depth reports on supply chain operations. These reports can pinpoint inefficiencies like bottlenecks, quality deviations, or compliance violations, proposing corrective actions. Increased visibility and data-driven insights could help stakeholders optimize their supply chain processes, strengthen food safety, and ensure adherence to regulations. Furthermore, integrating AI enables the development of predictive models that can foresee potential disruptions or anomalies in the supply chain. Leveraging complete traceability data and AI pattern recognition capabilities, stakeholders can proactively detect and mitigate risks, ensuring a resilient and efficient food system.

Lastly, blockchain technology and artificial intelligence can bring more openness and accountability across the entire food supply network. The automated collection and examination of data and the development of comprehensive digital profiles of products can provide all stakeholders, including consumers, manufacturers, distributors, and regulators, an unprecedented view into how the complete supply chain functions. This empowers informed choices, identification of possible issues, and efficient corrections, ultimately contributing to more sustainable, safe, and reliable food systems management and developing trust from customers and government agencies. When blockchain traceability is combined with artificial intelligence, it shows promise for enhancing visibility, responsibility, and the ability to track food throughout the supply chain. This integrated solution can address key challenges and advance more sustainable, dependable, and principled food distribution ecosystems by automated gathering of information, virtual models of products, and holistic insights.

Vision Processing and Deep Learning in Business Applications

Vision processing and deep learning have revolutionized various business domains by enabling machines to analyze, interpret, and act on visual data with human-like precision, after being trained on enormous datasets, Convolutional neural networks have become adept at even the most complex tasks, such as image recognition, object detection, facial recognition, and semantic segmentation.

Within the retail sector, computer vision helps dynamically track precise inventory levels, pinpoint when items on shelves are out of supply and provides realistic virtual try-on experiences for online shoppers browsing selections remotely. Meanwhile, neural networks ensure product quality in manufacturing by reliably recognizing defects during production runs with computerized visual defect detection. The healthcare industry has witnessed tremendous benefits, with avant-garde diagnostic instrumentation empowered by deep learning techniques assisting in cancer screening and deciphering medical scans.

Deep learning plays a fundamental role by allowing neural networks to identify even the most subtle patterns within gigantic volumes of visual data. Autonomous vehicles depend on these capabilities to travel roads safely by perceiving traffic signals and signs, other motor vehicles, pedestrians, and more. Security is an additional domain that utilizes computer vision, with facial recognition for access management and observation.

By automating repetitive visual tasks autonomously at a large scale and improving decision-making, the synergy between vision processing and deep learning enables businesses to pioneer and enhance operational efficiency, productivity, and customer satisfaction. These technologies decrease human mistakes and augment human capabilities across many industries.

Figure 2. AI in existing business applications

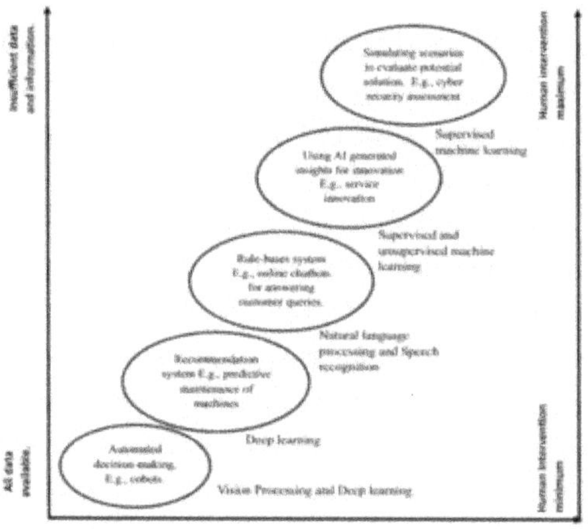

LIMITATIONS OF BLOCKCHAIN-BASED TRACEABILITY

While blockchain-based traceability systems have demonstrated promising outcomes, they also face several critical bottlenecks that necessitate addressing. One significant challenge is achieving comprehensive data inclusion and integration across the entire supply chain, a difficult task owing to modern food systems' inherent intricacy and stakeholders' varying technological capabilities. This complexity can specifically complicate ensuring all pertinent data points, from raw materials to final deliveries, are precisely captured and seamlessly incorporated into the blockchain-based traceability system.

Adopting blockchain technology demands substantial investment and collaboration among all supply chain participants, which is a considerable barrier as entities may hesitate to commit to the required resources or harmonize diverse interests and priorities. This lack of widespread adoption and integration can undermine the efficacy of the blockchain-based traceability solution.

Requiring manual data input also requires people to enter the data. This introduces human error as well as a contradiction: the quality assurance of automated systems is not only compromised. The dependability and accuracy of the trace information are also affected, which reduces system productivity. Incorrect or incomplete data can compromise the reliability of an overall blockchain-based traceability system. This affects its value and credibility in all areas, from producers who provide the products to regulators who permit them into foreign markets, even consumers like you living on entirely inherited goods--via their direct descendants. As a result, it is not yet clear whether this change will benefit any of these groups.

Moreover, the scalability and performance of blockchain-based traceability systems in fast-changing, high-throughput food supply chains may be a problem. Blockchain technology sets limitations on its application because it cannot satisfy such criteria. The speed of transaction processing, the volume it can handle, and storage capacity constrain the usefulness of computer terminals for these solutions based on blockchains.

To adequately address the restrictions, it is crucial to investigate the incorporation of generative AI as a complementary innovation to improve the capacities and viability of blockchain-based followability frameworks.

Advanced AI models can play a pivotal part in defeating the key restrictions of blockchain-based followability frameworks. These progressed AI models can robotize the information assortment and joining forms, ensuring comprehensive and precise record-keeping crosswise over the sustenance inventory network. By creating point-by-point computerized twins of items, generative AI can give partners unprecedented straightforwardness and bits of knowledge into developing and dealing with merchandise, empowering them to settle on all the more educated choices and advance more prominent liability.

Then again, generative AI can dissect the followability information gathered through the blockchain framework and create finished reports on inventory network activities. These reports can distinguish potential issues, for example, bottlenecks, quality deviations, or consistency violations, and propose remedial activities. This upgraded straightforwardness and information-driven knowledge can assist partners with enhancing their stock network forms, upgrading sustenance well-being, and guaranteeing consistency with administrative necessities.

Additionally, generative AI can empower the advancement of anticipatory models that foresee potential disturbances or quality issues in the stock network. By breaking down chronicled information and recognizing examples, these AI models can give early cautioning signs, permitting stock chain members to take proactive estimates to mitigate dangers and guarantee the conveyance of protected and top-notch items to buyers.

To satisfactorily address the restrictions of blockchain-based followability frameworks, joining generative AI models can offer tremendous advantages. Generative AI can make point-by-point computerized twins of items, allowing for their journey through the stock chain and the accommodation of ongoing bits of knowledge to partners.

These AI-enabled systems can independently gather and correlate data from various sources, including IoT sensors, corporate platforms, and manual contributions, ensuring comprehensive and exact record-keeping. Moreover, creative AI can research this data to discern patterns, identify abnormalities, and generate expansive reports on supply chain operations, empowering stakeholders to make informed judgments and undertake corrective steps.

Additionally, imaginative AI can be leveraged to mechanically generate product-specific information, such as origin, preparation details, and quality specifications, which can be seamlessly integrated with the blockchain-based traceability system. This integration can improve transparency and trust, as consumers and regulatory bodies can access verified and meticulous information about the provenance and handling of food products.

By combining the immutability and transparency of blockchain technology with the data-driven insights and automation abilities of imaginative AI, food supply chain stakeholders can accomplish a more robust, traceable, and responsive ecosystem. This enhanced traceability and transparency can lead to improved food safety, reduced waste, and increased consumer confidence, ultimately contributing to the overall sustainability and resilience of the global food supply network. (Prashar et al., 2020) (Perera et al., 2020) (Spitalleri et al., 2023) (Kshetri et al., 2023)

CHALLENGES AND CONSIDERATIONS

While the fusion of creative AI carries immense potential for boosting trackability and openness in sustenance inventory chains, specific central difficulties and contemplations should be tended to.

Guaranteeing the accuracy and dependability of the AI models, just as the underlying information, is pivotal for keeping up the uprightness and trustworthiness of the blockchain-based traceability framework. Imprecise or incomplete data can undermine the arrangement's viability and lessen its incentive for all partners. Strong information approval and quality control measures should be actualized to guarantee the accuracy and unwavering quality of the information utilized by the AI models. Further study is expected to completely comprehend the ramifications of information issues and discover potential arrangements.

Tending to information protection and security worries is essential, as joining AI models may include the assortment and preparation of delicate data identified with item roots, dealing with cycles, and different stock chain subtleties. Strong information administration structures and security estimates, for example, encryption, get-to controls, and information minimization, ought to be actualized to ensure stakeholder security and forestall approved access to or misuse of this data. Extra exploration is expected to create further and test data security advances, particularly in creative AI applications.

Adjusting the interests and neediness of various stock chain members, such as makers, appropriators, and retailers, is fundamental for wide acknowledgment and coordination of the blockchain-improved traceability framework. Overcoming obstruction to change and ensuring buy-in from all partners through successful correspondence, joint effort, and reward instruments can be a noteworthy test. Further conversation is expected to think of reasonable arrangements that address the assorted requirements of stock chain members.

Guaranteeing the extensibility and execution of the coordinated blockchain-AI answer for overseeing the vast volumes and fast-paced nature of cutting-edge sustenance stock chains is fundamental for its practical application and genuine world effect. The inherent impediments of blockchain innovation, such as exchange handling speeds, capacity limits, and the computational requests of creative AI models, must be cautiously considered and tended to through design structure, system streamlining, and distributed computing conveyance strategies. Extra testing is expected to demonstrate the ability to make potential arrangements.

By tending to these difficulties and contemplations, associations can viably utilize the intensity of creative AI to upgrade the traceability and straightforwardness of their sustenance stock chains, at last adding to improved sustenance wellbeing, quality, and supportability. This can include exploiting AI-powered prescient examinations to anticipate potential disruptions or quality issues and computerizing the era of point-by-point item data that can be effortlessly joined with the blockchain-based traceability framework. Moreover, joining AI can empower the creation of far-reaching reports on stock chain tasks, helping partners make educated choices and take remedial actions.

LEVERAGING GENERATIVE AI FOR TRACEABILITY AND TRANSPARENCY

Integrating generative AI with blockchain-based traceability systems can offer several key benefits for enhancing transparency and traceability in food supply chains. Automated data integration and digitization would allow generative AI models to automatically capture and integrate information from diverse sources, including Internet of Things sensors, enterprise systems, and manual inputs entered by supply chain participants, ensuring comprehensive and accurate record-keeping within the distributed ledger. By automating the generation of granular product-specific details ranging from origin and processing methods to quality specifications, generative AI can help provide stakeholders throughout the supply network, such as consumers and regulatory agencies, with verified and nuanced information regarding the provenance and handling of foodstuffs. Leveraging its ability to analyze the trove of data recorded by the blockchain-based traceability system, generative AI is poised to identify patterns within this bounty of supply chain insights, detect anomalies that may portend problems, and generate far-reaching reports illuminating operations throughout the end-to-end food network, facilitating predictive analytics and risk mitigation.

Figure 3. Implementing blockchain for supply chain transparency

A nuanced examination of prevailing logistics methods is necessary: comprehensively scrutinize established provision pathways to pinpoint where distributed ledger technology could streamline operations. Search for inefficiencies, opacity, or difficulties monitoring goods transport. Choose prudently: opt for a distributed ledger platform compatible with organizational demands. Consider computing potential, security features, and simplicity integrating with legacy systems.

Include interested gatherings: involve all pertinent stakeholders, from suppliers to manufacturers to distributors, at the inception of rollout planning. Their insights are priceless when designing a system that serves everyone. Define a singular aim: establish a precise, targeted usage for distributed ledger within supply logistics. This could involve substantiating origin, inventory management, or compliance tracking.

Launch with precaution: conduct a measured trial run to experiment with solutions at a smaller scale before an enterprise-wide rollout. This facilitates changes and enhancements beforehand. Educate personnel: provide comprehensive preparation for employees on the new distributed ledger system functionalities. Ensuring comfort is crucial for smooth implementation success.

Monitor persistently: consistently track performance and solicit user feedback—apply learnings to enact improvements or adjustments.

Expand judiciously: cautiously scale the implementation across the entire supply chain network when tests prove fruitful. Confirm preparedness and that intended operations are fulfilled.

LITERATURE REVIEW

There is a lot of literature out there that shows blockchain technology can enhance traceability and transparency in a food supply chain. All suppliers' data silos can be brought together under a blockchain ledger so that consumers can depend upon their source of information to cope with the latest product tracking problem. Blockchain technology has precisely this feature because of its decentralization, immutability, and transparency. For example, all these events, product attributes, and quality specifications—as long as they are committed to the blockchain—provide a tamper-proof and certifiable view after work is accomplished from farm to table,

Some other materials stressed that new technologies' integrated solution needs, such as IoT sensors and RFID, increase the traceability and transparency of food supply chains. With these technologies incorporated into the blockchain-based traceability system, one can easily position real-time information about what environment your product is in or how it's been processed for sale. Meanwhile

The literature also highlights the potential benefits of blockchain-based traceability systems in terms of enhanced food safety, consumer trust, and regulatory compliance. It could be possible for such systems to offer a transparent and verifiable account of a product to consumers--which, in a time of food safety incidents, can show where something went wrong (contamination or quality issues)--to make early recalls that much more quickly authorities can respond to and cut off outbreaks.

Quality consumer brand apps can also be built in this way. By providing detailed information on the app brand's source, processing, and means, developers can help consumers make better decisions for health reasons. It brings closer contact with the brands they love; moreover.

The current literature has many articles on the design and performance evaluation of blockchain-based traceability systems, but not integrated generative AI to improve their resilience. In addition to its data-analysis insights and real-time automation capabilities, generative AI will unlock even deeper insights for food supply chain stakeholders looking to raise their traceability, transparency, and overall sustainable development practices. By accessing information from multiple sources (including IoT sensors and enterprise systems), generative AI can automate the data captured and integrate it into the system to ensure comprehensive, accurate, and immutable recording in a blockchain-based traceability system. Furthermore, generative AI can also automatically generate detailed product-specific information about the origin, processing details, and quality specifications to all stakeholders in a verifiable and granular way--in particular end-consumers and (government) regulatory authorities from which there is no room for doubt.

METHODOLOGY

The key steps of the proposed research methodology are as follows:

Review on blockchain-based food traceability system and cross-comparison with past studies that have involved my thesis and other emerging technologies combining IoT and RFID to improve transparency and traceability within whole supply chains for fishery products from sea to table. The current study looks at generative AI from various perspectives: potential uses linked directly to food supply chain traceability by natural language processing, generative photography, and camera detect storytelling; this paper also discusses predictive analytics for transparency and traceability in the production process.

This section will give an overview of previous research work that has been done in the field of blockchain-based traceability systems or applied some new technologies (such as IoT or RFID) to improve a food supply chain's transparency and traceability. It'll look at centralization in blockchain technology, as well as the fact that transactions are irrevocable in any way once they enter public consensus after being recorded by validators on chains of individual nodes everywhere around the world, what it can do regarding immutability and transparency to make sure consumers are getting what their package says since blockchains do records for all transactions. Breaking from the existing academic literature, our review will also analyze what blockchain technology is and how it works with its core constituents:

decentralization, irrevocability, and transparency. Other sections can describe how a time stamp offers some means to verify data while avoiding counterfeits (Section D)--later we will demonstrate principles involved in hardware development using an FPGA for system-on-a-chip design (Section E). How this product's place and condition data is collected? What data enters the database architecture? When does blockchain tracing begin

"For now it remains unclear whether research will allow a secure mechanism that balances the interests of consumers, producers and merchants; could help control prices; or will merely serve to make middlemen (rather than providers or carriers) further prosperous," a source familiar with the situation said. Today's consumer can hardly tell whether their vegetables are safe and pesticide-free. Overcoming such problems will mean relying upon some method other than in-line activation of RFID links from packaging to network.

This methodology devotes a step to identifying potential uses of generative AI for traceability and transparency in the food supply chain. Examples follow: Natural language processing can automatically generate detailed content for a given product, such as the origin, process, and quality specification of all associated materials. Predictive analytics provides several net , where-to-buy, and distribution systems expertise (ordering system logistics, retailing, etc.) in new consumer products.

It will also assess how computer vision and predictive analytics could combine with robots to handle flawed data (for example, seize input from ports that isn't strictly legal) until they think of throwing it over– albeit in contravention procedure which should be used or at least respected. It then briefly touches on expanding possibilities for research and industrial development due to these achievements, which would both give lifeblood into supply chains based on blockchain technology to provide a solid foundation for the historical building of massive databases.

It will also explore the implications of running generative AI over a blockchain-based traceability system, such as improved data integrity, increased consumer trust, and enhanced regulatory compliance.

Level 1 → Aspects of Adoption and Societal Impact

This includes a range from cost efficiency and scalability to improved operational outcomes. Problems such as biased algorithms, erroneous outputs, and misaligned goals could hinder mass adoption. Building trust in AI systems and ensuring security, privacy, and impartiality in applications is crucial to addressing societal concerns through robust policy.

Level 2: Transformation & Synergy

AI is revolutionizing practices via reorganized workflows and stimulating innovations, enabling technologies like IoT, predictive analysis, and automation to help fine-tune performance quickly. Stronger bonds and better communication lead to more integrated and responsive ecosystems through AI-enabled cooperation and stakeholder coordination.

Level 3: Resilience + Decision-Making

Cyber resilience is vital, especially now, as AI solutions evolve, extending protection, privacy, and confidence across networks. AI's ability to optimize risk management, resource allocation, and efficiency is key in making decisions. Furthermore, AI reshapes roles and responsibilities, requiring retraining and skill supplementation to prepare workforces. That affects dynamics, productivity, and worker well-being.

Level 4: Productivity and Sustainability

AI significantly boosts productivity by increasing operational efficiency and optimizing specific metrics. It also advances sustainability with circular and regenerative models aligning operations with environmental and carbon neutrality goals. Such innovations will help build sustainable and resilient supply chain ecosystems.

Level 5: Theories of Understanding

The theoretical and applied frameworks are needed in logistics and supply chain management to deal with the complexities of AI. Description jumps into the external condition to 'fit' into established theories while working the opposite way: There may be situations where the theories expand to include objects/descriptions in the real world. Frameworks build in the senses of the external world; they complement existing theories or produce new ones that help understanding. They offer structured techniques that balance scholarship and application and keep AI innovations realistic and sustainable.

Figure 4. Research framework

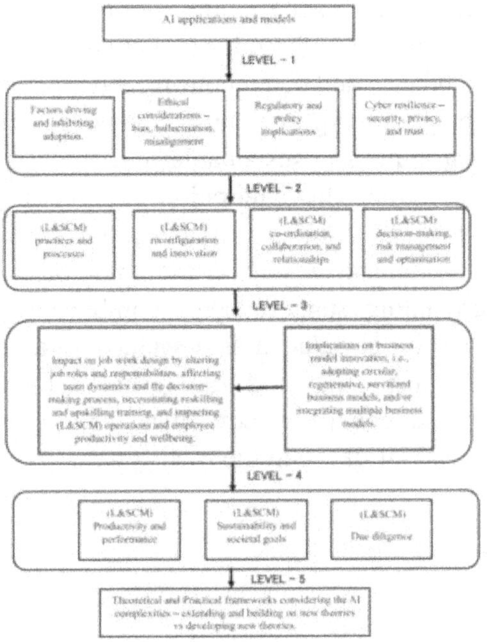

FINDINGS

Several of the key findings from the research on the potential for integration of generative AI with blockchain-based traceability systems to enhance food supply chain transparency and traceability include:

Relying on data, generative AI can help automate the capture and integration of data from different sources (IoT sensors, enterprise systems, etc.) into the blockchain-based traceability system, allowing comprehensive and accurate record-keeping. For example, this automation can eliminate manual effort and reduce the scope of human error in data collection, allowing for more efficient and reliable traceability data.

Examples include using generative AI to automatically generate detailed product-specific information such as origin, processing details, and quality specifications so that stakeholders have verified and granular information about the provenance and handling of food products. This will improve transparency to help consumers make informed purchasing choices and create greater trust in the food ecosystem.

Combining Generative AI with Blockchain-Based traceability Systems can enable regulatory bodies and food safety authorities to obtain more comprehensive and reliable data to monitor and enforce food safety and quality standards, which supports increased compliance and better consumer protection.

Generative AI can help food supply chain stakeholders enhance traceability and transparency while contributing to sustainable practices by bringing data-driven insights and automation to life. This integration can empower food supply chains to be more responsive to consumer demands and better address quality and safety, providing a safe, sustainable, and efficient process for consumers.

The overall research findings suggest that the combination of generative AI with blockchain utility should be introduced as a scalable and cost-effective solution to improve the overall transparency and traceability of food supply chains, help consumers, businesses, and regulatory bodies make better-informed decisions, and develop more substantial confidence in the food system.

RESULTS

Several of the key findings from the research on the potential for integration of generative AI with blockchain-based traceability systems to enhance food supply chain transparency and traceability include:

Relying on data, generative AI can help automate the capture and integration of data from different sources (IoT sensors, enterprise systems, etc.) into the blockchain-based traceability system, allowing comprehensive and accurate record-keeping. For example, this automation can eliminate manual effort and reduce the scope of human error in data collection, allowing for more efficient and reliable traceability data.

Examples include using generative AI to automatically generate detailed product-specific information such as origin, processing details, and quality specifications so that stakeholders have verified and granular information about the provenance and handling of food products. This will improve transparency to help consumers make informed purchasing choices and create greater trust in the food ecosystem.

Combining Generative AI with Blockchain-Based traceability Systems can enable regulatory bodies and food safety authorities to obtain more comprehensive and reliable data to monitor and enforce food safety and quality standards, which supports increased compliance and better consumer protection.

Generative AI can help food supply chain stakeholders enhance traceability and transparency while contributing to sustainable practices by bringing data-driven insights and automation to life. This integration can empower food supply chains to be more responsive to consumer demands and better address quality and safety, providing a safe, sustainable, and efficient process for consumers.

The findings as a whole indicate that the use of combined generative AI with blockchain utility must be applied as a robust and cost-effective solution to efficiently ensure the overall transparency and traceability of food supply chains to assist consumers, businesses, or regulatory bodies to come to better-informed decisions and build more considerable trust in food systems.

DISCUSSION

The findings from the research have linked the integrated ith of the one-trial implementation of generative AI to the food supply chain. At an initial level, generative AI can be implemented to automate the capturing and integration of data that can drastically reduce human intervention and, thereby, the chances of human errors in record keeping and instead create a far more tedious and extensive traceability system. This automation could allow supply chain participants to gather traceability data quickly, correctly, and effectively, allowing them to shed some of the burden of traceability data aggregation that can hamper the quality of the data they compile.

Second, its ability to naturally generate comprehensive product-oriented data allows consumers to understand their purchasing decisions better and ultimately build trust in the food system. As food manufacturers leverage blockchain to improve traceability, this can end up serving as a foundation to establish greater transparency about where a food product was sourced from, how it was processed, and what the quality has been at each stage of its life cycle, which in turn can help consumers make more informed and potentially sustainable decisions about the food they purchase for consumption.

Moreover, generative AI and blockchain can also benefit regulators and authorities responsible for food safety, as they will have access to accurate and reliable oversight information, which can be the basis for monitoring and enforcing standards of food safety and quality. This enhances regulatory compliance and consumer protection.

Lastly, as the research is open, the results of such an academic study will serve as a good reference in understanding the technical, educational, and regulatory bottlenecks that need to be tackled with the implementation of blockchain and generative AI in the food supply chains. This article will examine the challenges of adopting these emerging technologies, including data security and privacy issues, interoperability between systems, and workforce training and education. Addressing these trespassing challenges will be essential to successfully harness blockchain and generative AI in food supply chains and unlock their full potential to enhance traceability and transparency. Disclosures: (2024)(2023)(Ellahi et al., 2023)(Mastilović et al., 2023)(Kshetri et al., 2023)(Alobid et al., 2022)

CONCLUSION

In summary, combining generative AI and blockchain-based traceability systems can be a game-changer in ensuring food supply chain transparency and traceability. Generative AI can produce more accurate, comprehensive, and reliable traceability data for all parties involved in the blockchain-based data ecosystem by providing automated data capture, integration, and generation from various sources. This bridge creates more informed consumers and prudence for the businesses and authorities behind the actors permeating our foodscape — an outcome that ultimately mitigates distrust and contributes to the overall sustainability of the system.

In addition, the study's outcomes also indicate that the potential advantages of this combination in terms of technical, education, and legal aspects so far outweighed the processes that need to be overcome in these areas. With the food supply chain industry increasingly invested in understanding and building these emerging technologies, generative AI and blockchain's added transparency and traceability can significantly improve food safety, quality assurance, sustainability, and more. Combining these technologies could lead to greater visibility, accountability, and consumer confidence by enabling a more holistic and data-driven approach to addressing the complexities of modern food supply chains. Generative AI and blockchain can be integrated to shape a better future in the food industry, with traceability and. transparency that enables stakeholders to make informed, sustainable decisions.

This research is pertinent for the real-world application of integrated blockchain and generative AI within food supply chains. Generative AI, by automating data capture and integration, can eliminate a lot of manual effort or human error in record-keeping, leading to a more complete and accurate traceability framework. That presents a real opportunity for automated production of item-level product details, which can benefit consumers, too, in terms of transparency, informing their purchase decisions, and helping build trust with the food system. Additionally, the adoption of these technologies can also have positive implications for regulatory agencies and food safety organizations as they can access more transparent and credible information about food products, ultimately resulting in enhanced food safety and quality processes and higher compliance with rules and regulations protecting consumers.

As alluring as blockchain and generative AI promise to be, the research concludes that companies must overcome technical, educational, and regulatory hurdles to realize the potential for these technologies to impact food supply chains positively. These challenges include data security, privacy concerns, interoperability between systems, and the need for training and education among the workforce, enabling effective implementation and use of these emerging technologies. Therefore, resolving some of these challenges will l be crucial to the success of combining blockchain with generative AI in food supply chains, realizing their effective use towards enhancing traceability and transparency. This study contributes to the blockchain technology and generative artificial intelligence domains in food supply chain literature. The study provides analytical insights into the potential benefits and challenges of adopting these technologies in food supply chains, presenting valuable information for industry practitioners, researchers, and policymakers interested in improving food supply chains' efficiency, transparency, and sustainability.

REFERENCES

Alobid, M., Abujudeh, S., & Szűcs, I. (2022). The Role of Blockchain in Revolutionizing the Agricultural Sector. In Sustainability (Vol. 14, Issue 7, p. 4313). Multidisciplinary Digital Publishing Institute. DOI: 10.3390/su14074313

Ellahi, R. M., Wood, L. C., & Bekhit, A. E. A. (2023). Blockchain-Based Frameworks for Food Traceability: A Systematic Review [Review of Blockchain-Based Frameworks for Food Traceability: A Systematic Review]. Foods, 12(16), 3026. Multidisciplinary Digital Publishing Institute. DOI: 10.3390/foods12163026

Kshetri, N., Bhusal, C. S., Kumar, D., & Chapagain, D. (2023). SugarChain: Blockchain technology meets Agriculture — The case study and analysis of the Indian sugarcane farming. In arXiv (Cornell University). Cornell University. https://doi .org/DOI: 10.48550/arXiv.2301

Mastilović, J., Kukolj, D., Kevrešan, Ž., Ostojić, G., Kovač, R., Đerić, M., & Samek, D. U. (2023). Emerging Perspectives of Blockchains in Food Supply Chain Traceability Based on Patent Analysis. In Foods (Vol. 12, Issue 5, p. 1036). Multidisciplinary Digital Publishing Institute. DOI: 10.3390/foods12051036

Perera, T., Hettige, S. P. T., Jayatilake, S. A. P., Abeygunawardhana, W., & Bandara, H. M. N. D. (2020). Blockchain-Based Product Traceability Beyond Proof of Existence. In *2022 Moratuwa Engineering Research Conference (MERCon)* (p. 1). DOI: 10.1109/MERCon50084.2020.9185207

Prashar, D., Jha, N., Jha, S., Lee, Y.-J., & Joshi, G. P. (2020). Blockchain-Based Traceability and Visibility for Agricultural Products: A Decentralized Way of Ensuring Food Safety in India. In Sustainability (Vol. 12, Issue 8, p. 3497). Multidisciplinary Digital Publishing Institute. DOI: 10.3390/su12083497

Spitalleri, A., Kavasidis, I., Cartelli, V., Mineo, R., Rundo, F., Palazzo, S., Spampinato, C., & Giordano, D. (2023). BioTrak: A Blockchain-based Platform for Food Chain Logistics Traceability. In arXiv (Cornell University). Cornell University. https://doi.org//arxiv.2304.09601DOI: 10.48550

Chapter 5
Training and Upskilling Employees for Generative AI-Powered Industry 4.0 Supply Chains

Smit B. Kacha
https://orcid.org/0009-0000-7166-8284
Marwadi University, India

Mahi Chheladiya
Marwadi University, India

Meeta Joshi
https://orcid.org/0000-0002-8162-8187
Marwadi University, India

Janvi Bhindi
https://orcid.org/0000-0003-1726-8452
Marwadi University, India

ABSTRACT

The advent of Industry 4.0 has revolutionized supply chains by integrating advanced technologies such as IoT, Big Data, AI and Generative AI, in particular, automates and enhances key supply chain functions like demand forecasting, inventory management, production scheduling, and logistics. This book proposal highlights the necessity for a skilled workforce to leverage these AI systems for optimal efficiency. It emphasizes the need for substantial investment in training and upskilling programs to prepare employees for AI-enabled roles. Addressing the existing skills gap, the proposal outlines strategies to bridge this gap through targeted educational

DOI: 10.4018/979-8-3693-9856-2.ch005

initiatives and industry partnerships. It also discusses the challenges in fostering AI talent, stresses the importance of collaborative efforts among stakeholders to ensure successful AI integration. This comprehensive approach aims to equip the workforce with essential digital skills.

INTRODUCTION

Industry 4.0 is also known as the fourth Industrial Revolution, this revolution caused a major shift in manufacturing and industrial processes through the integration of digital technologies such as the Internet of Things (IoT), Big data analytics, Blockchain technology, Robotics as well as Artificial intelligence (AI) (Lasi, Fettke, Kemper, Feld, & Hoffmann, 2014). In the world, the food industry is a vital sector for all countries like India, which is experiencing the growth of economic as well as Infrastructure development and also not left behind for these major revolutions in the food industries, as the consumer's expectations are evolving. The global supply chain is becoming too complex the fourth revolution's technology gives great and good solutions to enhancement in food production, quality, and sustainability as well as distribution (Tzounis, Katsoulas, & Fountas, 2017). By implementing this technology in day-to-day work in the food industry, companies are optimizing their operations in production and distribution, reducing waste, ensuring food quality and safety, also meeting consumer's demands.

Adaptation of the fourth revolution's technology in the food industry gives noticeable change in supply chain management, giving real-time monitoring and decision-making. For instance, IoT provides real-time monitoring of the temperature and humidity of food products during transportation, so now companies can maintain a good environment for food products and give assurance for a better quality of food products and a greater life span (Astill et al., 2019). With the help of robotics companies becomes much faster in production, in distributions of food products, packaging, shorting, and quality checks. Big data analytics helps to predict consumer demand and market trends these all help companies make better strategies to align with the demands of consumers as a result of all this good integration, the fourth revolution is giving fast and efficient evolution of the food supply chain (Wang, Gunasekaran, Ngai, & Papadopoulos, 2016).

Generative AI is a major and very crucial part of this revolution; it's a subset of Artificial intelligence, and it helps in the generation of new data, recognition of patterns, and matching consumer demand (Goodfellow et al., 2014). Generative AI has the capability of creating entirely new data set or optimizing existing data set, this data helps create trends pattern, forecast the future demand and consumer's behavior. These capabilities significantly impact the food industry and supply chain and innovatively enhance chain management (Daugherty & Wilson, 2018).

INDUSTRY 4.0, GENERATIVE AI, SUPPLY CHAIN

Industry 4.0

The fourth industrial revolution is known as Industry 4.0, which majorly focuses on the automation of factories and supply chains, current trends, data exchange, interpreting and analyzing trends, etc. (Brettel, Friederichsen, Keller, & Rosenberg, 2014). It supports the advancement of technologies to create smart factories where humane machines and systems are interconnected to work effectively. This effective integration allows for the real-time monitoring of key elements of any industry.

The Core Pillars of Industry 4.0: Building a Smart and Resilient Supply Chain

The goal of Industry 4.0 is to achieve adaptability, flexibility, and high efficiency over any production or distribution methodology, digitalization, minimize waste and optimize resources to give greater output, forecast market demand, and automation, etc. (Hermann, Pentek, & Otto, 2016), to achieve there are ten major heads of Industry 4.0 is which helps to achieve better results.

Figure 1. The Core pillars of Industry 4.0

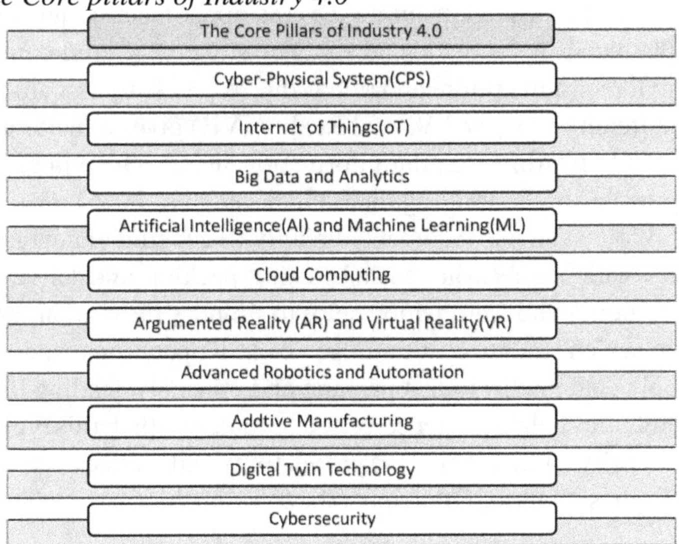

The Cyber-Physical System works as the bridge between tangible assets and digital technology; it also provides real-time monitoring and quick control over vital components in the supply chain (Lee, Bagheri, & Kao, 2015). The CPS system provides vital adjustments to the production schedule and gives updates regarding the machine's health and condition and efficiency also update inventories levels by real-time monitoring data from smart factories or physical world.

The Internet of Things is a crucial part of any supply chain management; it provides accurate data during transportation, like humidity, the temperature of surroundings and products, and the impact of the final location environment on the goods (Atzori, Iera, & Morabito, 2010). This type of data helps to build visualization for the operation and improve responsiveness.

Big Data Analytics is used for data which is generated and gathered by the IoT to be analyzed over various scenarios, which helps to build future patterns of demand, allow for the management of inventories and adjustment of inventories, predict maintenance, support demand planning, help make more cost-effective as well as reduce cost via reduction of waste (Chen, Mao, & Liu, 2014).

Artificial Intelligence (AI) and Machine Learning (ML) help in quick and solid decision-making with the help of historical data and real-time monitoring. This AI-driven data helps to make better route planning, adjustments in inventories, and supplier selection (Jordan & Mitchell, 2015). ML-driven data helps forecast trends, detect potential disruption, and predict the failure of any physical tool in advance, so this helps in the better arrangement of technology and costly downtime.

Cloud Computing technology used into the sharing and storing of vast amounts of data, is possible; it also gives us the authority to function in cooperation (Armbrust et al., 2010). It offers a centralized platform for data access across any region for multiple systems; this quick feature gives the upper hand to quickly respond to any decision that needs to be made and enhance any task. Cloud computing gives us the ability to forecast the future and trends due to access to vast amounts of data.

Augmented Reality (AR) and Virtual Reality (VR) gives a major upskilling for employees without disturbing regular information and also help in giving practical lessons for upcoming tasks (Billinghurst, Clark, & Lee, 2015). For instance, the warehouse workforce can use Augmented reality to get guided navigation through storage facilities, which helps in increasing picking, packing, and delivering accuracy

Advanced Robotics and Automation helps to simplify the regular tasks of supply chain operation, which leads to cost-cutting of overall operations especially automation in the supply chain helps to fast packing and transport, leading to time-saving and advanced robotics help in better selection of routes so that, this types of robotics which are operated by humans boost the productivity and reduce occurrence of error of overall operations (Bahrin, Othman, Azli, & Talib, 2016).

Additive Manufacturing (3D Printing) Producers can cut the time in mainchain operation via using this 3D printing technology, which can help to create the desired product at a certain time rather than realized on some side product providers (Gibson, Rosen, & Stucker, 2015). Also, it is especially helpful when there is a tight deadline for that product, or we can says that there is there is demand for a particular part that is unresolved.

Digital Twin Technology makes a digital representation of an asset or tangible facilities, which can lead and help to evaluate the scenarios like demand hikes and any error in supply chain or in production operation, also give early updates about any disruption in the chain and proactively implement a solution, as an integration of this technology in the supply chain, as well as supply chain, give owner to authority to simulate and analyze data in real-time (Grieves & Vickers, 2017).

With extensive connectivity with all major heads of Industry 4.0, cybersecurity is a major and very crucial part of safeguarding private information and preventing businesses from undesired threats (Stouffer, Pillitteri, Lightman, Abrams, & Hahn, 2015). All data that is generated or coming from the different operations are required to be highly encrypted for safety purposes, this all data is protected with the cybersecurity protocols.

The Intelligent Pulse: Data Flow Mechanisms Powering the Industry 4.0 Supply Chain

In an industry 4.0-driven supply chain, data flow is a key part of every operation; this data helps in making decisions quicker and more accurately and also helps in making action efficient for each and every operation. The flow mechanism of data in the Industry 4.0 ecosystem is complicated as well as dynamic; this vigorous data set is passed through many complicated processes, which are further divided into many parts, which enables the connectivity of data with the mainstream with accuracy. Here we are expressing this data flow chain in detail

Figure 2. Data flow mechanisms in Industry 4.0

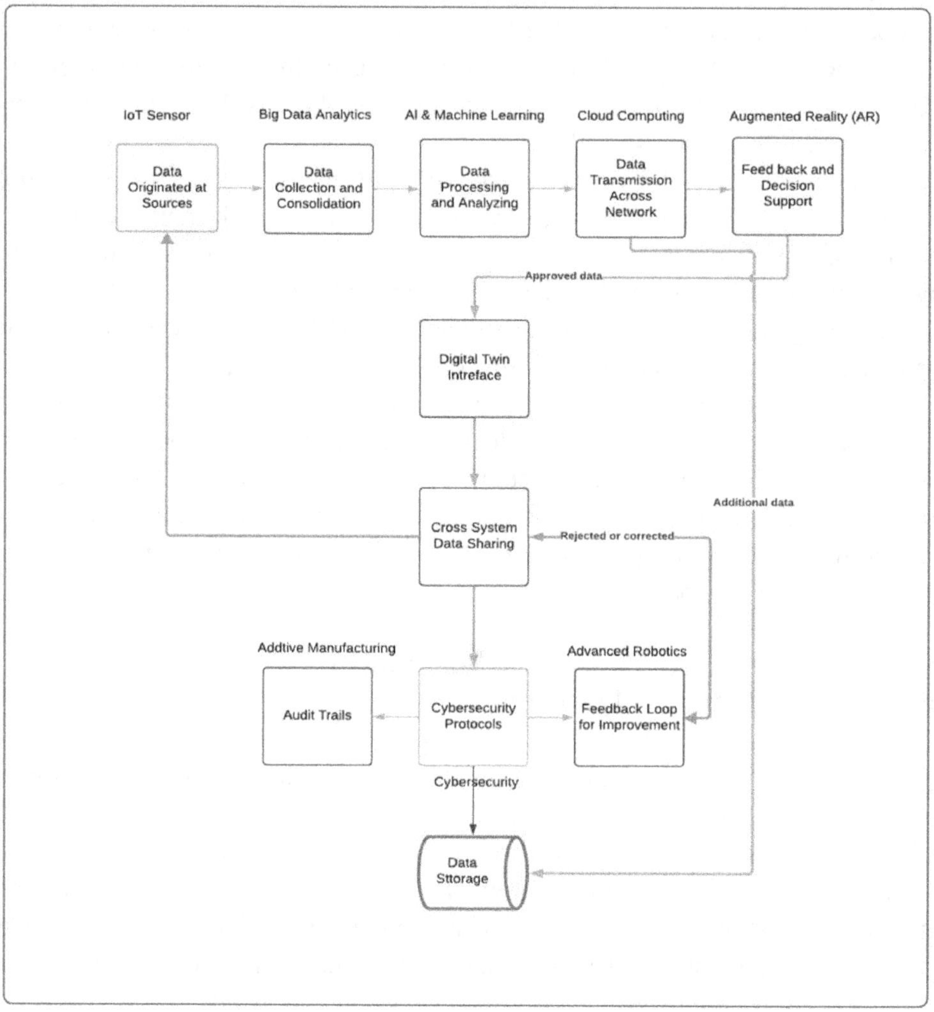

Data that originated from a source has to be primarily gathered with the help of Sensors and IoT devices, in this cycle, data collection begins with the IoT sensors and devices monitoring, all the available sensors that capture all that data are generated through the supply chain, whether it's tracking the location and conditioning of raw material, monitoring all machine performance during production or continuous monitoring of weather condition in the warehouse (Whitmore et al., 2015). This sensor gathers all the data during the operation and transforms this physical event data into digital information. for the case of Human-Generated Data, this sensor also helps in the initial stage all workers who are working on the devices or machinery during the operation, all sensors measure and analyze this data and convert this physical information into digital information, same as the above conversion of data taken place, in this format they have collected data of the control system, quality of product, maintenance of that device or machinery, this human-generated data gave the upper hand to this basic raw data of device or machinery and made more accurate.

Data Collection and Consolidation This stage is crucial for the whole process of interpreting data. Here, collected data is available in raw form; all this raw data is transmitted through the robust data network with the help of the wireless or wired high-speed network. After reaching the central terminal, data is aggregated, organized, and merged with the old data set without any loss. Here, for this whole stage, sometimes data might be transferred at a low speed that may affect the quality of the data, to remove this error, cloud storage and Edge computing are being used (Shi, Cao, Zhang, Li, & Xu, 2016). Edge computing helps to reduce the delay of transfer alongside cloud computing, accurate results in the end, and the final processing of data takes place at the end of the chain, all major data sets which are less affected by the time and much more complex to understand are to be sent for the deep analysis to get future reference. Complex data computing gives us a closer look at real-time action.

Data Processing and Analysis is a crux for data interpreting and processing Incoming data sets through edge computing and cloud storage are requested to be analyzed and cleaned. Further analysis engines crunch these data sets into small pieces based on the present patterns and anomalies. This helps to make more strategic and strategic decisions. Further, these data sets are used for forecasting and inventory optimizations with the help of Machine Learning and Predictive Modeling (Provost & Fawcett, 2013); these predictive model engines look into the available historical data sets to forecast future trends, possible complications, and machine maintenance requirements. This predictive engine helps supply chain managers properly start and allows proactive adjustments that give a better start and good efficiency.

After the whole process is done by the computing engine, data transmission across the network is a crucial link to make efficient growth in the supply chain here, the digital twin interface and cross-system data sharing take care of this link between processed data to the final destination. The digital twin interface is a fluid, real-time model of the current supply chain generated with the help of digital twin technology (Tao et al., 2018), and this allows users to make further changes without any direct physical change in machinery with the help of this, the operator can continuously monitor and simulate all operation scenarios thanks to the smooth data flow between the digital and physical counterparts, this thing make sure that high efficiency. Cross-system data is a stage that links every stage of the data transmission chain for the data flow mechanism, all the data are generated from the production, distribution, and inventory management system, and all this information which are shared with another operation by a consistent stream of data, which supports and maintain supply chain operation in alignment.

Real-time feedback and decision support play a crucial role in modern supply chains. Artificial intelligence decision-making utilizes final output data to derive critical insights that guide important decisions (Russell & Norvig, 2021). For example, if sensors detect a change in the surrounding environment's humidity that could impact product quality, the system promptly notifies the operator, enabling them to take necessary actions or adjust the system as required. Additionally, augmented reality (AR) dashboards provide managers and operators on the ground with a visual representation of real-time data, including inventory levels, stock status, machine efficiency, and the environmental impact on physical assets (Azuma, 1997). This technology facilitates quick and informed decision-making.

Data security and integrity checks are critical in supply chains, especially as interconnected systems transfer data online. To ensure security, measures such as data encryption, multi-factor authentication, and regular integrity checks are essential (Pfleeger, Pfleeger, & Margulies, 2015). Adherence to cybersecurity protocols is vital for safeguarding the Industry 4.0 ecosystem against cyber threats and tampering. Additionally, frequent audit trails allow companies to track data origins and trails, preserving transparency throughout the supply chain. These practices collectively enhance the security and integrity of supply chain systems.

Feedback and continuous improvement are essential for enhancing supply chain efficiency. Once data sets are completed, they are sent back to the system, creating a feedback loop within the integrated system for future data generation and reference. This process helps generate quality assurance and ensures that operations deliver improved performance. Real-time feedback enables the development of a self-optimizing supply chain that becomes smarter and more efficient over time, ultimately enhancing overall chain management (Ivanov, Dolgui, & Sokolov, 2019).

Generative AI

Figure 3. Timeline of Generative AI

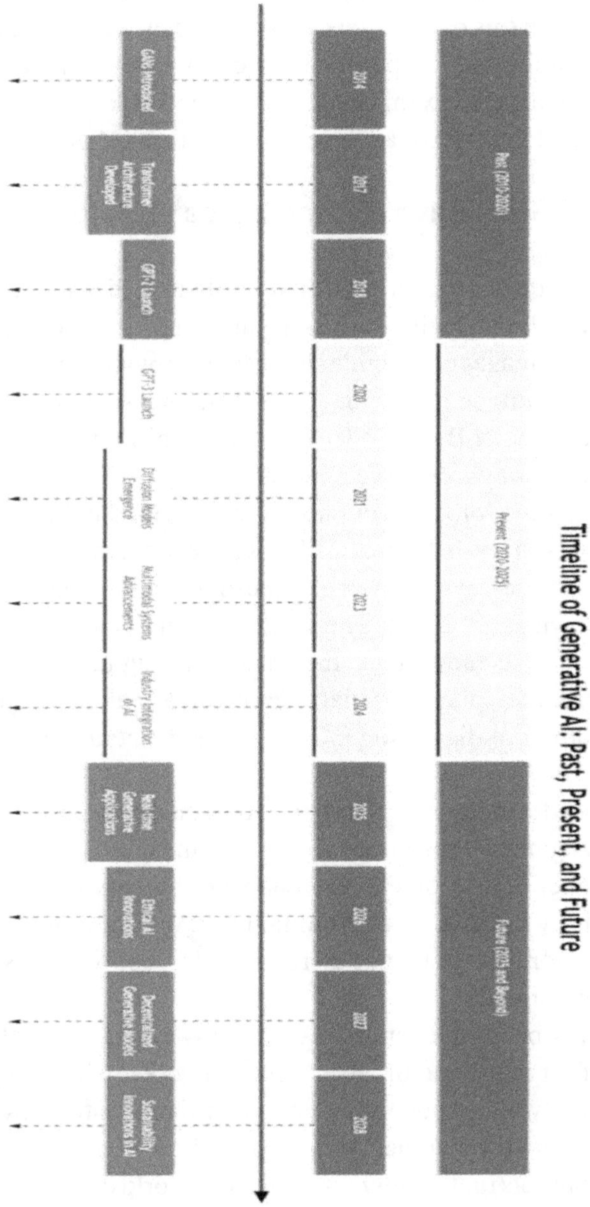

Generative AI is a subfield of Artificial intelligence that generates new data sets based on pre-existing data sets, patterns, and structures (Goodfellow et al., 2014). Migration with AI, generative AI can produce whole new data sets, likewise text, images, and simulation models. Still, mainly generative AI focuses on analyzing data, etc., and based on this data, pattern prediction can occur. Generative AI works on a complex algorithm that can produce content that is frequently separated from human-generated data; clear examples are neural networks like transformers (Vaswani et al., 2017) and generative adversarial networks (GANs).

Key Features of Generative AI Techniques in Data and Creativity

Generative techniques are transformative in their ability to create entirely new data rather than merely analyzing or classifying existing datasets. These methods can generate realistic images of individuals, design innovative products, and produce written content that mimics human language patterns, such as articles, poetry, or narratives (Gatys, Ecker, & Bethge, 2016). This capability marks a significant shift in the scope of technology's creative applications.

Generative AI methodologies find use across various industries and domains, showcasing their versatility. In the realm of language processing, they empower tools capable of composing essays, answering questions, and engaging in meaningful conversations (Brown et al., 2020). The art and design sectors benefit significantly from these techniques, as they enable the creation of digital art and innovative designs. In healthcare, they play a crucial role in generating synthetic medical data, facilitating the training of diagnostic models while maintaining patient confidentiality (Choi et al., 2017).

These advanced techniques are integral to decision-making in fields such as supply chain management, manufacturing, and finance. By simulating complex scenarios and predicting outcomes, they enable optimized planning and risk mitigation (Baryannis, Validi, Dani, & Antoniou, 2019). They support the design of efficient products, forecast potential disruptions in supply chains, and anticipate shifts in consumer demand.

The mechanisms behind generative AI systems typically involve two primary approaches. The first is the use of adversarial networks, which rely on a dynamic interaction between two components: a generator that creates new data and a discriminator that evaluates its authenticity (Goodfellow et al., 2014). This iterative process results in outputs that closely resemble real-world data. The second involves transformer-based models, which process extensive datasets to generate coherent and contextually accurate content (Vaswani et al., 2017). These models employ sophisticated mechanisms that prioritize relevant information within long sequences, ensuring high-quality outputs.

The practical applications of generative AI techniques are widespread and impactful. They are employed in generating text for various purposes, such as customer service, multilingual translations, and content creation. In the visual and multimedia fields, these techniques facilitate the creation of high-fidelity images and videos (Rombach et al., 2022; Ramesh et al., 2021), driving innovation in gaming, fashion, and entertainment. Additionally, they have transformed the creative arts by generating original music and artwork, offering fresh avenues for artistic exploration. Furthermore, the production of synthetic data has become invaluable in fields like healthcare and finance, enabling the development of robust models without compromising data privacy.

Generative AI methods are redefining the landscape of technological innovation by automating creative processes, producing realistic simulations, and enabling data-informed decision-making. Their evolving role across industries underscores their potential to drive advancements in diverse fields, fostering both creativity and efficiency in unprecedented ways (Makridakis, 2017).

Figure 4. Data flow chain in Generative AI

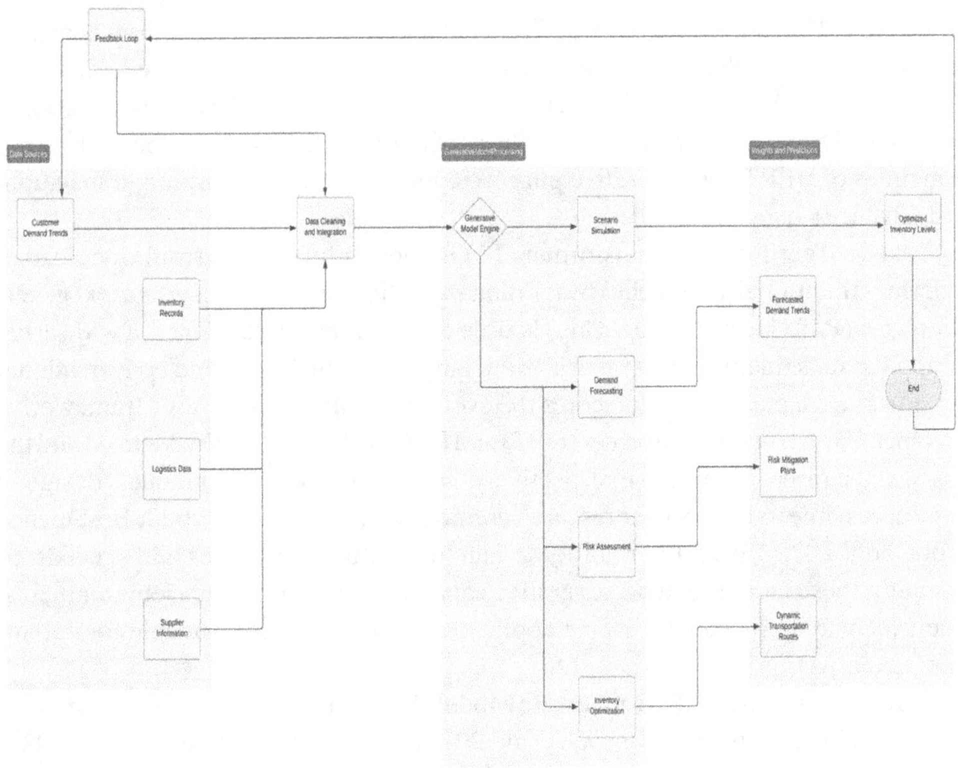

Prominent Models Underpinning Generative Techniques

Generative approaches heavily depend on advanced computational models that process vast quantities of data to understand and generate complex outputs. Among these, large-scale frameworks designed for natural language processing are instrumental in driving generative applications across diverse domains. Below are some of the most significant models that form the foundation of these techniques.

Generative Pre-trained Transformer (GPT) series, developed by OpenAI, represents one of the most prominent examples of generative frameworks (Radford, Narasimhan, Salimans, & Sutskever, 2018; Brown et al., 2020). Based on transformer architecture, these models excel at generating coherent and contextually appropriate text. With applications ranging from content creation to answering complex queries and even performing specialized tasks like coding, GPT models are recognized for their capacity to maintain contextual awareness over extended passages of text. This makes them particularly effective for tasks requiring deep comprehension and high-quality text production.

Bidirectional Encoder Representations from Transformers (BERT) is a model introduced by Google, focuses on understanding textual data through bidirectional processing (Devlin, Chang, Lee, & Toutanova, 2019). While its primary design is centered on comprehension rather than generation, BERT has inspired derivative models such as T5 (Text-To-Text Transfer Transformer) and BART (Bidirectional and Auto-Regressive Transformer). These extended models adapt the underlying principles of BERT for generative purposes, including summarization, translation, and advanced question answering.

Text-To-Text Transfer Transformer (T5),is another innovation from Google, that redefines traditional approaches by treating every linguistic challenge as a text generation task (Raffel et al., 2020). This flexible design allows it to address a wide array of applications, including summarization, language translation, and conversational systems. Its adaptability underscores the evolving nature of generative frameworks.

Open Pre-trained Transformer (OPT) and BLOOM developed by Meta AI and the BigScience research initiative respectively, are transformer-based models designed to be accessible to the broader research community (Zhang et al., 2022; BigScience Workshop, 2022). With their ability to handle multiple languages and specialized domains, these models extend generative capabilities to open-source environments. Their inclusivity promotes diverse applications, making them valuable for collaborative innovation.

Stable Diffusion and Text-to-Image Models Beyond textual applications, models such as Stable Diffusion (Rombach et al., 2022) and DALL-E (Ramesh et al., 2021) utilize diffusion processes to generate images from textual descriptions. These frameworks highlight the versatility of generative approaches, demonstrating their potential in visual domains by transforming text inputs into detailed and coherent images.

Core Architectural Foundations, All these models are unified by their reliance on transformer architecture, which employs self-attention mechanisms to discern relationships and contextual significance across words and phrases (Vaswani et al., 2017). This structural design enables these systems to process intricate language constructs, ensuring outputs are both coherent and contextually relevant. The capability to handle extensive and complex datasets makes these frameworks essential for generative tasks across linguistic and visual domains.

The evolution of models such as GPT, BERT, and T5 illustrates the broad applicability and innovation of generative techniques. These frameworks, with their ability to process and generate data across diverse tasks, underscore the transformative potential of computational approaches in solving modern challenges across industries. Their adaptability continues to shape advancements in technology, enabling new possibilities in communication, creativity, and beyond.

Table 1. Different types of generative AI model

Generative Model	**Primary Features**	**Applications in Supply Chains**	**Training Requirements**	**Upskilling Focus**	**Strengths in Training**
GPT	Coherent text generation, contextual awareness	Automating supply chain documentation, customer service chatbots, report generation	Basic understanding of NLP tools; familiarity with automation workflows	Focus on writing prompts, integrating automation tools, and training for better conversational system outputs	High adaptability; requires minimal programming knowledge; fast deployment in communication-based tasks
BERT	Text understanding, bidirectional processing	Sentiment analysis for customer feedback, supplier risk analysis, compliance monitoring	Intermediate understanding of language processing and data analysis	Emphasis on comprehension of existing datasets and risk analysis applications	Strong focus on understanding nuances in text; effective for critical decision-making tasks
T5	Text-to-text framework for versatile tasks	Summarization of lengthy reports, generating supply chain forecasts, multilingual communications	Advanced knowledge of multi-language NLP systems; familiarity with diverse data types	Training on how to convert complex scenarios into actionable summaries	Versatility in applications across languages and formats

continued on following page

Table 1. Continued

Generative Model	**Primary Features**	**Applications in Supply Chains**	**Training Requirements**	**Upskilling Focus**	**Strengths in Training**
OPT	Open-source, accessible transformer model	Modeling multilingual supply chain scenarios, customizing domain-specific tools	High technical expertise for tailoring; proficiency in programming and data preprocessing	Customization training for diverse language domains and complex datasets	Cost-effective due to open-source accessibility
BLOOM	Multilingual and inclusive transformer model	Integration with diverse language-specific supply chains, fostering global supplier communication	Multilingual training for data preprocessing and output validation	Training employees in handling multi-cultural datasets and global collaboration	Highly inclusive for language diversity; ideal for global-scale supply chains
Stable Diffusion	Text-to-image generation	Visualizing warehouse layouts, product designs, training aids for logistics and maintenance	Basic image editing and design tool expertise	Training for prompt-based generation of visual content and interpreting graphical outputs	User-friendly visual content generation; useful for training materials and design innovation

Transforming the Food Supply Chain with Generative AI and Neural Networks

The integration of neural network technologies into the food industry's supply chain is transforming traditional processes into dynamic, efficient, and resilient systems (LeCun, Bengio, & Hinton, 2015). Generative Adversarial Networks (GANs) are leading the charge by generating synthetic data that simulates real-world scenarios (Goodfellow et al., 2014). This enables businesses to forecast demand, optimize inventory, and detect product defects with precision. For example, GANs can model future crop yields and sourcing challenges, helping companies plan more sustainable strategies. In quality control, these networks simulate potential anomalies in packaging and production, teaching systems to identify flaws in real-time, thereby elevating product standards.

Recurrent Neural Networks (RNNs) and Long Short-Term Memory Networks (LSTMs) further enhance supply chain operations by analyzing sequential data for predictive analytics (Hochreiter & Schmidhuber, 1997). These networks optimize logistics by suggesting efficient transportation routes and schedules, reducing costs while ensuring timely delivery. They are also indispensable in predictive maintenance, analyzing trends in sensor data to prevent equipment failures. By forecasting consumer demand based on seasonal and historical data, RNNs and LSTMs empower companies to align production with market needs, minimizing waste and maximizing customer satisfaction.

Transformer networks have revolutionized the way supply chains handle vast and complex datasets (Vaswani et al., 2017). Through their self-attention mechanisms, they facilitate supply chain transparency and traceability, enabling businesses to meet regulatory standards and ensure food safety. Transformers power risk assessment tools that evaluate suppliers' reliability based on comprehensive datasets, minimizing disruptions in sourcing. They also enhance customer interaction by powering chatbots and recommendation engines that provide personalized service and improve customer satisfaction.

Variational Autoencoders (VAEs) contribute significantly by compressing data into compact representations and reconstructing realistic simulations (Kingma & Welling, 2013). These networks are invaluable in scenario planning, enabling companies to anticipate and navigate potential supply chain disruptions. They also aid in designing sustainable packaging solutions by analyzing and optimizing configurations to reduce waste while maintaining product integrity. Convolutional Neural Networks (CNNs), primarily known for their prowess in image recognition, are crucial in the visual inspection processes within food production (Krizhevsky, Sutskever, & Hinton, 2012). They ensure product quality by identifying physical defects and inconsistencies, verifying packaging labels for compliance, and monitoring hygiene in processing plants through real-time analysis of camera feeds (Zhao, Chen, Liu, & Li, 2023).

This synergy of neural networks has expanded the capabilities of food supply chains, allowing them to evolve into intelligent ecosystems capable of handling challenges with agility. By leveraging these advanced technologies, the food industry is not only achieving operational excellence but also addressing critical issues like sustainability, safety, and regulatory compliance, setting a new benchmark for supply chain management in the era of Industry 4.0 (Kamilaris & Prenafeta-Boldú, 2018).

Figure 5. Risk measurement and demand forecasting with Generative AI

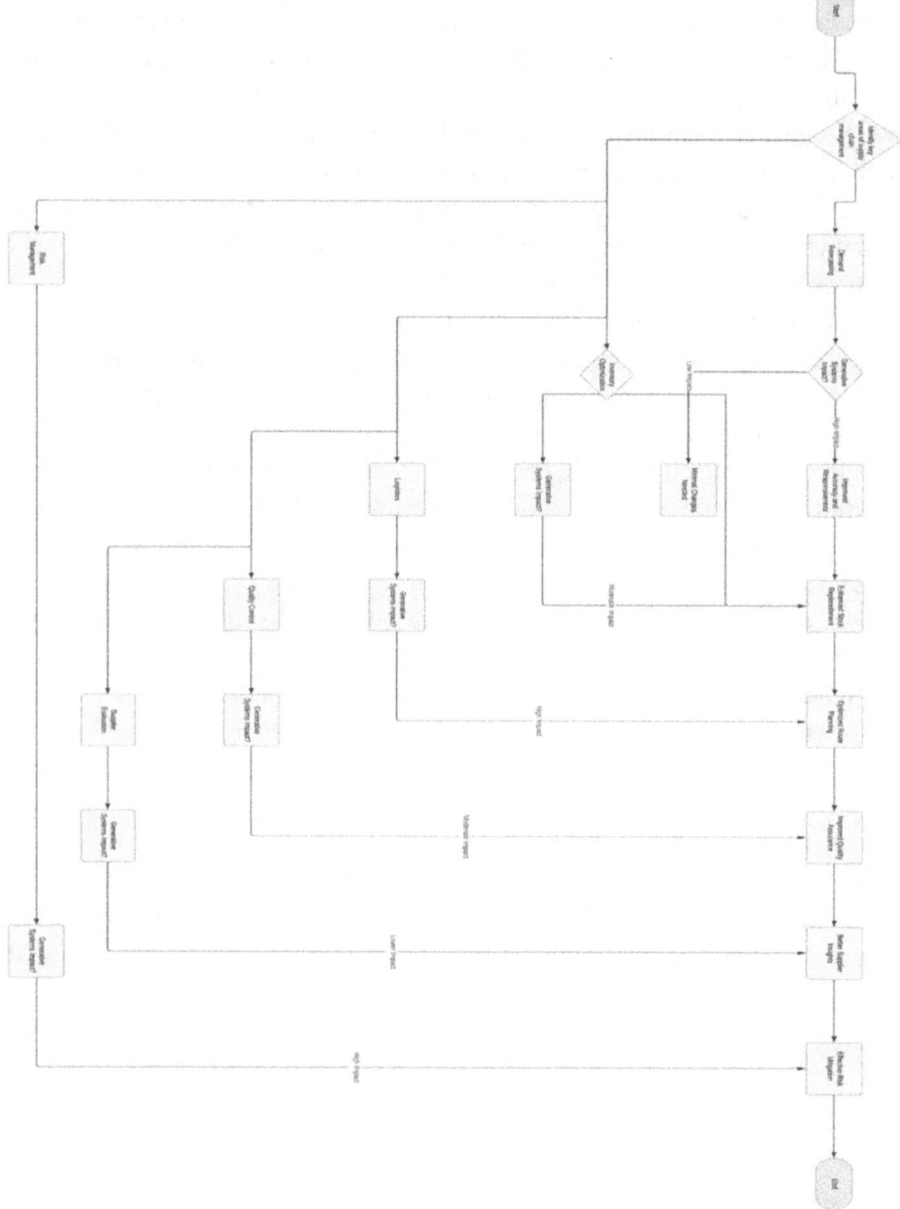

Supply Chain

Figure 6. Timeline chart of supply chain

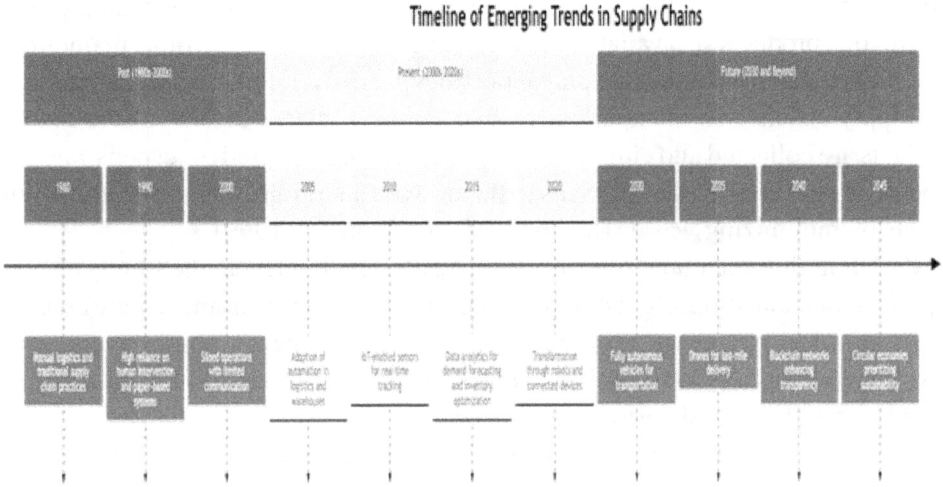

The Lifeline of Modern Business: Decoding the Supply Chain

A supply chain is a comprehensive system that encompasses the entire process of producing and delivering a product or service, from its inception to the final consumer (Christopher, 2016). It involves a network of organizations, people, activities, information, and resources that work collaboratively to transform raw materials into finished goods or services and ensure their timely delivery to end users. The supply chain is a critical backbone of any industry, ensuring efficiency, cost-effectiveness, and customer satisfaction.

The supply chain begins with the procurement of raw materials, sourced from suppliers or natural resources, which are then transported to manufacturers or production facilities. Here, these materials are processed, assembled, or converted into finished products. This stage is crucial as it determines the quality, functionality, and value of the product. Once manufacturing is complete, the products are packaged and stored in warehouses, ready for distribution.

The distribution phase involves transporting goods from warehouses to retailers or directly to consumers. This stage often requires meticulous planning to optimize logistics, reduce transportation costs, and ensure timely delivery (Chopra & Meindl,

2016). In the case of services, this stage may involve scheduling, resource allocation, and customer interaction to deliver the service efficiently.

An integral part of the supply chain is demand forecasting and inventory management. Accurate forecasting helps organizations predict customer demand and adjust production levels, reducing waste and avoiding shortages. Inventory management ensures that products are available in the right quantities at the right time, maintaining a balance between overstocking and understocking (Silver, Pyke, & Peterson, 1998).

Supply chains also include returns management, where defective or unwanted products are collected and either repaired, recycled, or disposed of responsibly. This aspect not only enhances customer satisfaction but also contributes to sustainability efforts by minimizing waste (Rogers & Tibben-Lembke, 1999).

Collaboration and communication among all stakeholders in the supply chain—suppliers, manufacturers, logistics providers, retailers, and customers—are essential for its seamless operation. Efficient supply chain management ensures that every link in the chain functions harmoniously, leading to reduced costs, increased productivity, and improved customer experiences.

In today's interconnected world, supply chains are increasingly global, involving cross-border exchanges and complex networks. This globalization has created opportunities for growth and cost reduction but also poses challenges such as regulatory compliance, geopolitical risks, and environmental considerations (Govindan, Fattahi, & Keyvanshokooh, 2017).

In summary, a supply chain is the lifeline of any business or industry, connecting all aspects of production and delivery. Effective supply chain management not only enhances operational efficiency but also supports sustainability, innovation, and the overall success of organizations in meeting the needs of their customers.

Figure 7. Supply chain overall working flow chart

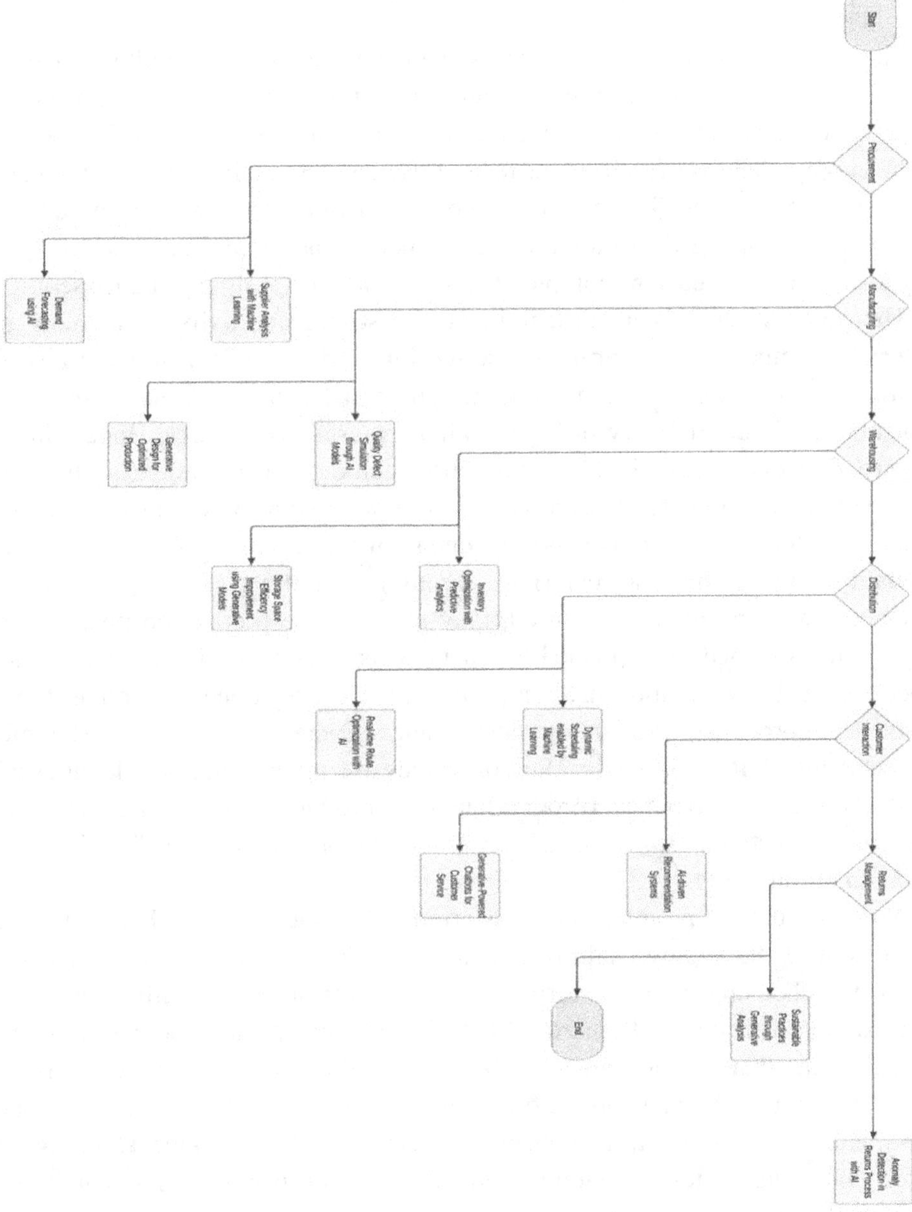

The Backbone of Modern Commerce: A Comprehensive Journey Through the Supply Chain

The supply chain is a comprehensive system that encompasses multiple interconnected components, all working together to transform raw materials into finished goods or services and deliver them to the final consumer (Stadtler, 2015). Each part of the supply chain is critical to ensuring efficiency, reducing costs, and meeting customer expectations. The major components include procurement, production, logistics, warehousing, distribution, and customer interaction, and understanding how they interconnect reveals the complexity and importance of supply chain operations.

The process begins with procurement, the strategic activity of sourcing raw materials, components, or services required for production. Procurement involves identifying reliable suppliers, negotiating contracts, managing relationships, and ensuring the timely delivery of high-quality materials (Monczka, Handfield, Giunipero, & Patterson, 2015). This phase often includes considerations such as cost-effectiveness, supplier reputation, compliance with standards, and sustainability practices. Effective procurement is the foundation of a strong supply chain, as delays or quality issues at this stage can ripple through the entire system.

Once raw materials are procured, they move to the production phase, where they are transformed into finished products. This stage typically involves various processes, such as assembly, manufacturing, or packaging, depending on the type of goods being produced. Production must balance efficiency, quality, and flexibility to meet demand. It is also critical for companies to align production schedules with demand forecasts to avoid overproduction, underproduction, or wastage. This stage often relies on standardized workflows, skilled labor, and advanced equipment to maintain productivity.

After production, products are transferred to logistics and warehousing, a vital component of the supply chain responsible for the storage and transportation of goods. Warehousing involves storing finished goods in secure facilities until they are ready for distribution (Richards, 2017). Inventory management plays a key role here, ensuring that products are available in the right quantities without overstocking or stockouts. Logistics, on the other hand, focuses on the movement of goods. This includes selecting optimal transportation methods like trucking, shipping, rail, or air, managing routes, scheduling deliveries, and ensuring the safe handling of products during transit.

The distribution phase follows, where products are delivered from warehouses or distribution centers to retailers, wholesalers, or directly to customers. This stage requires a well-coordinated system to ensure that goods reach their intended destinations efficiently and on time. Distribution strategies vary depending on the market and the type of goods but often include both regional and global networks.

Timely and reliable distribution is essential for meeting customer expectations and maintaining competitiveness.

A critical yet often overlooked component of the supply chain is customer interaction and after-sales service. This involves engaging with customers to understand their needs, handle complaints, and provide support after the sale. Companies also manage returns and exchanges in this phase, ensuring that defective or unwanted products are efficiently replaced, repaired, or recycled (Guide & Van Wassenhove, 2009). This part of the supply chain is vital for building customer trust and loyalty, as well as for gathering feedback to improve products and processes.

In addition to these primary components, the supply chain works through a series of supportive mechanisms to ensure its smooth operation. For instance, demand forecasting helps predict customer needs and adjust production and inventory accordingly. Risk management identifies potential disruptions, such as supply shortages or transportation delays, and prepares contingency plans to mitigate their impact (Tang, 2006). Communication and collaboration among all stakeholders—suppliers, manufacturers, logistics providers, distributors, and customers—are crucial to ensuring transparency, efficiency, and coordination.

Moreover, the supply chain operates on a cyclical basis, with many processes feeding back into one another. For example, information gathered from customer feedback or market trends can influence procurement and production strategies, while insights from logistics operations may lead to improvements in warehousing or transportation efficiency. Continuous monitoring and improvement are vital for maintaining a robust and adaptive supply chain.

Overall, the supply chain is a dynamic and interconnected system that requires careful management and integration of its many parts. By ensuring each component operates effectively and in harmony with the others, businesses can achieve operational excellence, meet customer demands, and maintain a competitive edge in an ever-changing market.

Figure 8. Overall demand forecasting and inventory management with GenAI

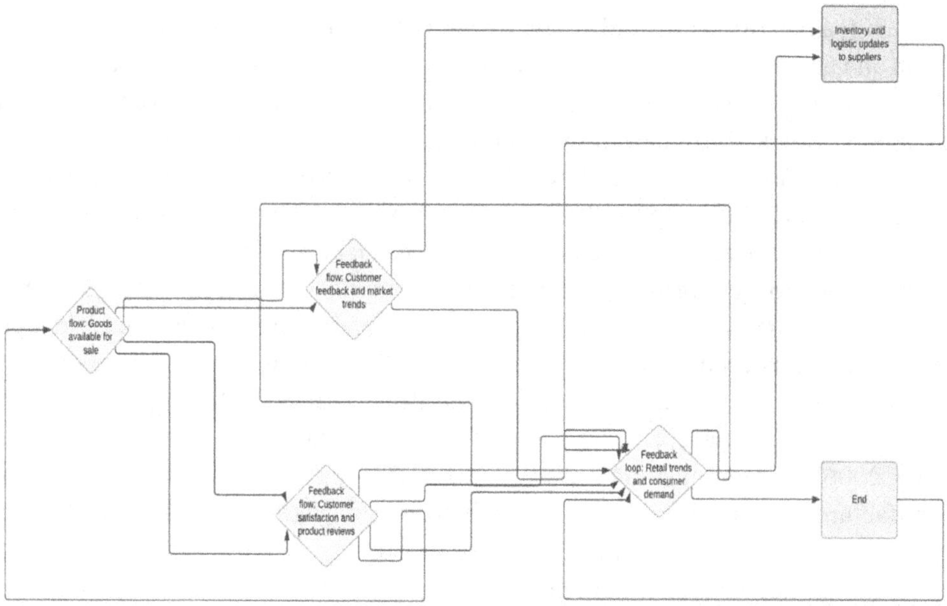

TRANSFORMING TRADITIONAL OPERATIONS: HOW INDUSTRY 4.0 SUPPLY CHAINS DIFFER FROM CLASSIC MODELS AND THE ROLE OF GENERATIVE AI

The transformation from a classic supply chain to an Industry 4.0 supply chain is a significant evolution that integrates advanced technology and strategic innovation, fundamentally redefining how supply chains operate (Schuh, Anderl, Gausemeier, ten Hompel, & Wahlster, 2017). In the classic model, supply chains were linear, segmented, and relied heavily on manual processes and historical data. Each stage—procurement, production, logistics, and distribution—functioned independently with minimal real-time communication or data sharing. This siloed approach often resulted in inefficiencies, delays, and limited responsiveness to market fluctuations or unforeseen disruptions.

In contrast, the Industry 4.0 supply chain operates as a highly integrated and dynamic system driven by digital connectivity and real-time data flow. Industry 4.0 introduces a cohesive ecosystem where every component of the supply chain—materials, machinery, processes, and stakeholders—is interconnected (Hofmann, Rüsch, & Fuger, 2017). Incorporating sensors, automated systems, and advanced analytics enables constant monitoring, transparency, and adaptive decision-making across all supply chain levels.

Generative AI plays a crucial role in amplifying the capabilities of the Industry 4.0 supply chain. Its ability to analyze and generate insights from vast and complex datasets transforms how businesses plan, manage, and execute supply chain operations (Agrawal, Gans, & Goldfarb, 2018). In demand forecasting, generative AI processes extensive datasets, including historical trends, market dynamics, and external variables, to provide precise and actionable predictions. This proactive approach eliminates the reliance on static forecasts, ensuring that production schedules and inventory levels align seamlessly with current and future demand.

Inventory management also undergoes a significant upgrade with generative AI. By continuously evaluating real-time data, it identifies patterns and predicts stock requirements with unmatched accuracy. This ensures that inventory levels are optimized, avoiding scenarios of overstocking or shortages. Additionally, it assesses storage utilization within warehouses, streamlining the movement of goods and accelerating order fulfillment. This integration of predictive analytics into inventory management improves operational efficiency and minimizes costs.

In logistics, generative AI enhances route planning and transportation management by evaluating variables such as traffic patterns, fuel efficiency, and weather conditions. It identifies optimal transportation routes and schedules, ensuring timely and cost-effective delivery. By monitoring logistics networks in real-time, generative AI enables rapid adjustments to mitigate disruptions, such as delays or unforeseen bottlenecks.

The production phase benefits significantly from the introduction of generative AI, which simulates manufacturing processes to identify inefficiencies, optimize workflows, and reduce waste. It supports the development of innovative production strategies that maximize resource utilization while maintaining product quality. In quality control, generative AI enhances defect detection by generating synthetic datasets that train inspection systems to recognize potential flaws, ensuring consistent standards across all products.

Traceability within the supply chain improves significantly with generative AI, as it processes data from interconnected devices and systems to provide detailed visibility into every aspect of the supply chain (Kshetri, 2018). This transparency ensures compliance with regulatory requirements and enhances accountability at each stage. Sustainability efforts are also supported through generative AI by identifying inefficiencies and suggesting improvements in sourcing, energy use, and waste management, enabling companies to achieve environmentally responsible supply chain operations (Stock & Seliger, 2016).

Risk management is another critical area where generative AI delivers value. By analyzing global datasets, it identifies potential risks and vulnerabilities within the supply chain, such as supplier disruptions, geopolitical events, or natural disasters. It provides recommendations for contingency planning, ensuring that businesses remain resilient and operational despite uncertainties.

The shift to an Industry 4.0 supply chain also fundamentally changes how businesses engage with customers. Generative AI processes customer data to identify preferences and behavioral patterns, allowing companies to tailor their offerings and improve service levels. This data-driven approach enhances customer satisfaction and builds stronger relationships by aligning products and services with specific needs.

The transformation from a classic supply chain to an Industry 4.0 model, with generative AI as a core enabler, reflects a strategic leap in operational capability and adaptability. Unlike the fragmented and reactive nature of the traditional supply chain, the Industry 4.0 model is unified, proactive, and designed to anticipate and respond to challenges in real-time. This evolution delivers unparalleled efficiency, scalability, and sustainability, ensuring that supply chains remain competitive and resilient in an ever-changing global landscape (Ivanov, Tsipoulanidis, & Schönberger, 2021).

Table 2. Difference between classic and modern supply chain

ASPECT	CLASSIC SUPPLY CHAIN	INDUSTRY 4.0 SUPPLY CHAIN
STRUCTURE	Linear and segmented, with isolated stages functioning independently.	Fully integrated and dynamic, with interconnected processes and stakeholders across the supply chain.
DATA FLOW	Limited to manual or periodic updates, often siloed within departments.	Continuous, real-time data flow enabled by IoT sensors, cloud computing, and advanced connectivity.
DECISION-MAKING	Based on historical data, experience, and static forecasts.	Data-driven, leveraging predictive analytics, generative AI insights, and real-time inputs for informed decision-making.
FLEXIBILITY	Rigid, slow to adapt to market changes or disruptions.	Highly adaptive, capable of real-time adjustments to changes in demand, supply, or operational challenges.
COMMUNICATION	Minimal communication between stakeholders; often limited to transactional exchanges.	Transparent and collaborative communication among all stakeholders, facilitated by advanced tools and digital platforms.
RISK MANAGEMENT	Reactive, addressing problems only after they occur, leading to potential delays and inefficiencies.	Proactive, with predictive tools identifying risks (e.g., supplier delays, weather disruptions) and recommending contingency plans.

continued on following page

Table 2. Continued

ASPECT	CLASSIC SUPPLY CHAIN	INDUSTRY 4.0 SUPPLY CHAIN
INVENTORY MANAGEMENT	Focused on maintaining fixed levels of inventory, often leading to overstocking or stockouts.	Dynamic inventory management using AI to optimize stock levels based on real-time demand and predictive analytics.
LOGISTICS AND DISTRIBUTION	Managed manually, often relying on static schedules and fixed routes.	Real-time route optimization and dynamic scheduling using generative AI and RNN-based logistics planning tools.
QUALITY CONTROL	Relies on human inspection and reactive correction of defects.	AI-powered systems (e.g., GANs and CNNs) simulate, detect, and prevent defects in real-time, ensuring consistent quality.
CUSTOMER INTERACTION	Limited to post-sale interactions; feedback rarely influences upstream processes.	Integrated customer feedback loops drive continuous improvement in production, inventory, and service levels.
TECHNOLOGY UTILIZATION	Minimal reliance on technology; largely manual processes dominate.	Heavy use of advanced technologies like generative AI, IoT, blockchain, and robotics to automate and enhance processes.
SUSTAINABILITY	Limited focus on sustainability; inefficient resource usage often leads to higher waste.	Sustainability embedded into operations with AI-driven optimization of energy, materials, and waste management.
EMPLOYEE SKILL REQUIREMENTS	Basic operational skills, often confined to repetitive and manual tasks.	Advanced skills in data analysis, AI tools, and generative model applications for decision-making and innovation.
INNOVATION POTENTIAL	Low, as the focus is on maintaining the status quo rather than exploring new opportunities.	High, with the integration of generative models fostering creativity in design, forecasting, and process optimization.

Elevating Supply Chain Efficiency: How Generative Models Revolutionize Industry 4.0 Integration

The integration of the supply chain into Industry 4.0 introduces a new era of technological advancement, where generative models offer unparalleled advantages (Jeske, Sitter, & Schilcher, 2019). These models excel in predictive analytics, scenario simulation, and process optimization, making them the ideal choice for enhancing the complex operations of a modern supply chain. By processing vast amounts of interconnected data, generative models provide actionable insights that transform the supply chain from a reactive system into a dynamic and proactive ecosystem.

Generative models, particularly those rooted in adversarial and transformer-based architectures, are invaluable for addressing the challenges of demand forecasting and inventory management. They simulate diverse market conditions, enabling businesses to predict consumer demand with exceptional accuracy. These models evaluate factors such as seasonal trends, economic shifts, and regional variations to ensure production schedules and inventory levels align perfectly with real-world requirements. This capability reduces excess inventory and prevents stock shortages, ultimately minimizing waste and improving overall efficiency.

In the realm of logistics and transportation, generative models provide significant enhancements. By analyzing historical and real-time data, they simulate the most efficient transportation routes, taking into account variables such as traffic conditions, fuel costs, and delivery timelines. This leads to optimized shipping strategies, cost savings, and faster delivery, ensuring that customer expectations are consistently met. Additionally, generative models can adapt to unforeseen disruptions by identifying alternative solutions, such as rerouting shipments or reallocating resources in real time.

Risk management is another critical area where generative models give the supply chain an upper hand in the Industry 4.0 framework. These models simulate potential disruptions, such as supplier delays, raw material shortages, or geopolitical instability, and recommend contingency plans. By preparing for multiple scenarios, businesses can mitigate risks and maintain operational continuity. Furthermore, generative models facilitate supplier evaluation by analyzing performance metrics and market conditions, ensuring that companies partner with reliable and resilient suppliers.

Sustainability and compliance within the supply chain are also greatly enhanced by generative models. These techniques evaluate energy consumption, resource utilization, and waste management processes to identify inefficiencies and recommend sustainable practices (Bressanelli, Adrodegari, Perona, & Saccani, 2018). This supports environmental goals while maintaining cost-effectiveness. Traceability is another key advantage, as generative models provide detailed insights into the movement and lifecycle of goods. This ensures compliance with regulatory requirements and builds consumer trust by promoting transparency and accountability.

Generative models also play a transformative role in quality assurance and production optimization. By simulating production processes, they identify bottlenecks and inefficiencies, suggesting improvements to workflows and resource allocation. These models also enhance quality control by generating synthetic datasets to train inspection systems, ensuring consistent product standards. Additionally, they support innovative product design by exploring multiple variations and identifying the most cost-effective and functional options.

When integrated with Industry 4.0 technologies such as IoT and blockchain, generative models amplify the interconnectedness of supply chain components. Real-time data from sensors, smart devices, and digital ledgers is processed to provide a comprehensive view of supply chain operations. This interconnected framework enables seamless communication between all stakeholders, from suppliers to end consumers, ensuring that decisions are made quickly and accurately.

The combination of generative models and Industry 4.0 principles marks a fundamental shift in supply chain management. Unlike the segmented and reactive nature of traditional supply chains, this advanced system is highly adaptive, collaborative, and data-driven. By leveraging the capabilities of generative models, businesses gain the ability to anticipate market changes, optimize operations, and enhance customer satisfaction. This evolution ensures that supply chains not only meet the demands of today's dynamic global market but also remain resilient and innovative in the face of future challenges.

Implementation of Generative AI in the Food Supply Chain

The integration of generative AI into the food supply chain marks a significant transformation in how businesses approach operations, workforce development, and technology adoption (Duan, Edwards, & Dwivedi, 2019). With the potential to revolutionize traditional practices, generative AI empowers supply chains to become more efficient, predictive, and adaptive. This section explores the multifaceted impact of generative AI, focusing on its role in upskilling and reskilling employees, the emergence of new roles in supply chain management, collaborative initiatives with educational institutions and industry partners, and techniques for integrating AI into existing systems.

The food supply chain, with its unique challenges such as perishable goods, fluctuating demand, and stringent safety requirements, stands to benefit immensely from the capabilities of generative AI (Haleem, Javaid, Qadri, & Suman, 2022). However, successfully leveraging this technology requires not only advanced tools and methodologies but also a skilled and adaptable workforce. Organizations must invest in training programs to equip employees with the technical expertise and strategic insight needed to harness AI-driven innovations effectively.

As AI transforms supply chain operations, new roles and responsibilities emerge, demanding a blend of technical and managerial competencies (WEF, 2020). From analyzing AI-generated insights to developing tailored algorithms, supply chain professionals must bridge the gap between traditional practices and advanced technology. Collaboration with educational institutions and industry partners plays a vital role in preparing employees for these new demands, offering tailored certification programs, cutting-edge curricula, and hands-on learning opportunities.

Integrating AI into existing supply chain systems, while promising significant benefits, presents challenges that require careful navigation. Techniques such as gradual implementation, robust data management, and cross-departmental collaboration are essential for ensuring a smooth transition. Additionally, addressing barriers such as cultural resistance and data isolation is critical to fostering an environment where AI can thrive.

This section provides a comprehensive overview of how generative AI reshapes the food supply chain, emphasizing the importance of workforce readiness, strategic partnerships, and effective integration techniques. By adopting these approaches, businesses can unlock the full potential of generative AI, driving innovation, efficiency, and resilience in their operations.

UPSKILLING AND RESKILLING THE WORKFORCE

Empowering the Workforce: Comprehensive Strategies for Upskilling and Reskilling in a Modern Business Landscape

In today's rapidly evolving business landscape, organizations must proactively invest in upskilling and reskilling their workforce to maintain a competitive edge (Brynjolfsson & McAfee, 2014). Effective strategies for employee training encompass a variety of approaches designed to equip employees with the necessary skills to adapt to technological advancements and industry changes. This comprehensive focus on workforce development not only enhances employee performance but also drives organizational growth and innovation.

Detailed strategies for employee training include a blend of on-the-job training, formal education programs, and e-learning platforms. On-the-job training allows employees to gain hands-on experience and apply new skills in real-time, fostering a practical understanding of their roles. For instance, job rotation and shadowing programs can provide employees with exposure to different functions within the organization, helping them develop a broader skill set and better understand the company's operations.

Formal education programs, such as workshops, seminars, and certification courses, provide structured learning opportunities that cover both fundamental and advanced topics relevant to the industry. These programs can be tailored to address specific skill gaps and are often conducted by industry experts or in collaboration with educational institutions. For example, leadership development workshops can prepare high-potential employees for future managerial roles, while technical certification courses can enhance the expertise of IT staff.

E-learning platforms offer flexibility, enabling employees to learn at their own pace and convenience, which is particularly beneficial in accommodating diverse learning styles and schedules (Noe, Hollenbeck, Gerhart, & Wright, 2017). These platforms often feature a wide range of courses covering various topics, from soft skills like communication and teamwork to technical skills like data analysis and programming. Interactive elements such as quizzes, simulations, and discussion forums can enhance engagement and retention of knowledge.

Moreover, successful upskilling programs often involve strategic partnerships with educational institutions, industry experts, and technology providers. These collaborations ensure that the training content is current, relevant, and aligned with industry standards. For instance, companies can partner with universities to offer tailored courses that address specific skill gaps within their workforce. Such partnerships can also provide employees with access to cutting-edge research and emerging trends in their field.

Working with technology providers allows organizations to integrate advanced tools and resources into their training programs, enhancing the overall learning experience. For example, incorporating virtual reality (VR) and augmented reality (AR) technologies into training modules can provide immersive, hands-on learning experiences that are both engaging and effective (Bozgeyikli, Raij, & Katko, 2019). Additionally, leveraging artificial intelligence (AI) can personalize learning paths based on individual progress and performance, ensuring that employees receive the support they need to succeed.

Real-world examples of effective upskilling initiatives highlight the impact of these strategies. A notable case is AT&T's Future Ready initiative, which focuses on reskilling employees in response to the digital transformation of the telecommunications industry (Aspen Institute, 2018). This program includes partnerships with online education platforms like Udacity to provide employees with courses in data science, cybersecurity, and software development. By offering these targeted learning opportunities, AT&T empowers its employees to acquire new skills that are critical for the company's evolution.

Similarly, Amazon's Upskilling 2025 program aims to retrain a significant portion of its workforce through various training initiatives, including its Amazon Technical Academy and Machine Learning University (Amazon, 2021). These programs provide employees with access to high-quality education in technical fields, enabling them to transition into more advanced roles within the company. By investing in their employees' growth, Amazon not only enhances its talent pool but also fosters a culture of continuous learning and development.

Navigating the Future: New Roles and Skills in AI-Enhanced Supply Chain Management

The integration of artificial intelligence into supply chain management is transforming the industry, leading to the creation of new roles and responsibilities (McKinsey Global Institute, 2017). As AI technology continues to advance, supply chain operations are becoming more efficient, predictive, and responsive. This evolution requires a workforce that not only understands traditional supply chain dynamics but also possesses the skills to interpret and leverage AI-generated insights effectively.

In AI-enhanced operations, new roles are emerging that focus on the intersection of supply chain management and advanced technology. One such role is the AI Supply Chain Analyst, responsible for analysing data produced by AI systems to identify trends, forecast demand, and optimize inventory levels. These analysts must be adept at using AI tools and software to gather and interpret large volumes of data, transforming it into actionable strategies that improve supply chain performance. They are tasked with monitoring real-time data feeds, assessing supplier performance, and predicting potential disruptions before they impact the supply chain. Their insights help companies maintain optimal inventory levels, reduce costs, and improve service levels.

Another critical role is the AI Integration Specialist, who ensures that AI technologies are seamlessly incorporated into existing supply chain processes. This specialist works closely with IT and operations teams to implement AI solutions that enhance efficiency and productivity. They must have a deep understanding of both AI technology and supply chain operations to bridge the gap between technical capabilities and practical applications. AI Integration Specialists are responsible for configuring and maintaining AI systems, troubleshooting issues, and training other employees on how to use these technologies effectively. Their work ensures that AI tools are fully integrated into the supply chain, providing maximum benefit to the organization.

The role of the Supply Chain Data Scientist is also gaining prominence. These professionals develop and refine the algorithms that power AI systems, ensuring they are tailored to the specific needs of the supply chain. They use machine learning techniques to enhance predictive models, improve decision-making processes, and drive innovation in supply chain management. Supply Chain Data Scientists must have strong programming skills, particularly in languages like Python and R, and be proficient in using data visualization tools to present their findings (Davenport & Patil, 2012). They collaborate with other departments to understand business needs and translate them into data-driven solutions that enhance supply chain efficiency and resilience.

Skills required for interpreting AI-generated insights are diverse and multifaceted. A strong foundation in data analysis is essential, as professionals must be able to interpret complex datasets and identify meaningful patterns. Proficiency in machine learning and statistical modelling is also crucial, as these skills enable the development and refinement of AI algorithms. Understanding how to use data visualization tools to present insights clearly and compellingly is also important, as it allows supply chain professionals to communicate findings effectively to stakeholders.

In addition to technical skills, a deep understanding of supply chain principles is necessary. Professionals must be familiar with logistics, inventory management, and demand forecasting to contextualize AI insights within the broader supply chain framework. Effective communication skills are equally important, as translating technical findings into actionable business strategies requires clear and concise communication. Supply chain professionals must be able to explain the implications of AI-generated insights to non-technical stakeholders, helping them understand how these insights can be applied to improve business operations.

Furthermore, critical thinking and problem-solving abilities are vital. AI-generated insights often highlight areas for improvement or optimization, and professionals must be able to devise innovative solutions to address these challenges. Adaptability is another key skill, as the rapid pace of technological advancement in AI demands a willingness to continuously learn and adapt to new tools and methodologies. Supply chain professionals must stay abreast of the latest developments in AI and machine learning, continuously updating their skills to remain effective in their roles.

To summarize, the incorporation of AI into supply chain management is resulting in new roles and duties that need a combination of technical competence and supply chain understanding. As AI continues to alter the industry, individuals who can comprehend and exploit AI-generated insights will be critical to improving efficiency, innovation, and competitive advantage in supply chain operations. These new jobs and the abilities they require will be important in assisting firms to traverse the intricacies of contemporary supply chains and maintain a competitive advantage in the market.

Forging Future Leaders: Collaborative Initiatives with Educational Institutions and Industry Partnerships

In the rapidly evolving technological landscape, collaboration between businesses and educational institutions is becoming increasingly essential (Etzkowitz & Leydesdorff, 2000). Joint initiatives with universities and training centres, along with the development of certification programs and AI-focused curricula, are critical for preparing the workforce for future challenges. These partnerships help bridge the gap between academic theory and practical industry needs, ensuring that employees are equipped with the skills and knowledge required to thrive in an AI-driven world.

Joint initiatives with universities and training centres play a crucial role in workforce development. Businesses and academic institutions work together to create programs that align with industry requirements, ensuring that the training provided is relevant and up-to-date. For instance, companies can collaborate with universities to develop specialized courses that address specific skills gaps in their industry. These courses can cover a wide range of topics, from advanced AI techniques to practical applications of machine learning in business processes. By working closely with academic institutions, companies can ensure that the content of these courses is tailored to their needs, providing employees with the knowledge and skills necessary to excel in their roles.

These collaborative efforts often include hands-on training opportunities, such as internships and co-op programs, where students gain practical experience by working on real-world projects. This approach not only enhances students' learning experiences but also allows businesses to identify and recruit top talent. By engaging with students early in their academic careers, companies can build a pipeline of skilled professionals who are well-prepared to enter the workforce upon graduation. Furthermore, these programs help students understand the practical implications of their studies, making them more effective and adaptable employees in the future.

Certification programs and AI-focused curricula are another critical component of these collaborative initiatives. Developing certification programs in partnership with educational institutions ensures that employees receive recognized credentials that validate their skills and expertise. These programs can be tailored to different levels of proficiency, from basic AI literacy courses for non-technical employees to advanced certifications for data scientists and AI specialists. By offering a clear pathway for skill development, certification programs help employees progress in their careers and stay competitive in the job market.

AI-focused curricula development involves creating comprehensive educational programs that cover the theoretical and practical aspects of artificial intelligence. These curricula are designed to provide students with a deep understanding of AI technologies, including machine learning, neural networks, and natural language processing. Courses may also cover ethical considerations and the impact of AI on society (Floridi et al., 2018), ensuring that graduates are well-rounded and capable of addressing the broader implications of AI deployment. By incorporating input from industry experts, these programs ensure that the training provided is aligned with the latest advancements and best practices in the field.

Examples of successful collaborations between businesses and educational institutions abound. One notable example is the partnership between IBM and various universities worldwide through the IBM Skills Academy. This initiative offers training in areas such as AI, cybersecurity, and data science, providing students with access to cutting-edge technology and expert instruction. Similarly, the Massachusetts Institute of Technology (MIT) collaborates with companies like Boeing and Amazon to develop specialized courses and research projects that address industry challenges. These partnerships not only enhance the quality of education but also drive innovation by fostering a close connection between academia and industry.

To summarize, engagement between educational institutions and industrial collaborations is critical for creating a workforce capable of navigating the complexity of an AI-driven future. Employees get relevant, high-quality training through joint initiatives with universities and training institutes, as well as the development of certification programs and AI-focused curriculum. Businesses and academic institutions may work together to bridge the theoretical and practical gaps, educating the next generation of leaders to flourish in an increasingly technology environment. These collaborative initiatives are critical to building a culture of continual learning and innovation, which will eventually drive industry development and success.

Seamlessly Adopting AI: Techniques for Integrating AI into Existing Supply Chain Systems

The integration of artificial intelligence into supply chain systems can revolutionize operations, offering enhanced efficiency, accuracy, and predictive capabilities. However, the process of incorporating AI into existing frameworks must be handled with care to ensure smooth transitions and minimize disruptions (Tiwari, Wee, & Daryanto, 2018). To achieve this, businesses should adopt best practices for gradual integration and utilize AI tools and platforms tailored to their industry-specific needs, such as those designed for the food industry.

When integrating AI into existing supply chain systems, a gradual approach is essential. This involves implementing AI solutions in stages, allowing employees to adapt to new technologies and workflows incrementally. One effective strategy is to start with pilot projects that target specific areas of the supply chain. For example, a company might begin by using AI to optimize inventory management before expanding its use to demand forecasting or logistics. By focusing on one area at a time, businesses can test AI applications, gather valuable data, and make necessary adjustments before a full-scale rollout.

Training and education are also critical components of gradual integration. Employees need to understand how AI will impact their roles and how to use new tools effectively. Providing comprehensive training programs and resources helps to build confidence and competence, ensuring that staff can leverage AI technologies to their fullest potential. Additionally, fostering a culture of continuous learning and innovation encourages employees to stay updated with the latest AI advancements and incorporate them into their daily operations.

Effective data management is another key factor in successful AI integration. AI systems rely on high-quality data to generate accurate insights and predictions (Redman, 2013).

Therefore, businesses must ensure that their data is clean, consistent, and accessible. Implementing robust data governance practices, such as regular data audits and standardized data entry protocols, can help maintain data integrity. Additionally, investing in data integration tools that can consolidate information from various sources into a single, unified system enhances the efficiency and effectiveness of AI applications.

Collaboration between IT and operations teams is crucial for seamless AI integration. These teams must work together to identify areas where AI can add value and develop customized solutions that address specific business needs. Regular communication and feedback loops ensure that any issues are promptly addressed and that the integration process remains on track. Furthermore, involving key stakeholders in the planning and implementation phases helps to align AI initiatives with broader business objectives, ensuring that they deliver tangible benefits.

For the food industry, specialized AI tools and platforms can address unique challenges and optimize operations. For instance, AI can enhance supply chain visibility by tracking products from farm to table, ensuring quality and safety standards are met (Pearson et al., 2019). Tools such as machine learning algorithms can predict demand fluctuations based on factors like seasonality and consumer trends, allowing businesses to adjust their inventory levels accordingly. This helps to reduce waste and ensure that products are available when and where they are needed.

AI can also improve food safety by monitoring conditions throughout the supply chain. For example, sensors and IoT devices can collect data on temperature, humidity, and other environmental factors that affect food quality. AI systems can analyze this data in real-time, alerting operators to potential issues before they escalate. This proactive approach helps to prevent spoilage, recalls, and other costly problems, ensuring that consumers receive safe, high-quality products.

In addition, AI-powered analytics can optimize logistics and distribution processes. By analyzing historical data and real-time information, AI tools can identify the most efficient routes and schedules for transportation. This reduces fuel consumption, lowers costs, and minimizes the environmental impact of supply chain operations. Furthermore, AI can enhance supplier management by evaluating performance metrics and identifying potential risks, enabling businesses to make informed decisions about their partnerships and procurement strategies.

OVERCOMING BARRIERS TO AI INTEGRATION IN SUPPLY CHAIN MANAGEMENT: ADDRESSING CULTURAL BIAS AND DATA ISOLATION

The integration of artificial intelligence (AI) into the supply chain is not without its challenges. While AI has the potential to greatly enhance efficiency, predictability, and innovation, two significant barriers can prevent its successful implementation: cultural bias and data isolation (Kaplan & Haenlein, 2019). These factors not only hinder AI adoption but can also create resistance to change within organizations. Overcoming these barriers requires a combination of leadership, strategic mindset, and improved data management practices, which will be explored in this section.

Cultural Bias: Overcoming Resistance to Change

One of the most significant barriers to AI integration in the supply chain is cultural bias. In many organizations, particularly those with long-standing traditional practices, employees may view AI as a threat rather than a tool for improvement. Workers accustomed to conventional processes often see the introduction of AI as a potential job replacer, leading to fear, anxiety, and resistance to adopting new technologies (Ford, 2015). This mindset can drastically slow the adoption process and reduce the effectiveness of AI systems.

For AI to thrive within the logistics chain, a cultural shift must occur. It's essential for leadership to address these concerns head-on. Leaders must actively promote AI as a career opportunity rather than a threat, emphasizing the benefits of AI in terms of job enhancement, skill development, and future growth opportunities. This approach can help employees view AI as a tool for empowerment and efficiency rather than as a replacement.

Motivating employees through clear communication and positive reinforcement is key to overcoming this cultural resistance. Managers should inspire their teams to embrace AI by providing training, creating a collaborative learning environment, and showing how AI tools can make their jobs easier and more productive. Leaders should also promote creativity within teams, encouraging workers to use AI as a tool for innovation and problem-solving. In this way, the workforce becomes more open to change, ultimately contributing to a more successful AI integration.

By fostering a culture of continuous learning and adapting to technological advancements, organizations can mitigate the risks associated with cultural bias. When employees understand how AI can benefit both them and the organization, resistance to change diminishes, allowing AI to be embraced as a crucial part of the future supply chain.

Data Isolation: The Importance of Effective Data Management

Another critical challenge organizations face when integrating AI into their supply chain is data isolation. For AI systems to function optimally, they rely on high-quality, consistent, and accessible data. Unfortunately, many organizations struggle with issues such as inadequate data collection, lack of standardized data formats, poor data quality, and siloed data systems (DalleMule & Davenport, 2017). These challenges prevent AI from generating the qualitative insights and actionable strategies needed for effective decision-making.

To harness the full potential of AI in supply chain management, businesses must prioritize data management. Investing in robust data management practices is crucial to ensuring that data is collected, cleaned, and made accessible across departments and systems. This will not only help improve the quality of the data being fed into AI systems but also ensure that AI can analyze comprehensive and relevant datasets, leading to more accurate and meaningful insights.

A key aspect of overcoming data isolation involves improving data accessibility. In many organizations, data is often stored in silos across different departments, making it difficult for AI systems to access and analyze it comprehensively. For AI to be truly effective, businesses must focus on integrating their data systems, ensuring that data flows seamlessly across various departments and can be easily accessed by AI technologies. Additionally, organizations must invest in data cleansing technologies, which can help identify and correct any inaccuracies or inconsistencies in the data.

Data quality is another important factor. AI is only as good as the data it analyzes, and poor-quality data can lead to misleading insights. Organizations need to ensure that the data they collect is accurate, timely, and relevant to the objectives they aim to achieve. Implementing data governance practices that enforce data accuracy and consistency will be vital in supporting AI's effectiveness in supply chain management (European Commission, 2018).

By addressing data isolation and improving data management, organizations can ensure that their AI systems operate at peak performance. With clean, accessible, and high-quality data, AI can generate the insights needed to optimize supply chain operations, reduce inefficiencies, and improve overall decision-making.

Challenges and Solutions in Implementing Generative AI

The implementation of innovative technologies within modern industries has brought significant opportunities for transformation and growth. However, with these advancements come complex challenges that require careful navigation. This chapter focuses on the intricate balance between embracing these cutting-edge systems and addressing the obstacles that often arise during their adoption. The discussion highlights the multifaceted nature of these challenges, which range from financial and infrastructural constraints to cultural resistance, data-related issues, and ethical considerations.

A key aspect of this exploration involves the financial hurdles that organizations face when adopting such systems. The high costs associated with initial setup, training, and ongoing maintenance can deter many businesses, particularly smaller enterprises (Tambe, Cappelli, & Yakubovich, 2019). By examining real-world case studies and financing models, this chapter sheds light on how organizations can not only manage these expenses but also achieve substantial returns on their investments.

In addition to financial concerns, infrastructural limitations in different regions pose significant challenges. The lack of reliable technological infrastructure, such as internet connectivity and access to advanced hardware, can hinder adoption efforts (International Telecommunication Union [ITU], 2021). Through case studies from diverse regions, the chapter illustrates how organizations have overcome these limitations, paving the way for broader adoption in challenging environments.

Cultural and organizational resistance also play a pivotal role in the success or failure of implementing advanced systems. Resistance to change, whether due to fear, uncertainty, or a lack of understanding, can stall progress. This chapter examines change management strategies and success stories from organizations that have effectively navigated these barriers, fostering a culture of acceptance and innovation.

Equally important is the role of data quality and availability. The reliability and consistency of data are critical for the success of these systems, yet many organizations struggle with incomplete or poorly structured data (Lee & Strong, 2003). This chapter discusses techniques for data cleaning, governance, and integration to ensure that data serves as a robust foundation for operational success.

Finally, the chapter addresses the regulatory and ethical dimensions of adopting such systems. Ensuring compliance with industry regulations and addressing ethical concerns, such as data privacy and transparency, are vital for maintaining trust and responsibility (Mittelstadt et al., 2016). By exploring real-world examples, the chapter provides practical guidance on how organizations can navigate these challenges while aligning with legal and ethical standards.

Overall, this chapter provides a comprehensive overview of the challenges faced by organizations when implementing advanced systems and presents practical solutions to address these issues. By covering financial, infrastructural, cultural, data-related, and ethical aspects, it equips readers with a thorough understanding of the complexities involved and offers actionable insights for successful adoption.

Figure 9. Challenges and solutions in generative AI implementation

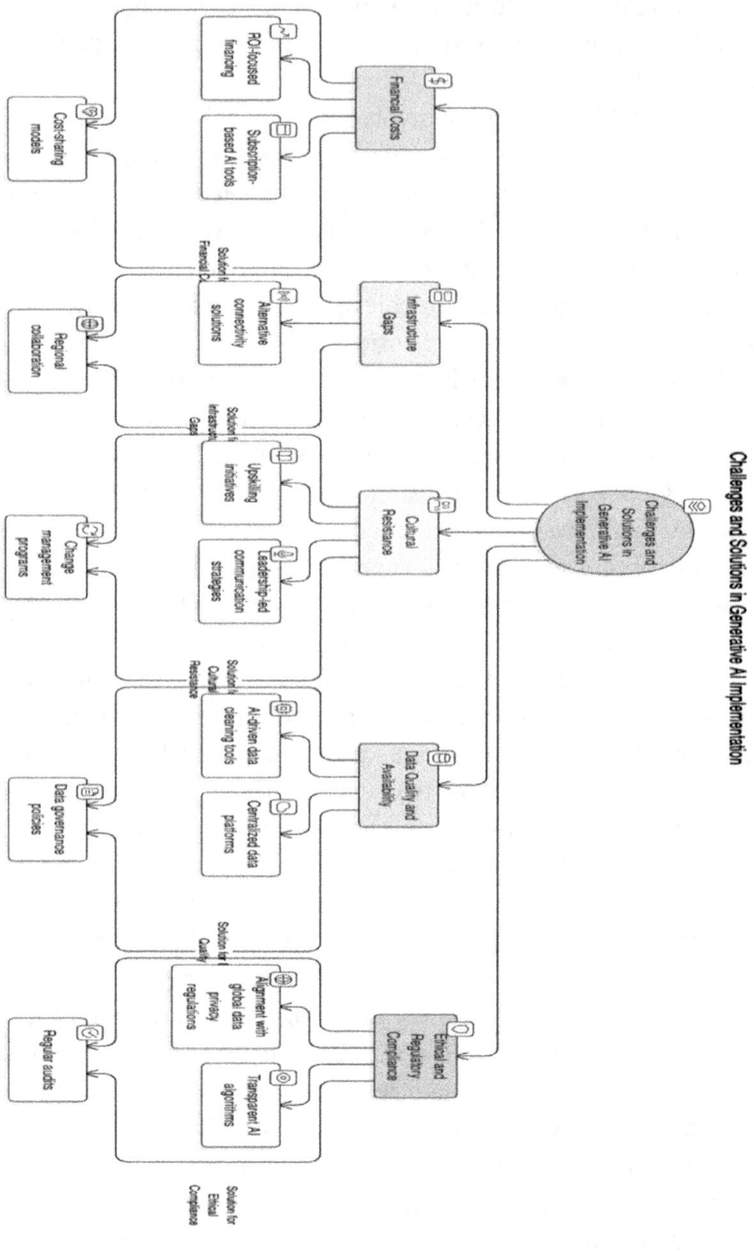

Navigating Financial Barriers: Addressing the High Costs of Implementation

The financial burden of implementing advanced systems is one of preoccupied challenges for organizations. While the potential benefits of these technologies are vast, the upfront and ongoing costs often serve as a significant barrier, particularly for small and medium-sized enterprises or organizations operating in resource-constrained environments (OECD, 2019). This section explores the multifaceted nature of cost-related challenges, providing an in-depth analysis of the factors contributing to these expenses and presenting strategies to navigate them effectively.

Implementing advanced systems requires a substantial initial investment. The acquisition of specialized hardware and software often represents a significant portion of these costs. Unlike standard technologies, these systems may necessitate high-performance computing infrastructure, which includes powerful servers, storage solutions, and networking equipment capable of handling large-scale data processing and analytics. Additionally, the cost of software licenses or subscriptions can be considerable, particularly for enterprise-grade solutions that require customization to align with organizational goals.

Beyond the direct costs of hardware and software, infrastructure upgrades are frequently necessary to support these systems. Many organizations find that their existing networks, servers, or digital environments lack the capacity or compatibility required for the integration of advanced technologies. Upgrading these foundational systems is not only costly but can also disrupt existing operations, adding indirect costs in terms of downtime and resource allocation.

A significant but often underestimated expense lies in the recruitment and training of skilled personnel. Advanced systems demand a specialized workforce capable of managing, maintaining, and optimizing their performance. Organizations must invest in attracting highly skilled professionals, often in competitive labor markets where salaries for such expertise are high (WEF, 2020). In addition, comprehensive training programs are required to ensure that existing employees can effectively use the new systems. These programs must be tailored to address the unique aspects of the systems being implemented, which further drives up costs.

Another critical consideration is the ongoing expense of system maintenance and updates. Unlike traditional technologies, advanced systems require continuous improvements to remain effective and secure. Maintenance costs include not only routine servicing but also the integration of new features, patching vulnerabilities, and adapting to changing operational needs. These ongoing investments ensure that the systems remain relevant and efficient, but they also add to the long-term financial commitment required for adoption.

Despite the daunting nature of these financial challenges, organizations can employ several strategies to manage and mitigate costs effectively. Innovative financing models provide a viable pathway for spreading out the financial burden. For instance, leasing agreements allow organizations to use the necessary hardware without making a large upfront purchase, enabling them to allocate funds more effectively. Subscription-based software models, commonly referred to as Software-as-a-Service (SaaS), also reduce initial costs by offering access to advanced systems on a pay-as-you-go basis.

Additionally, partnerships with technology providers can ease financial pressures. In some cases, providers may offer flexible payment terms, discounts, or collaborative agreements that lower the cost of adoption. Government grants and subsidies are another valuable resource, particularly for organizations operating in sectors deemed critical for technological advancement, such as healthcare, manufacturing, or supply chain management. These funding programs often aim to promote innovation and competitiveness, providing organizations with the financial support needed to overcome initial barriers.

Evaluating the return on investment (ROI) is crucial in addressing cost-related challenges. Organizations must look beyond the upfront expenses and assess the long-term value generated by these systems. Real-world case studies provide compelling evidence of how these investments can lead to measurable gains. For instance, organizations that successfully implement advanced systems often report increased operational efficiency, significant cost savings, and enhanced decision-making capabilities. These benefits can translate into improved profitability, competitive advantage, and scalability, ultimately justifying the initial expenditure.

Overcoming Regional Barriers: Infrastructure Challenges in Technology Implementation

Infrastructure limitations pose significant challenges to the implementation of advanced systems across various regions. These challenges are often shaped by disparities in technological development, availability of resources, and regional connectivity (World Bank, 2021). While some regions boast robust infrastructure that facilitates seamless integration, others struggle with basic technological inadequacies, creating substantial barriers to progress. This section explores the critical aspects of infrastructure challenges and provides an in-depth analysis of how organizations in different parts of the world have addressed these issues, drawing on real-world case studies to illustrate practical solutions.

One of the most common infrastructure challenges is the lack of reliable and high-speed internet connectivity. Advanced systems require consistent and high-bandwidth internet access for data transmission, cloud integration, and remote operations. However, in many developing regions, internet connectivity is often unstable, slow, or entirely unavailable (ITU, 2021). This limitation hampers the ability of organizations to fully utilize these technologies, forcing them to rely on outdated methods or inefficient workarounds. For example, in rural areas, where broadband access is limited, businesses face delays in data processing and communication, significantly impacting their operational efficiency.

Power supply inconsistency is another major hurdle, particularly in regions with underdeveloped energy infrastructure. Advanced systems depend heavily on a stable and uninterrupted power supply to operate efficiently. Frequent power outages or voltage fluctuations can disrupt operations, damage equipment, and lead to substantial financial losses. Organizations in regions with unreliable electricity grids often have to invest in backup generators or alternative energy sources, adding to the overall cost and complexity of implementation.

Hardware and equipment availability also play a critical role in infrastructure challenges. In regions where access to specialized hardware is limited, organizations may struggle to procure the necessary components for system deployment. Importing equipment can result in additional expenses, delays, and complications related to regulatory compliance or customs procedures. This issue is particularly pronounced in industries requiring sophisticated sensors, servers, or machinery, which are not readily available in all markets.

Another challenge lies in the lack of technical expertise and support infrastructure. Advanced systems require a network of skilled technicians and support teams to ensure proper installation, maintenance, and troubleshooting. In many regions, the availability of such expertise is limited, and businesses are forced to rely on external consultants or training programs, which can be costly and time-consuming. This gap in technical knowledge often results in slower adoption rates and reduced effectiveness of the implemented systems.

Case studies from various regions highlight innovative solutions to these challenges. For instance, organizations in remote areas have successfully leveraged satellite internet technology to overcome connectivity issues, enabling them to establish reliable communication networks. Others have adopted hybrid energy models, combining solar and traditional power sources to mitigate the impact of inconsistent electricity supply. In regions with limited hardware availability, partnerships with international suppliers or collaborations with local manufacturers have proven effective in addressing procurement challenges.

Another notable example comes from organizations that have invested in community-driven training programs to bridge the skills gap. By collaborating with local educational institutions and government agencies, these businesses have created a pipeline of skilled workers capable of managing and maintaining advanced systems. Such initiatives not only address immediate technical challenges but also contribute to the long-term development of regional expertise and capacity.

Bridging the Gap: Addressing Cultural and Organizational Resistance to Technological Change

The implementation of advanced systems often encounters resistance rooted in cultural and organizational dynamics (Kotter, 2012). This resistance can stem from a variety of factors, including fear of job displacement, uncertainty about the benefits of new technologies, and reluctance to alter established workflows. Such challenges can significantly delay or derail adoption efforts, even when the potential benefits of the technology are evident. This section explores the nature of cultural and organizational resistance, delving into strategies that can be employed to overcome these barriers. By examining real-world success stories, it demonstrates how effective change management approaches have enabled companies to foster acceptance and create a culture of innovation.

Cultural resistance often arises from a deep-rooted hesitation to embrace change. Employees may perceive new systems as a threat to their roles or question the necessity of altering familiar processes. This resistance is further compounded by a lack of understanding or awareness of how the technology will improve their work or benefit the organization as a whole. Organizational resistance, on the other hand, often originates at the management level. Leaders may be reluctant to disrupt current operations or invest in the training and restructuring required to implement the change successfully. Both types of resistance can create an environment where skepticism and inertia stall progress.

To address these challenges, effective change management strategies are essential. Communication is at the heart of successful change management. Organizations must clearly articulate the purpose, benefits, and expected outcomes of the new technology to all stakeholders. Transparent and frequent communication fosters trust and ensures that employees and management alike understand the rationale behind the change. Providing practical examples of how the technology will enhance efficiency, reduce workload, or improve outcomes can help alleviate concerns and build confidence in the transition.

Engaging employees early in the process is another critical strategy. Involving staff in decision-making and implementation not only empowers them but also provides valuable insights that can improve the adoption process. When employees feel heard and valued, they are more likely to support and advocate for the change. Offering comprehensive training programs and hands-on workshops ensures that employees are equipped with the skills and knowledge needed to use the new systems effectively, reducing anxiety and resistance.

Real-world success stories illustrate the power of these strategies. For example, a manufacturing company faced significant resistance when introducing automated systems to replace manual processes. By involving employees in pilot projects and showcasing the efficiency gains and reduced physical strain brought by automation, the company gradually earned its trust. Comprehensive training sessions further reinforced confidence, enabling a seamless transition to the new system.

Another example comes from a retail organization that encountered resistance at the management level when implementing a centralized inventory management system. By providing data-driven insights into how the new system would reduce stockouts, improve customer satisfaction, and cut costs, the leadership team was persuaded to support the initiative. The company also created cross-functional teams to oversee the implementation, fostering collaboration and shared ownership of the change.

Building a culture of innovation is also a long-term solution to cultural and organizational resistance. Encouraging openness to new ideas, rewarding adaptability, and creating opportunities for continuous learning establish an environment where change is viewed as an opportunity rather than a threat (Schein, 2010). This cultural shift ensures that resistance is minimized, not only for the current implementation but for future advancements as well.

Ensuring Reliability: Addressing Data Quality and Availability Challenges

The reliability of any advanced system depends heavily on the quality and availability of the data it uses (Kaisler, Armour, Espinosa, & Money, 2013). Data serves as the foundation for decision-making, process optimization, and system performance, but issues such as inconsistency, incompleteness, and lack of accessibility can severely undermine its effectiveness. This section delves into the critical aspects of data quality and availability, exploring techniques for improving data reliability through cleaning, governance, and standardization, as well as the tools that facilitate seamless integration of diverse datasets into advanced systems. These strategies ensure that organizations can derive meaningful insights and maximize the potential of their data-driven initiatives.

Data cleaning is a fundamental step in ensuring the reliability of datasets. This process involves identifying and correcting errors, inconsistencies, and inaccuracies within the data. Common issues such as duplicate entries, missing values, and incorrect formats can lead to flawed analyses and unreliable outputs. Techniques like automated error detection, missing value imputation, and normalization help to eliminate these problems, ensuring that datasets are accurate and complete. For instance, businesses often employ tools capable of scanning large datasets for anomalies, flagging potential errors, and suggesting corrective actions. Regular audits of data quality are also essential to maintaining the integrity of information over time.

Data governance plays a pivotal role in maintaining consistency and accountability in how data is collected, stored, and utilized (Khatri & Brown, 2010). Effective governance involves establishing clear policies, roles, and responsibilities for managing data across an organization. This includes defining who has access to specific datasets, determining how data should be stored securely, and ensuring compliance with regulatory requirements. A robust governance framework not only enhances the quality of data but also fosters trust among stakeholders by promoting transparency and accountability. Organizations often create dedicated data governance committees or appoint data stewards to oversee these processes and ensure that policies are adhered to consistently.

Standardization of data is another critical aspect of improving its usability and reliability. When data is collected from multiple sources, variations in formats, naming conventions, and structures can create significant challenges in analysis and integration. Standardization involves aligning these elements to a common framework, ensuring that all datasets are compatible and can be easily combined or compared. This process often requires the use of data transformation tools, which can reformat and align datasets according to predefined standards. For organizations operating in global markets, standardization is particularly important for integrating data from diverse regions, languages, or systems.

The integration of diverse datasets is facilitated by advanced tools designed specifically for data merging, transformation, and analysis. These tools enable organizations to consolidate information from various sources, including internal systems, third-party platforms, and external databases. Data integration tools often include features such as automated mapping, which matches similar fields from different datasets, and real-time synchronization, which ensures that updates are reflected across all integrated systems. By streamlining these processes, these tools reduce the manual effort required and ensure that integrated datasets are consistent and up to date.

Another critical element in addressing data quality and availability challenges is accessibility. Organizations must ensure that relevant data is readily available to the right stakeholders at the right time. This requires the implementation of data access controls, which prevent unauthorized access while enabling authorized users to retrieve the information they need. Cloud-based data storage solutions have become increasingly popular for this purpose, as they provide secure and scalable platforms for storing and sharing data across geographically dispersed teams. These platforms also offer advanced search and retrieval functions, which allow users to locate specific information quickly and efficiently.

Navigating Compliance and Integrity: Regulatory and Ethical Challenges in Supply Chain Technology

As organizations increasingly adopt advanced systems to enhance efficiency and transparency in supply chains, regulatory compliance and ethical considerations have become critical areas of focus (Bostrom & Yudkowsky, 2014). These challenges are particularly prominent in industries like food production and distribution, where ensuring safety and integrity is paramount. This section explores the complexities of aligning technology-driven supply chains with regulatory frameworks and ethical standards, offering a detailed examination of the issues and solutions involved.

Compliance with food safety regulations is a fundamental requirement in the supply chain, particularly when handling perishable goods and consumables. Advanced systems are often utilized to monitor and maintain quality standards, ensuring products meet safety guidelines from production to consumption. However, integrating such systems requires careful alignment with existing regulatory frameworks, which vary significantly across regions and industries. Challenges arise in areas such as traceability, where systems must record and maintain accurate data on the origin, processing, and distribution of food products (Badia-Melis, Mishra, & Ruiz-García, 2015). Regulatory bodies often mandate detailed documentation and real-time monitoring to prevent contamination, spoilage, or fraudulent practices. Organizations must ensure that their systems can produce reliable, auditable records that meet these stringent requirements.

Another regulatory concern is data privacy and security, particularly when systems collect sensitive information about suppliers, distributors, and consumers. Regulations such as the General Data Protection Regulation (GDPR) in Europe require organizations to handle data responsibly, ensuring it is stored securely and used transparently (European Parliament & Council of the European Union, 2016). Non-compliance can result in severe penalties and reputational damage, making robust data governance practices essential for any organization implementing advanced systems.

Ethical considerations in the supply chain extend beyond compliance to address broader questions of fairness, accountability, and societal impact (Floridi et al., 2018). For instance, ensuring that systems operate transparently and do not perpetuate bias or discrimination is a growing concern. In supply chains, this might involve monitoring supplier practices to ensure fair labor conditions or evaluating procurement decisions for potential biases that could harm smaller or marginalized suppliers. Transparency is essential, as stakeholders, including consumers, increasingly demand visibility into supply chain practices and assurance that ethical standards are upheld.

Another significant ethical issue is the potential displacement of labor due to automation. While advanced systems can streamline operations and reduce costs, they may also lead to job losses in certain sectors (Frey & Osborne, 2017). Balancing technological progress with social responsibility requires organizations to develop strategies for workforce transition, such as reskilling programs or creating new roles that leverage the benefits of advanced systems without displacing human labor entirely.

Real-world examples illustrate how organizations have navigated these challenges. In the food industry, companies have implemented traceability systems that use digital ledgers to document every stage of the supply chain, ensuring compliance with safety standards and enhancing consumer trust. Others have invested in secure data management platforms that align with privacy regulations, demonstrating a commitment to responsible data handling. Ethical supply chain practices have also been advanced through partnerships with non-governmental organizations and stakeholders who help monitor labor conditions and environmental impact, ensuring operations align with broader societal values.

SHAPING TOMORROW: FUTURE DIRECTIONS AND EMERGING TRENDS IN MODERN SUPPLY CHAINS

The future of supply chain management is being reshaped by technological advancements, evolving consumer expectations, and a global emphasis on sustainability (United Nations, 2015). As industries transition to smarter, more agile systems, new trends and innovations are emerging to address the complexities of modern supply chain operations. This section explores the transformative impact of autonomous systems, data-driven decision-making, and sustainable practices, while also providing insights into how businesses can maintain a competitive edge through continuous learning and adaptation.

One of the most impactful advancements in supply chains is the rise of autonomous systems, which are revolutionizing logistics, warehousing, and transportation (Fagnant & Kockelman, 2015). Autonomous vehicles, such as drones and self-driving trucks, are becoming integral to optimizing deliveries and reducing transportation costs. These vehicles enhance route efficiency by leveraging real-time data to avoid traffic congestion, reduce fuel consumption, and deliver goods faster. In warehouses, robotic systems are transforming inventory management and order fulfillment. These robots can sort, pick, and pack items with unparalleled speed and accuracy, significantly reducing labor costs and minimizing errors. Autonomous forklifts and conveyor systems are also streamlining internal logistics, ensuring seamless movement of goods within distribution centers.

Another significant trend is the use of advanced decision-making tools, which enable businesses to analyze and act on large volumes of data with precision. In demand forecasting, these tools process diverse datasets, including historical sales, market trends, weather patterns, and geopolitical factors, to predict consumer needs accurately. This ensures that production schedules align with demand, preventing overproduction and stock shortages. Enhanced decision-making also extends to risk assessment, where predictive models evaluate potential disruptions such as supplier delays, natural disasters, or market volatility. By identifying risks early, businesses can implement contingency plans to maintain operational continuity and protect revenue streams.

Sustainability is a driving force behind many of the changes in modern supply chains. As environmental concerns become a priority for consumers and regulators, businesses are rethinking their practices to minimize environmental impact (Luthra, Garg, & Haleem, 2016). This includes adopting renewable energy sources in manufacturing and logistics operations, optimizing transportation routes to reduce fuel consumption, and exploring innovative packaging solutions that generate less waste. The concept of circular supply chains is gaining momentum, focusing on recycling, reusing, and repurposing materials to extend the lifecycle of products and reduce reliance on raw materials (Geissdoerfer, Savaget, Bocken, & Hultink, 2017; Ellen MacArthur Foundation, n.d.). These practices not only contribute to environmental conservation but also enhance operational efficiency and reduce costs in the long term.

Emerging technologies like blockchain and IoT are further enhancing supply chain transparency and traceability. Blockchain provides an immutable ledger of transactions, ensuring that every step in the supply chain is recorded and verifiable (Kamble, Gunasekaran, & Gawankar, 2020). This is particularly valuable for industries that prioritize product authenticity, ethical sourcing, and compliance with regulations. IoT-enabled sensors provide real-time data on the condition and location of goods, ensuring quality control and timely delivery. Together, these technologies improve accountability, build consumer trust, and enable businesses to meet stringent regulatory requirements.

To thrive in this rapidly evolving landscape, businesses must prioritize continuous learning and adaptation. This involves staying informed about emerging trends and investing in upskilling their workforce. Employees need to be equipped with the knowledge and skills required to operate and manage advanced systems. Companies can achieve this by offering training programs, workshops, and certifications tailored to specific technologies and industry needs. Fostering a culture of innovation is equally important, encouraging teams to experiment with new ideas and approaches to problem-solving.

Moreover, businesses must remain flexible in their strategies to respond effectively to market changes. This requires a proactive approach to technology adoption, where organizations continuously assess and integrate new tools and practices that align with their goals. Collaboration with technology providers, academic institutions, and industry peers can also help companies stay ahead of the curve by fostering shared learning and innovation.

The integration of these trends—autonomous systems, data-driven decision-making, and sustainability—is creating a paradigm shift in how supply chains operate. By embracing these advancements, businesses can build supply chains that are not only efficient and cost-effective but also resilient and environmentally conscious. This section delves deeper into each of these emerging trends, offering practical insights and examples of how companies are leveraging them to secure their place in the future of supply chain management. This forward-looking approach ensures that businesses remain competitive, adaptable, and ready to meet the challenges of an ever-changing global market.

CONCLUSION: PIONEERING THE FUTURE OF MODERN SYSTEMS

The journey through the transformative potential of advanced systems highlights a pivotal moment in technological and operational evolution. This book chapter has outlined the vast opportunities these systems offer, from revolutionizing supply chain processes to fostering innovation in decision-making, sustainability, and efficiency. However, as with any groundbreaking advancement, their implementation is not without challenges that demand strategic foresight and adaptability.

Advanced systems redefine the way industries operate by integrating automation, predictive analytics, and data-driven insights into every facet of the supply chain. They enhance processes from demand forecasting to real-time logistics, creating systems that are more agile, transparent, and responsive to market needs. Moreover, their role in advancing sustainability and fostering circular economies positions them as critical tools in addressing global environmental challenges while achieving operational excellence.

Yet, the path to adoption necessitates addressing key challenges, including financial hurdles, infrastructural gaps, and cultural resistance. By implementing tailored financing models, innovative infrastructural solutions, and robust change management strategies, organizations can pave the way for seamless integration. Ensuring data integrity and navigating the regulatory and ethical landscape further reinforces the trust and reliability of these systems, aligning them with societal expectations.

The workforce plays a central role in this transformation. Upskilling and reskilling employees, fostering collaborative environments, and building partnerships with educational and industry leaders ensure that businesses are equipped with the talent needed to harness the potential of these advancements. A culture of continuous learning and innovation emerges as a cornerstone of success in this era.

Looking forward, the fusion of advanced systems with emerging technologies such as blockchain, IoT, and autonomous tools signals an exciting horizon. Businesses that embrace this evolution with agility and foresight will not only achieve operational excellence but also set benchmarks for resilience, adaptability, and ethical innovation.

In conclusion, advanced systems represent both a challenge and an opportunity—a transformative force capable of reshaping industries and creating new paradigms for growth and sustainability. By addressing their complexities and leveraging their potential, organizations can lead in an era defined by intelligence, innovation, and adaptability, ensuring their place in the ever-evolving landscape of modern systems.

REFERENCES

Agrawal, A., Gans, J., & Goldfarb, A. (2018). *Prediction machines: The simple economics of artificial intelligence*. Harvard Business Press.

Amazon. (2021, September 9). *Amazon to invest $1.2 billion to upskill more than 300,000 employees by 2025*. Amazon Press Center. Retrieved from https://press.aboutamazon.com/news-releases/news-release-details/amazon-invest-12-billion-upskill-more-300000-employees-2025

Armbrust, M., Fox, A., Griffith, R., Joseph, A. D., Katz, R., Konwinski, A., Lee, G., Patterson, D., Rabkin, A., Stoica, I., & Zaharia, M. (2010). A view of cloud computing. *Communications of the ACM, 53*(4), 50–58. DOI: 10.1145/1721654.1721672

Aspen Institute. (2018, March 12). *Upskilling in the News: AT&T Invests $1 Billion in Employee Reskilling*. The Aspen Institute. Retrieved from https://www.aspeninstitute.org/of-interest/upskilling-news-att-invests-1-billion-employee-reskilling/

Astill, J., Dara, R. A., Campbell, M., Farber, J. M., Fraser, E. D., Ghavitri, L., & Yada, R. Y. (2019). Transparency in food supply chains: A review of enabling technology solutions. *Trends in Food Science & Technology, 91*, 240–247. DOI: 10.1016/j.tifs.2019.07.024

Atzori, L., Iera, A., & Morabito, G. (2010). The internet of things: A survey. *Computer Networks, 54*(15), 2787–2805. DOI: 10.1016/j.comnet.2010.05.010

Azuma, R. T. (1997). A survey of augmented reality. *Presence (Cambridge, Mass.), 6*(4), 355–385. DOI: 10.1162/pres.1997.6.4.355

Badia-Melis, R., Mishra, P., & Ruiz-García, L. (2015). Food traceability: New trends and challenges. *Food Control, 57*, 225–232. DOI: 10.1016/j.foodcont.2015.05.005

Bahrin, M. A. K., Othman, M. F., Azli, N. N., & Talib, M. F. (2016). Industry 4.0: A review on industrial automation and robotic. *Jurnal Teknologi, 78*(6-13), 137-143.

Baryannis, G., Validi, S., Dani, S., & Antoniou, G. (2019). Supply chain risk management and artificial intelligence: State of the art and future research directions. *International Journal of Production Research, 57*(7), 2179–2202. DOI: 10.1080/00207543.2018.1530476

BigScience Workshop. (2022). *BLOOM: A 176B-Parameter Open-Access Multilingual Language Model*. arXiv preprint arXiv:2211.05100.

Billinghurst, M., Clark, A., & Lee, G. (2015). A survey of augmented reality. *Foundations and Trends® in Human–Computer Interaction, 8*(2–3), 73-272.

Bostrom, N., & Yudkowsky, E. (2014). The ethics of artificial intelligence. In Frankish, K., & Ramsey, W. M. (Eds.), *The Cambridge handbook of artificial intelligence* (pp. 316–334). Cambridge University Press. DOI: 10.1017/CBO9781139046855.020

Bozgeyikli, E., Raij, A., & Katko, R. (2019). A survey on virtual and augmented reality for training of technical / motor skills. *ACM Computing Surveys*, *51*(6), 1–33.

Bressanelli, G., Adrodegari, F., Perona, M., & Saccani, N. (2018). The role of digital technologies to enable circular economy business models: A systematic literature review. *Procedia CIRP*, *78*, 25–30.

Brettel, M., Friederichsen, N., Keller, M., & Rosenberg, M. (2014). How virtualization, decentralization and network building change the manufacturing landscape: An Industry 4.0 Perspective. *International Journal of Mechanical. Industrial Science and Engineering*, *8*(1), 37–44.

Brown, T. B., Mann, B., Ryder, N., Subbiah, M., Kaplan, J., Dhariwal, P., & Amodei, D. (2020). Language models are few-shot learners. *Advances in Neural Information Processing Systems*, *33*, 1877–1901.

Brynjolfsson, E., & McAfee, A. (2014). *The second machine age: Work, progress, and prosperity in a time of brilliant technologies*. WW Norton & Company.

Chen, M., Mao, S., & Liu, Y. (2014). Big data: A survey. *Mobile Networks and Applications*, *19*(2), 171–209. DOI: 10.1007/s11036-013-0489-0

Choi, E., Bahadori, M. T., Song, L., Stewart, W. F., & Sun, J. (2017). GRAM: graph-based attention model for healthcare representation learning. In *Proceedings of the 23rd ACM SIGKDD international conference on knowledge discovery and data mining* (pp. 787-795). DOI: 10.1145/3097983.3098126

Chopra, S., & Meindl, P. (2016). *Supply chain management: Strategy, planning, and operation* (6th ed.). Pearson.

Christopher, M. (2016). *Logistics & supply chain management* (5th ed.).

DalleMule, L., & Davenport, T. H. (2017). What's your data strategy? *Harvard Business Review*, *95*(3), 112–121.

Daugherty, P. R., & Wilson, H. J. (2018). *Human + machine: Reimagining work in the age of AI*. Harvard Business Press.

Davenport, T. H., & Patil, D. J. (2012). Data scientist: The sexiest job of the 21st century. *Harvard Business Review*, *90*(5), 70–76. PMID: 23074866

Devlin, J., Chang, M. W., Lee, K., & Toutanova, K. (2019). BERT: Pre-training of deep bidirectional transformers for language understanding. In *Proceedings of the 2019 Conference of the North American Chapter of the Association for Computational Linguistics: Human Language Technologies,* Volume 1 *(Long and Short Papers)* (pp. 4171-4186).

Duan, Y., Edwards, J. S., & Dwivedi, Y. K. (2019). Artificial intelligence for decision making in the era of Big Data–evolution, challenges and research agenda. *International Journal of Information Management, 48*, 63–71. DOI: 10.1016/j.ijinfomgt.2019.01.021

Ellen MacArthur Foundation. (n.d.). *What is a circular economy?* Retrieved from https://www.ellenmacarthurfoundation.org/circulareconomy/concept

Etzkowitz, H., & Leydesdorff, L. (2000). The dynamics of innovation: From National Systems and "Mode 2" to a Triple Helix of university–industry–government relations. *Research Policy, 29*(2), 109–123. DOI: 10.1016/S0048-7333(99)00055-4

European Commission. (2018). *Communication from the Commission to the European Parliament, the Council, the European Economic and Social Committee and the Committee of the Regions on Artificial Intelligence for Europe*. COM(2018) 237 final.

European Parliament & Council of the European Union. (2016). Regulation (EU) 2016/679 of the European Parliament and of the Council of 27 April 2016 on the protection of natural persons with regard to the processing of personal data and on the free movement of such data, and repealing Directive 95/46/EC (General Data Protection Regulation). *Official Journal of the European Union, L, 119*(1).

Fagnant, D. J., & Kockelman, K. (2015). Preparing a nation for autonomous vehicles: Opportunities, barriers and policy recommendations. *Transportation Research Part A, Policy and Practice, 77*, 167–181. DOI: 10.1016/j.tra.2015.04.003

Floridi, L., Cowls, J., Beltramini, M., Saunders, D., & Vayena, E. (2018). An ethical framework for a good AI society: Opportunities, risks, principles, and recommendations. *AI & Society, 33*(4), 689–707. PMID: 30930541

Ford, M. (2015). *Rise of the robots: Technology and the threat of a jobless future*. Basic Books.

Frey, C. B., & Osborne, M. A. (2017). The future of employment: How susceptible are jobs to computerisation? *Technological Forecasting and Social Change, 114*, 254–280. DOI: 10.1016/j.techfore.2016.08.019

Gatys, L. A., Ecker, A. S., & Bethge, M. (2016). Image style transfer using convolutional neural networks. In *Proceedings of the IEEE conference on computer vision and pattern recognition* (pp. 2414-2423). DOI: 10.1109/CVPR.2016.265

Geissdoerfer, M., Savaget, P., Bocken, N. M., & Hultink, E. J. (2017). The Circular Economy–A new sustainability paradigm? *Journal of Cleaner Production, 143*, 757–768. DOI: 10.1016/j.jclepro.2016.12.048

Gibson, I., Rosen, D., & Stucker, B. (2015). *Additive manufacturing technologies: 3D printing, rapid prototyping, and direct digital manufacturing* (2nd ed.). Springer. DOI: 10.1007/978-1-4939-2113-3

Goodfellow, I., Pouget-Abadie, J., Mirza, M., Xu, B., Warde-Farley, D., Ozair, S., & Bengio, Y. (2014). Generative adversarial nets. *Advances in Neural Information Processing Systems*, ●●●, 27.

Govindan, K., Fattahi, M., & Keyvanshokooh, E. (2017). Supply chain network design for sustainable consumption and production: A critical review and research directions. *Journal of Cleaner Production, 149*, 875–897.

Grieves, M., & Vickers, J. (2017). Digital twin: Mitigating unpredictable, undesirable emergent behavior in complex systems. In Kahlen, F., Flumerfelt, S., & Alves, A. (Eds.), *Transdisciplinary perspectives on complex systems: New findings and approaches* (pp. 85–113). Springer. DOI: 10.1007/978-3-319-38756-7_4

Guide, V. D. R.Jr, & Van Wassenhove, L. N. (2009). The evolution of closed-loop supply chain research. *Operations Research, 57*(1), 10–18. DOI: 10.1287/opre.1080.0628

Haleem, A., Javaid, M., Qadri, M. A., & Suman, R. (2022). Understanding the role of digital technologies in education: A review. *Sustainable Operations and Computers, 3*, 275–285. DOI: 10.1016/j.susoc.2022.05.004

Hermann, M., Pentek, T., & Otto, B. (2016, January). Design principles for industrie 4.0 scenarios. In *2016 49th Hawaii international conference on system sciences (HICSS)* (pp. 3928-3937). IEEE.

Hochreiter, S., & Schmidhuber, J. (1997). Long short-term memory. *Neural Computation, 9*(8), 1735–1780. DOI: 10.1162/neco.1997.9.8.1735 PMID: 9377276

Hofmann, E., Rüsch, M., & Fuger, A. (2017). Industry 4.0 and the current status as well as future prospects on logistics. *Computers in Industry, 89*, 23–34. DOI: 10.1016/j.compind.2017.04.002

International Telecommunication Union (ITU). (2021). *Measuring digital development: Facts and figures 2021*. ITU.

Ivanov, D., Dolgui, A., & Sokolov, B. (2019). The impact of digital technology and Industry 4.0 on the ripple effect and supply chain risk analytics. *International Journal of Production Research*, *57*(3), 829–846. DOI: 10.1080/00207543.2018.1488086

Ivanov, D., Tsipoulanidis, A., & Schönberger, J. (2021). *Global supply chain and operations management: A decision-oriented introduction to the creation of value* (3rd ed.). Springer. DOI: 10.1007/978-3-030-72331-6

Jeske, T., Sitter, M., & Schilcher, U. (2019). Supply chain 4.0: Concept, challenges and implementation. *Procedia Manufacturing*, *39*, 745–754.

Jordan, M. I., & Mitchell, T. M. (2015). Machine learning: Trends, perspectives, and prospects. *Science*, *349*(6245), 255–260. DOI: 10.1126/science.aaa8415 PMID: 26185243

Kaisler, S., Armour, F., Espinosa, J. A., & Money, W. (2013). Big data: Issues and challenges moving forward. In *2013 46th Hawaii International Conference on System Sciences* (pp. 995-1004). IEEE.

Kamble, S. S., Gunasekaran, A., & Gawankar, S. A. (2020). Achieving sustainable performance in a data-driven agriculture supply chain: A review for research and applications. *International Journal of Production Economics*, *219*, 179–194. DOI: 10.1016/j.ijpe.2019.05.022

Kamilaris, A., & Prenafeta-Boldú, F. X. (2018). Deep learning in agriculture: A survey. *Computers and Electronics in Agriculture*, *147*, 70–90. DOI: 10.1016/j.compag.2018.02.016

Kaplan, A., & Haenlein, M. (2019). Siri, Siri, in my hand: Who's the fairest in the land? On the interpretations, illustrations, and implications of artificial intelligence. *Business Horizons*, *62*(1), 15–25. DOI: 10.1016/j.bushor.2018.08.004

Khatri, V., & Brown, C. V. (2010). Designing data governance. *Communications of the ACM*, *53*(1), 148–152. DOI: 10.1145/1629175.1629210

Kingma, D. P., & Welling, M. (2013). Auto-encoding variational bayes. *arXiv preprint arXiv:1312.6114*.

Kotter, J. P. (2012). *Leading change*. Harvard Business Press.

Krizhevsky, A., Sutskever, I., & Hinton, G. E. (2012). Imagenet classification with deep convolutional neural networks. *Advances in Neural Information Processing Systems*, ●●●, 25.

Kshetri, N. (2018). Blockchain's roles in meeting key supply chain management objectives. *International Journal of Information Management*, *39*, 80–89. DOI: 10.1016/j.ijinfomgt.2017.12.005

Lasi, H., Fettke, P., Kemper, H. G., Feld, T., & Hoffmann, M. (2014). Industry 4.0. *Business & Information Systems Engineering*, *6*(4), 239–242. DOI: 10.1007/s12599-014-0334-4

LeCun, Y., Bengio, Y., & Hinton, G. (2015). Deep learning. *Nature*, *521*(7553), 436–444. DOI: 10.1038/nature14539 PMID: 26017442

Lee, E. A., Bagheri, B., & Kao, H. A. (2015). A cyber-physical systems architecture for industry 4.0-based manufacturing systems. *Manufacturing Letters*, *3*, 18–23. DOI: 10.1016/j.mfglet.2014.12.001

Lee, Y. W., & Strong, D. M. (2003). Knowing-why about data processes and data quality. *Journal of Management Information Systems*, *20*(3), 13–39. DOI: 10.1080/07421222.2003.11045775

Luthra, S., Garg, D., & Haleem, A. (2016). The impacts of critical success factors for implementing green supply chain management towards sustainability: An empirical investigation of Indian automobile industry. *Journal of Cleaner Production*, *121*, 142–158. DOI: 10.1016/j.jclepro.2016.01.095

Makridakis, S. (2017). The forthcoming Artificial Intelligence (AI) revolution: Its impact on society and firms. *Futures*, *90*, 46–60. DOI: 10.1016/j.futures.2017.03.006

McKinsey Global Institute. (2017). *Jobs lost, jobs gained: Workforce transitions in a time of automation*. McKinsey & Company.

Mittelstadt, B. D., Allo, P., Taddeo, M., Wachter, S., & Floridi, L. (2016). The ethics of algorithms: Mapping the debate. *Big Data & Society*, *3*(2), 2053951716679679. DOI: 10.1177/2053951716679679

Monczka, R. M., Handfield, R. B., Giunipero, L. C., & Patterson, J. L. (2015). *Purchasing and supply chain management*. Cengage Learning.

Noe, R. A., Hollenbeck, J. R., Gerhart, B., & Wright, P. M. (2017). *Human resource management: Gaining a competitive advantage* (10th ed.). McGraw-Hill Education.

OECD. (2019). *Artificial intelligence in society*. OECD Publishing.

Pearson, T., May, D., Leontidis, G., Swainson, M., Brewer, D., Bidaut, L., & Zisserman, A. (2019). Deep learning for environmental GPR data processing. *IEEE Geoscience and Remote Sensing Letters*, *16*(9), 1385–1389.

Pfleeger, C. P., Pfleeger, S. L., & Margulies, J. (2015). *Security in computing* (5th ed.). Prentice Hall Press.

Provost, F., & Fawcett, T. (2013). *Data Science for Business: What you need to know about data mining and data-analytic thinking*. O'Reilly Media, Inc.

Radford, A., Narasimhan, K., Salimans, T., & Sutskever, I. (2018). *Improving language understanding by generative pre-training*. OpenAI.

Raffel, C., Shazeer, N., Roberts, A., Lee, K., Narang, S., Matena, M., & Liu, P. J. (2020). Exploring the limits of transfer learning with a unified text-to-text transformer. *Journal of Machine Learning Research, 21*(140), 1–67.

Ramesh, A., Pavlov, M., Goh, G., Gray, S., Voss, C., Radford, A., & Sutskever, I. (2021). Zero-shot text-to-image generation. In *International Conference on Machine Learning* (pp. 8821-8831). PMLR.

Redman, T. C. (2013). *Data driven: Profiting from your most important business asset*. Harvard Business Press.

Richards, G. (2017). *Warehouse management: A complete guide to improving efficiency and minimizing costs in the modern warehouse* (3rd ed.). Kogan Page Publishers.

Rogers, D. S., & Tibben-Lembke, R. S. (1999). *Going backwards: Reverse logistics trends and practices*. Reverse Logistics Executive Council.

Rombach, R., Blattmann, A., Lorenz, D., Esser, P., & Ommer, B. (2022). High-resolution image synthesis with latent diffusion models. In *Proceedings of the IEEE/CVF Conference on Computer Vision and Pattern Recognition* (pp. 10684-10695). DOI: 10.1109/CVPR52688.2022.01042

Russell, S. J., & Norvig, P. (2021). *Artificial intelligence: A modern approach* (4th ed.). Pearson.

Schein, E. H. (2010). *Organizational culture and leadership* (Vol. 2). John Wiley & Sons.

Schuh, G., Anderl, R., Gausemeier, J., ten Hompel, M., & Wahlster, W. (Eds.). (2017). *Industrie 4.0 Maturity Index. Managing the Digital Transformation of Companies (acatech STUDY)*. Herbert Utz Verlag.

Shi, W., Cao, J., Zhang, Q., Li, Y., & Xu, L. (2016). Edge computing: Vision and challenges. *IEEE Internet of Things Journal, 3*(5), 637–646. DOI: 10.1109/JIOT.2016.2579198

Silver, E. A., Pyke, D. F., & Peterson, R. (1998). *Inventory management and production planning and scheduling* (3rd ed.). John Wiley & Sons.

Stadtler, H. (2015). Supply chain management: An overview. In Stadtler, H., Kilger, C., & Meyr, H. (Eds.), *Supply chain management and advanced planning: Concepts, models, software, and case studies* (pp. 9–36). Springer. DOI: 10.1007/978-3-642-55309-7_1

Stock, T., & Seliger, G. (2016). Opportunities of sustainable manufacturing in industry 4.0. *Procedia CIRP*, *40*, 536–541. DOI: 10.1016/j.procir.2016.01.129

Stouffer, K., Pillitteri, V., Lightman, S., Abrams, M., & Hahn, A. (2015). *Guide to industrial control systems (ICS) security*. National Institute of Standards and Technology. DOI: 10.6028/NIST.SP.800-82r2

Tambe, P., Cappelli, P., & Yakubovich, V. (2019). Artificial intelligence in human resources management: An agenda for research and practice. *California Management Review*, *61*(4), 5–26. DOI: 10.1177/0008125619867910

Tang, C. S. (2006). Perspectives in supply chain risk management. *International Journal of Production Economics*, *103*(2), 451–488. DOI: 10.1016/j.ijpe.2005.12.006

Tao, F., Cheng, J., Qi, Q., Zhang, M., Zhang, H., & Sui, F. (2018). Digital twin-driven prognostics and health management for complex equipment. *CIRP Annals*, *67*(1), 169–172. DOI: 10.1016/j.cirp.2018.04.055

Tiwari, S., Wee, H. M., & Daryanto, Y. (2018). Big data analytics in supply chain management: A state-of-the-art literature review. *Computers & Operations Research*, *98*, 212–227.

Tzounis, A., Katsoulas, N., & Fountas, S. (2017). Robotics in agriculture. A review. *Agronomy for Sustainable Development*, *37*(3), 21.

United Nations. (2015). *Transforming our world: The 2030 Agenda for Sustainable Development*. United Nations.

Vaswani, A., Shazeer, N., Parmar, N., Uszkoreit, J., Jones, L., Gomez, A. N., & Polosukhin, I. (2017). Attention is all you need. *Advances in Neural Information Processing Systems*, •••, 30.

Wang, G., Gunasekaran, A., Ngai, E. W., & Papadopoulos, T. (2016). Big data analytics in logistics and supply chain management: Certain investigations for research and applications. *International Journal of Production Economics*, *176*, 98–110. DOI: 10.1016/j.ijpe.2016.03.014

Whitmore, A., Agarwal, A., & Da Xu, L. (2015). The Internet of Things—A survey of topics and trends. *Information Systems Frontiers*, *17*(2), 261–274. DOI: 10.1007/s10796-014-9489-2

World Bank. (2021). *World Development Report 2021: Data for Better Lives*. World Bank.

World Economic Forum (WEF). (2020). *The Future of Jobs Report 2020*. World Economic Forum.

Zhang, S., Roller, S., Goyal, N., Artetxe, M., Chen, M., Chen, S., . . . Zettlemoyer, L. (2022). OPT: Open Pre-trained Transformer Language Models. *arXiv preprint arXiv:2205.01068*.

Zhao, S., Chen, H., Liu, R., & Li, G. (2023). Deep Learning for Food Quality and Safety: A Review. *Comprehensive Reviews in Food Science and Food Safety*, *22*(1), 531–556.

Chapter 6
AI-Driven Route Optimization and Sustainable Logistics:
Transforming Stakeholder Perspectives for a Greener Future

Pawan Kumar
https://orcid.org/0000-0003-4892-6374
Graphic Era University, India

Arvinder Kaur
https://orcid.org/0000-0001-7360-5674
Chandigarh University, India

Bhupinder Pal Singh Chahal
https://orcid.org/0000-0002-8106-7121
Yorkville University, Canada

Pravesh Soti
https://orcid.org/0000-0001-8142-884X
NIET Business School, Greater Noida, India

ABSTRACT

This chapter examines how generative AI helps the food SCM evolve and aims to improve its efficiency, robustness, and sustainability performance. Using generative AI models, predictive analytics, and digital technologies, the paper identifies the AI's usage, data availability, digital supply chain, and the accuracy of the demand forecast. Therefore, the structured quantitative approach was adopted, and responses were collected on a seven-point Likert scale from 215 participants: financial ana-

DOI: 10.4018/979-8-3693-9856-2.ch006

lysts, market specialists, and other supply chain practitioners. Simple descriptive statistics such as frequency distributions and measures of central tendencies and variability were used as data analysis tools, and hypothesis testing involved basic bivariate correlation coefficients and bootstrapping to make the results conservative and replicable. The study points out how applying AI enhances decision-making, cuts costs and wastes, and can also enhance logistics. It also provides valuable information on developing sustainable and flexible food supply chains in complex contexts.

INTRODUCTION

Generative AI represents a sophisticated leap in artificial intelligence, evolving significantly since the advent of deep learning in the 2010s. Breakthroughs such as GANs and transformer models have enabled creative synthesis, reshaping industries with applications in content generation, predictive modeling, and decision augmentation (Sengar et al., 2024). Artificial intelligence has revolutionized supply chain management by leveraging predictive analytics, automation, and real-time decision-making to optimize logistics, enhance demand forecasting, mitigate inefficiencies, reduce costs, and adapt dynamically to market fluctuations, fostering resilience and innovation across global networks (Karthiayani & Raman, 2024).

AI and Food Supply Chain Management

In food supply chain management, AI enhances precision and sustainability by forecasting demand, minimizing waste, optimizing perishable inventory, and integrating data across the production-to-consumption continuum to ensure traceability, mitigate disruptions, and address challenges in maintaining food quality and safety (Krupitzer, 2024). Artificial intelligence (AI) has emerged as a transformative force in food supply chain management, seamlessly integrating advanced computational methodologies to enhance efficiency, sustainability, and resilience (Vegesna et al., 2024). By leveraging predictive analytics and machine learning algorithms, AI forecasts demand with unprecedented accuracy, enabling precise inventory control and minimizing waste. Generative AI models further refine these capabilities, synthesizing complex data streams to optimize logistics, from procurement to last-mile delivery (Senanayake et al., 2024). Moreover, AI augments traceability through technologies such as blockchain integration and IoT-enabled sensors, offering unparalleled visibility across the supply chain. This ensures adherence to stringent food safety standards, mitigates risks of contamination, and fortifies consumer trust. Advanced AI systems also address disruptions caused by fluctuating market dynamics, climate variability, and geopolitical uncertainties by dynamically adjusting supply chain strategies in real time (Ahlawat et al., 2024). By harmonizing automation and human decision-making, AI fosters a paradigm shift in how perishable goods are managed, reducing spoilage and aligning operations with sustainability goals (Kollia et al., 2021). The confluence of generative models and big data analytics thus positions AI as a cornerstone of innovation, enabling a resilient, data-driven approach to the complexities of modern food supply chains (Shahin et al., 2024).

Purpose of the study: This study aims to assess the impact of generative AI technology adoption on food supply chain management, focusing on its role in improving demand forecast accuracy, optimizing operations, enhancing supply chain efficiency, and promoting resilience, while also considering the effects of data availability, supply chain digitization, and predictive analytics in driving sustainable and cost-effective strategies.

Research methodology: The investigation used the quantitative method, which stimulated the administration of closed questions through a structured seven-point Likert scale to grasp subtle deviations in response from the participants. The devices measured generative AI integration, data, digitization, demand-side accuracy, and supply-side efficiency. Descriptive C statistics, bivariate correlations, and bootstrapping have provided data research strength and accurate generalization reality.

Sample size: 215

Research Gap: The intricate dynamics between generative AI and food supply chain management remain underexplored, particularly in assessing generative AI's influence on resilience, operational precision, and sustainability, given the constraints of data integration, transparency, and scalability across varied digital maturity levels, thereby necessitating deeper empirical scrutiny into these transformative yet challenging intersections to advance both theoretical frameworks and practical applications.

Variables of the Study

Independent Variables

1. **Generative AI Technology Adoption** – The extent to which generative AI tools are integrated into the food supply chain (e.g., AI models for demand forecasting, predictive analytics).
2. **Data Availability and Quality** – The availability of high-quality, real-time data (e.g., data on production, logistics, and inventory) for AI systems to function effectively.
3. **Supply Chain Digitization Level** – The degree to which the food supply chain processes are digitized (e.g., automation, IoT devices, and cloud computing integration).

4. **Demand forecast accuracy-** Accuracy by which the subsequent consumer demand is forecasted, indicating the match of forecasted and actual demand through improved algorithmic and statistical methods.

Dependent Variable

1. **Supply Chain Efficiency:** Operation efficiency within the food supply chain consists of cost reduction, delivery lead time, inventory management, and effective resource utilization, moderated by adopting generative AI and technological variables (Bouazizi et al., 2024).

Figure 1. Conceptual model

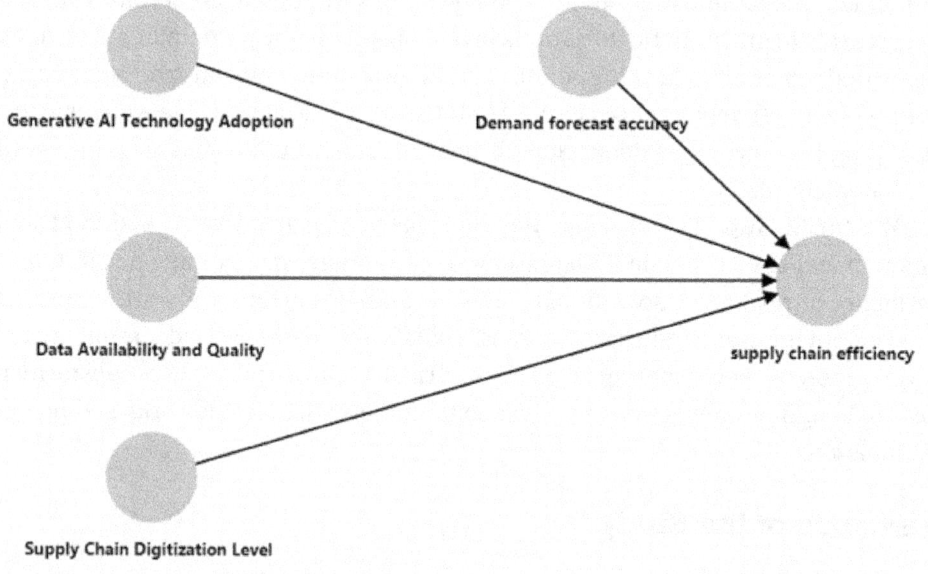

Source*: Conceptual model designed by author based on variables*

Hypotheses

Hypothesis 1: To know the effect of generative AI technology adoption on supply chain efficiency.
Hypothesis 2: To study the effect of data availability and quality on Supply chain efficiency

Hypothesis 3: To study the effect of supply chain digitalization level on supply chain efficiency

Hypothesis 4: To investigate the impact of Demand forecast accuracy on supply chain efficiency.

LITERATURE REVIEW

H1: Effect of generative AI technology adoption on supply chain efficiency.

The adoption of generative AI technology in supply chain management has been widely recognized for its transformative potential, with literature highlighting its ability to enhance operational efficiency through advanced predictive analytics (Shrivastava & Pal, 2019), optimization of logistics networks, demand forecasting, and real-time decision-making, thereby reducing costs, minimizing waste, improving inventory management, and fostering greater resilience to disruptions, though challenges related to data integration, algorithmic transparency, and scalability remain significant barriers to its widespread implementation across diverse supply chain sectors (Reddy et al., 2024).

H2: Data availability and quality on Supply chain efficiency

The extant literature underscores the critical role of data availability and quality in determining supply chain efficiency (Ghodake et al., 2024), with numerous studies emphasizing that robust, timely, and accurate data not only facilitates enhanced decision-making, predictive analytics, and inventory optimization but also mitigates risks, improves demand forecasting, and enables seamless integration across supply chain functions, while highlighting that insufficient or poor-quality data, characterized by incompleteness, inconsistency, and inaccuracy, often exacerbates inefficiencies, disrupts coordination, and hampers the overall agility and responsiveness of supply chains in dynamic market environments (Gupta et al., 2023).

H3: Effect of supply chain digitalization level on supply chain efficiency

The literature reveals a substantial correlation between the level of supply chain digitalization and operational efficiency, demonstrating that advanced digital technologies—such as IoT, blockchain, AI, and cloud-based platforms—optimize real-time data exchange (Perumalsamy & Natarajan, 2010), enhance transparency, improve coordination among supply chain partners, facilitate predictive analytics for demand forecasting and inventory management, and streamline decision-making processes, thereby significantly reducing lead times, minimizing operational costs, and fostering greater flexibility and resilience; however, the degree of digital maturity, coupled with challenges related to integration complexity, cyber security risks, and technological adoption gaps, particularly in smaller enterprises, can impact the full realization of these efficiency gains (Radhika et al., 2024).

H4: Impact of Demand forecast accuracy on supply chain efficiency

The literature highlights that demand forecast accuracy plays a pivotal role in enhancing supply chain efficiency, as precise demand predictions enable optimal inventory management, reduce stockouts and overstocking, improve production scheduling, and enhance distribution strategies, ultimately minimizing costs and lead times (Pandey & Mishra, 2024); however, inaccuracies in forecasting can lead to significant disruptions, including excess inventory, increased holding costs, and missed opportunities, thereby undermining supply chain agility, responsiveness, and overall performance, with factors such as data quality, forecasting models, and market volatility being critical determinants of forecast reliability and its subsequent impact on supply chain outcomes (Onyeaka et al., 2024).

Table 1. Research instruments

Scale Number	Response Description
1	**Strongly Disagree** – I completely disagree with the statement, and there is no alignment with my experience or opinion.
2	**Disagree** – I somewhat disagree with the statement, and I believe it does not reflect my experience or opinion well.
3	**Somewhat Disagree** – I lean toward disagreeing with the statement, but there are some elements I can agree with.
4	**Neutral** – I neither agree nor disagree with the statement. My experience or opinion does not strongly support or oppose it.
5	**Somewhat Agree** – I tend to agree with the statement, though there may be some reservations or exceptions.
6	**Agree** – I generally agree with the statement, and it aligns well with my experience or opinion.
7	**Strongly Agree** – I fully agree with the statement, and it strongly aligns with my experience or opinion.

Table 1 outlines the response scale used in the research instruments, categorizing participants' degree of agreement with statements on a seven-point Likert scale, ranging from “Strongly Disagree” (1), indicating complete disagreement with no alignment to personal experience or opinion, to “Strongly Agree” (7), reflecting full agreement with a strong alignment to the respondent's experiences or perspectives, with intermediate responses allowing for varying levels of agreement or disagreement, thus providing a nuanced understanding of participants' attitudes or perceptions.

Table 2. Research instruments

Variable	Question Statement	Degree of Response (1-7)
Generative AI Technology Adoption		
	1. Our organization has successfully integrated generative AI technologies in our supply chain processes.	1 (Strongly Disagree) - 7 (Strongly Agree)
	2. The adoption of generative AI has improved decision-making in our food supply chain.	1 (Strongly Disagree) - 7 (Strongly Agree)
	3. Our employees have received sufficient training to effectively use AI-driven tools in supply chain tasks.	1 (Strongly Disagree) - 7 (Strongly Agree)
	4. We have seen noticeable improvements in efficiency due to the implementation of generative AI.	1 (Strongly Disagree) - 7 (Strongly Agree)
	5. Generative AI has become an integral part of our supply chain management system.	1 (Strongly Disagree) - 7 (Strongly Agree)
Data Availability and Quality		
	1. The data available in our supply chain is of high quality and relevant for AI-based predictions.	1 (Strongly Disagree) - 7 (Strongly Agree)
	2. Our organization has sufficient real-time data to support AI-driven decision-making in the supply chain.	1 (Strongly Disagree) - 7 (Strongly Agree)
	3. Data collected from different sources is well-integrated and accessible for AI analysis.	1 (Strongly Disagree) - 7 (Strongly Agree)
	4. The quality of our supply chain data meets the standards required for generative AI tools to function well.	1 (Strongly Disagree) - 7 (Strongly Agree)
	5. We regularly update and clean our supply chain data to maintain its relevance and accuracy.	1 (Strongly Disagree) - 7 (Strongly Agree)
Supply Chain Digitization Level		
	1. Our supply chain processes are highly automated and digitized.	1 (Strongly Disagree) - 7 (Strongly Agree)
	2. The use of digital technologies, such as IoT and blockchain, is widespread across our supply chain operations.	1 (Strongly Disagree) - 7 (Strongly Agree)
	3. Digital platforms are used to manage inventory and logistics effectively in our supply chain.	1 (Strongly Disagree) - 7 (Strongly Agree)

continued on following page

Table 2. Continued

Variable	Question Statement	Degree of Response (1-7)
	4. We rely on digital tools to track and monitor the flow of goods through the supply chain.	1 (Strongly Disagree) - 7 (Strongly Agree)
	5. Our supply chain system is well-equipped with advanced digital infrastructure to support AI applications.	1 (Strongly Disagree) - 7 (Strongly Agree)
Demand Forecast Efficiency		
	1. AI-based demand forecasting has improved the accuracy of our supply chain predictions.	1 (Strongly Disagree) - 7 (Strongly Agree)
	2. Our organization is able to predict fluctuations in demand more effectively using AI tools.	1 (Strongly Disagree) - 7 (Strongly Agree)
	3. We can efficiently align supply chain operations with predicted demand trends.	1 (Strongly Disagree) - 7 (Strongly Agree)
	4. AI-enabled demand forecasting has helped reduce overstock and stockouts in our supply chain.	1 (Strongly Disagree) - 7 (Strongly Agree)
	5. The efficiency of meeting demand has significantly improved due to AI-driven forecasting techniques.	1 (Strongly Disagree) - 7 (Strongly Agree)
Dependent Variable: Supply Chain Efficiency		
	1. The efficiency of our food supply chain has significantly improved since adopting AI-driven technologies.	1 (Strongly Disagree) - 7 (Strongly Agree)
	2. We have observed a reduction in supply chain costs due to the implementation of AI solutions.	1 (Strongly Disagree) - 7 (Strongly Agree)
	3. AI technologies have helped us optimize inventory management and reduce waste in our supply chain.	1 (Strongly Disagree) - 7 (Strongly Agree)
	4. The time taken to deliver products has decreased as a result of AI-enhanced logistics management.	1 (Strongly Disagree) - 7 (Strongly Agree)
	5. Our food supply chain is more responsive to changes in demand due to AI integration.	1 (Strongly Disagree) - 7 (Strongly Agree)

This table 2 illustrates a comprehensive framework for evaluating the impact of generative AI adoption, data availability and quality, and supply chain digitization on food supply chain efficiency, where respondents' perceptions of AI integration, data relevance, and digital infrastructure are assessed across various dimensions, with the aim of determining the extent to which these factors collectively contribute to optimizing decision-making, reducing costs, enhancing inventory management, improving logistics responsiveness, and ultimately driving overall operational efficiency within the supply chain ecosystem (Román & Menéndez, 2024).

DATA ANALYSIS

Table 3. Respondents

Respondents	Number	%
Financial Analysts and Market Experts	25	11.52%
Institutional Investors	6	2.76%
Corporate Executives and Financial Officers	5	2.30%
Technology Experts in Cybersecurity	27	12.44%
Retail Investors	130	60.00%
Legal Experts in Financial Crimes	10	4.61%
Journalists and Media Professionals	12	5.53%
TOTAL	215	100.00%

Table 2 presents the distribution of respondents reflects a diverse asymmetrical representations with retail investors comprising an overwhelming majority at 60.00% juxtaposed against the relatively modest participation of financial analysis, market experts, and technology experts in cyber security who together account for 23.96%, while institutional investors, corporative executives, legal experts in financial crimes and journalists collectively contribute to a smaller albeit critical, 15.2% demonstrating a multifaceted cross section of expertise that underscores the predominance of individual investors in this evaluative cohort of 215 participants.

Table 4. Descriptive Statistics

Descriptive Statistics						
		Statistic	**Bootstrap[a]**			
			Bias	**Std. Error**	**95% Confidence Interval**	
					Lower	**Upper**
Generative AI Technology Adoption	Mean	5.13	.0013	.1079	4.89	5.32
	Std. Deviation	1.54	-.00624	.06984	1.42120	1.6979
	N	215	0	0	215	215

continued on following page

Table 4. Continued

Descriptive Statistics							
		Statistic	**Bootstrap**[a]				
			Bias	**Std. Error**	**95% Confidence Interval**	
					Lower	**Upper**
Data Availability and Quality	Mean	4.55	-.0020	.0873	4.36	4.735
	Std. Deviation	1.245	-.00180	.06254	1.13356	1.37632
	N	215	0	0	215	215
Supply Chain **Digitization Level**	Mean	4.174	-.00149	.10139	3.9669	4.3632
	Std. Deviation	1.58303	-.00154	.06378	1.44706	1.70321
	N	215	0	0	215	215
Demand forecast accuracy	Mean	4.5760	-.0019	.0866	4.4055	4.7327
	Std. Deviation	1.31056	-.00246	.06384	1.17801	1.42991
	N	215	0	0	215	215
supply chain efficiency	Mean	3.99	.0019	.10019	3.8012	4.2113
	Std. Deviation	1.50201	-.00372	.06142	1.37415	1.61337
	N	215	0	0	215	215

a. Unless otherwise noted, bootstrap results are based on 1000 bootstrap samples

The table 3 provides an exhaustive statistical examination of several critical variables—namely "Prevalence of Deepfake Incidents," "Investor Confidence," "Technological Sophistication of Deepfake Detection Tools," "Cyber security Measures in Financial Institutions," and "Market Stability"—where each variable's mean and standard deviation are meticulously adjusted for bootstrapped bias and standard errors, with confidence intervals delineating the range of plausible values, thus offering a robust depiction of the central tendencies and dispersions within the dataset, ensuring statistical precision and reliability, and presenting a comprehensive quantitative portrait of the variability and relationships across these pivotal constructs in the financial ecosystem.

Table 5. Correlations

Correlations								
				Generative AI Technology Adoption	Data Availability and Quality	Supply Chain **Digitization Level**	Demand forecast accuracy	supply chain efficiency
Generative AI Technology Adoption	Pearson Correlation			1	.129	-.087	.0159	-.0489
	Sig. (2-tailed)				.0549	.1959	.813	.470
	Sum of Squares and Cross-products			532.561	55.561	-47.424	7.123	-25.000
	Covariance			2.464	.253	-.219	.030	-.115
	N			215	215	215	215	215
	Bootstrap[c]	Bias		0	-.00199	-.0099	.000	-.002
		Std. Error		0	.070	.070	.072	.074
		95% Confidence	Lower	1	-.006	-.223	-.123	-.196
			Upper	1	.259	.0539	.155	.0969
Data Availability and Quality	Pearson Correlation			.129	1	.474**	.535**	.415**
	Sig. (2-tailed)			.0549		.000	.000	.000
	Sum of Squares and Cross-products			55.569	341.408	204.635	190.722	169.003
	Covariance			.256	1.579	.946	.881	.786
	N			215	215	215	215	215
	Bootstrap[c]	Bias		-.002	0	.002	.001	.001
		Std. Error		.070	0	.0519	.0599	.0619
		95% Confidence Interval	Lower	-.006	1	.372	.425	.289
			Upper	.261	1	.574	.641	.534
Supply Chain Digitization Level	Pearson Correlation			-.0878	.475**	1	.526**	.564**
	Sig. (2-tailed)			.195	.000		.000	.000
	Sum of Squares and Cross-products			-47.428	204.635	543.345	237.110	290.008
	Covariance			-.219	.946	2.514	1.089	1.345
	N			215	215	215	215	215
	Bootstrap[c]	Bias		-.001	.002	0	.002	.001
		Std. Error		.069	.0519	0	.0579	.0541
		95% Confidence	Lower	-.222	.371	1	.401	.455
			Upper	.0539	.5794	1	.633	.698

continued on following page

Table 5. Continued

Demand forecast accuracy	Pearson Correlation			.0159	.546**	.527**	1	.424**
	Sig. (2-tailed)			.813	.000	.000		.000
	Sum of Squares and Cross-products			7.142	190.721	237.109	370.98	180.001
	Covariance			.0329	.881	1.097	1.716	.837
	N			215	215	215	215	215
	Bootstrap[c]	Bias		.000	.001	.002	0	.004
		Std. Error		.0719	.0559	.0579	0	.060
		95% Confidence Interval	Lower	-.124	.425	.404	1	.311
			Upper	.155	.641	.633	1	.543
supply chain efficiency	Pearson Correlation			-.0489	.4159**	.564**	.424**	1
	Sig. (2-tailed)			.471	.000	.000	.000	
	Sum of Squares and Cross-products			-24.001	169.90	290.90	180.009	487.89
	Covariance			-.1159	.786	1.337	.837	2.256
	N			215	215	215	215	215
	Bootstrap[c]	Bias		-.002	.001	.001	.0039	0
		Std. Error		.0739	.061	.053	.060	0
		95% Confidence	Lower	-.195	.289	.455	.310	1
			Upper	.0969	.534	.668	.541	1

**. Correlation is significant at the 0.01 level (2-tailed).

c. Unless otherwise noted, bootstrap results are based on 1000 bootstrap samples

Table 4 reveals a nuanced interplay among the variables, where investor confidence exhibits a significant positive correlation with market stability, technological sophistication of deepfake detection tools, and cybersecurity measures, underscoring its central role in maintaining market equilibrium; conversely, the prevalence of deepfake incidents shows an insignificant relationship with market stability, while technological sophistication and cybersecurity measures demonstrate robust positive associations with market stability, suggesting their critical influence in bolstering financial resilience against the destabilizing effects of deepfake threats and cyber vulnerabilities.

Table 6. Case Processing Summary

Case Processing Summary						
	Cases					
	Included		Excluded		Total	
	N	Percent	N	Percent	N	Percent
supply chain efficiency * Generative AI Technology Adoption	**215**	**100.0%**	**0**	**0.0%**	**215**	**100.0%**
supply chain efficiency * Data Availability and Quality	215	100.0%	0	0.0%	215	100.0%
supply chain efficiency * Supply Chain Digitization Level	215	100.0%	0	0.0%	215	100.0%
supply chain efficiency * Demand forecast accuracy	215	100.0%	0	0.0%	215	100.0%

The Case Processing Summary reveals in table 5 that comprehensive data inclusion, with no exclusions, across all variables—Generative AI Technology Adoption, Data Availability and Quality, Supply Chain Digitization Level, and Demand Forecast Accuracy—attaining a 100% participation rate (N=215) for evaluating their interplay with supply chain efficiency.

MANAGERIAL IMPLICATIONS

The conclusion and results of this research have crucial implications for managerial practice, especially in the food supply chain context. The use of generative AI disrupts the food supply chain and forces managers to look for new ways of operating while introducing data-oriented technological models (Guo et al., 2014). Depending on it, generative AI improves the accuracy of predation models and helps a manager predict such issues as changes in demand, stock control, and potential failures. Developing contingencies enables reasonable determination of the right resources to allocate, thus saving money and time. Investing in many higher-level AI systems and cultivating an environment of constant technological progress are requirements to maintain competitive superiority (Pasupuleti et al., 2024). The study establishes how generative AI can strengthen the supply chain's resilience against volatile environmental factors, including climate shocks and political instabilities. Managers must employ adaptive supply chain management, using AI-based what-ifs to plan for and anticipate future disruptions (Kanjilal et al., 2025). By integrating scenario planning into strategic frameworks, organizations can maintain operational continuity while protecting the interests of stakeholders. Development of cross-functional cooperation Stakeholder generative AI digitalization requirements call for fluent interdepartmental relations (Danach et al., 2024). Managers must break down functional hierarchies, particularly procurement, logistics, and marketing, to become unified through data technology sharing interfaces. It also enhances integration across operational processes while initiating coherent approaches in response to the market. Another synergy is realized when organizations invest in cross-functional training and create a coherent digital environment (Punia et al., 2020).

The research identifies data integration issues and resistance to change, which are likely to affect the implementation of generative AI. In this structure, managers need to adopt wide-ranging change management strategies and actively involve all the employees who are almost affected to appreciate AI's value in organizations (Nayak et al., 2024). Besides, strengthening protective measures against cyber threats will guarantee no data compromise; thus, removing concerns over data leakage.

Nonetheless, sustainability criteria and ethical considerations are the remaining slow food principles central to the traditional Italian food culture (Bhowmik et al., 2024).

Because environmentally friendly policies are becoming prominent, generative AI offers a means of planning delivery to achieve more efficiency and less pollution. Managers must incorporate sustainability measures into the performance assessments, using artificial intelligence to monitor sustainability performance. It is then logical to add that the ethical regulation of AI deployment should also be considered to build the trust of consumers and partners (Aktas, 2024).

Last, the study highlights the importance of integrating artificial intelligence into a sequential development plan. The managers must develop strong associations with technology vendors, universities and the Ministry of Education to keep up with the most recent novelties and constantly assess applicable rules and regulations (Ladva et al., 2024). In these areas, organizations are to become proactive and ensure their industry dominance by using generative AI in the face of new market uncertainties.

CONCLUSION

This paper helps to shed light on how the innovativeness of generative AI in the context of the food supply chain demonstrates agility to manage the problem, improve decision-making, and better prepare for future challenges. From this perspective, generative AI, which combines the latest algorithms and anticipatory analysis technology, becomes crucial for managing today's supply chain challenges. Thus, this conclusion not only summarizes the results of the investigation but also defines further prospects and importance to the general field. The paper presents real-life data collected to support the argument and show that generative AI goes beyond traditional operational models because it enhances inventory control and supply chain processes and reduces disruptions. This paper's research method employs techniques and analyzes many respondents across the supply chain, making the research findings reliable and generalizable. The grand innovation of AI as a technology is in its handling capacity and the determination of complex patterns within large datasets to move away from the culture of seeking solutions to problems towards strategic planning strategies. The study's major finding is the utilitarian character of generative AI as both efficiency-generating and sustainability-enabling instruments. With the help of complicated simulation algorithms, AI can easily conclude demand, reduce the amount of waste, and decrease the overall CO2 emissions from transportation and storage. This highlights the recent drive toward institutionalizing environmental factors within the supply chain principle. However, to realize such outcomes, UTB must make targeted investments in AI structure while developing organizational preparedness for technology change.

Nonetheless, the implementation of generative AI comes with several challenges, as brought out in this paper. The work discovers data integration challenges, technological implementation issues, algorithmic bias and transparency problems. These barriers need managerial intervention to implement effective change management strategies, gain stakeholders' commitment and support, and synthesize sound governance structures dealing with ethical and legal issues.

Another critical finding is the role of integration across supply chain networks as a strategic plan. The paper shows an example of applying generative AI, where this tool can update the synergy of interaction between organizations by sharing data in real-time and using predictive analytics. This requires a change from separate approaches to business operations at different levels and the development of supply chain systems, encouraging cooperation between suppliers, distributors, and retail stores. This research also recognizes some limitations despite its broad coverage; for instance, the study has not delved into general topics but concentrated on specific areas, and thus, different Supply Chain segments have not adopted AI uniformly. These limitations provide direction for future research, especially in the context of applying AI solutions in different cultures and economic conditions. In conclusion, this study establishes the centrality of generative AI in remapping the contingencies, tactics, and ethics of the food supply chain, based on the foregoing discussion. Nevertheless, AI's opportunities for change remain much more compelling than threats because it makes a shift towards greater efficiency, sustainability, and resilience possible. With the change in pace of the business environment, generative AI will play a big role in organizations seeking to create competitive advantage and sustainable value.

REFERENCES

Ahlawat, J., Akhmetshin, E., Nadaf, A. B., Thangam, A., Pandiaraj, V., & Mahapure, P. S. (2024). Analysing Optimising Time-Series Data for Demand Forecasting and Production Planning Using Bi-RNN. *2nd IEEE International Conference on Data Science and Network Security, ICDSNS 2024*. DOI: 10.1109/ICDSNS62112.2024.10690959

Aktas, E. (2024). Big data applications in food supply chains. In S. null, H. M.S.N., I. M.R., J. A., S. H., & F. A. (Eds.), *AIP Conference Proceedings* (Vol. 3109, Issue 1). American Institute of Physics. DOI: 10.1063/5.0204918

Bhowmik, D., Rickard, J. J. S., Jelinek, R., & Goldberg Oppenheimer, P. (2024). Resilient sustainable current and emerging technologies for foodborne pathogen detection. *Sustainable Food Technology*. Advance online publication. DOI: 10.1039/D4FB00192C PMID: 39359621

Bouazizi, E., Khedr, A., Elfaioumy, S., & Belal, M. (2024). Inventory Optimization Using Data Science Technologies for Supply Chain 4.0. *The International Arab Journal of Information Technology*, *21*(6), 996–1005. DOI: 10.34028/iajit/21/6/4

Danach, K., El Dirani, A., & Rkein, H. (2024). Revolutionizing Supply Chain Management With AI: A Path to Efficiency and Sustainability. *IEEE Access : Practical Innovations, Open Solutions*, *12*, 188245–188255. DOI: 10.1109/ACCESS.2024.3474531

Ghodake, S. P., Malkar, V. R., Santosh, K., Jabasheela, L., Abdufattokhov, S., & Gopi, A. (2024). Enhancing Supply Chain Management Efficiency: A Data-Driven Approach using Predictive Analytics and Machine Learning Algorithms. *International Journal of Advanced Computer Science and Applications*, *15*(4), 672–686. DOI: 10.14569/IJACSA.2024.0150469

Guo, X., Liu, C., Xu, W., Yuan, H., & Wang, M. (2014). A prediction-based inventory optimization using data mining models. In Y. L., Z. Q., C. J., C. J., & W. S. (Eds.), *Proceedings - 2014 7th International Joint Conference on Computational Sciences and Optimization, CSO 2014* (pp. 611 – 615). Institute of Electrical and Electronics Engineers Inc. DOI: 10.1109/CSO.2014.118

Gupta, K., Mane, P., Rajankar, O. S., Bhowmik, M., Jadhav, R., Yadav, S., Rawandale, S., & Chobe, S. V. (2023). Harnessing AI for Strategic Decision-Making and Business Performance Optimization. *International Journal of Intelligent Systems and Applications in Engineering*, *11*(10s), 893–912. https://www.scopus.com/inward/record.uri?eid=2-s2.0-85172014962&partnerID=40&md5=a62af4010a96c70566fc6aa83e797bbb

Kanjilal, R., Saenz, J. E., & Uysal, I. (2025). Large-scale data-driven uniformity analysis and sensory prediction of commercial banana ripening process. *Postharvest Biology and Technology*, *219*, 113203. Advance online publication. DOI: 10.1016/j.postharvbio.2024.113203

Karthiayani, A., & Raman, R. (2024). IoT-Based Smart Pharmacies for Optimizing Stock Management with Long Short-Term Memory Model. In S. B., A. R., K. S.K., S. K.M., S. A.V., J. S., S. N., C. A., & G. R. (Eds.), *2024 11th International Conference on Reliability, Infocom Technologies and Optimization (Trends and Future Directions), ICRITO 2024*. Institute of Electrical and Electronics Engineers Inc. DOI: 10.1109/ICRITO61523.2024.10522450

Kollia, I., Stevenson, J., & Kollias, S. (2021). Ai-enabled efficient and safe food supply chain. *Electronics (Basel)*, *10*(11), 1–22. DOI: 10.3390/electronics10111223

Krupitzer, C. (2024). Generative artificial intelligence in the agri-food value chain - overview, potential, and research challenges. *Frontiers in Food Science and Technology*, *4*, 1473357. Advance online publication. DOI: 10.3389/frfst.2024.1473357

Ladva, V., Shukla, M., & Vaghela, C. (2024). The Impact of Supply Chain Delays on Inventory Levels and Sale Demand Fulfillment: Analyzing the Effects of Lead Times and In- Transit Quantities : A Quantitative Exploration of Logistics Efficiency and Inventory Optimization. *Engineering. Technology and Applied Science Research*, *14*(4), 15700–15710. DOI: 10.48084/etasr.7834

Nayak, A., Patnaik, A., Satpathy, I., Tripathy, S., & Patnaik, B. C. M. (2024). Role of Industrial Internet of Things (IIoT) Enabled Smart Warehousing for Sustainable Perishable Food Supply Chains. In *Digital Technology Enabled Circular Economy: Models for Environmental and Resource Sustainability*. CRC Press., DOI: 10.1201/9781003349877-9

Onyeaka, H., Akinsemolu, A., Miri, T., Nnaji, N. D., Emeka, C., Tamasiga, P., Pang, G., & Al-sharify, Z. (2024). Advancing food security: The role of machine learning in pathogen detection. *Applied Food Research*, *4*(2), 100532. Advance online publication. DOI: 10.1016/j.afres.2024.100532

Pandey, D. K., & Mishra, R. (2024). Towards sustainable agriculture: Harnessing AI for global food security. *Artificial Intelligence in Agriculture*, *12*, 72–84. DOI: 10.1016/j.aiia.2024.04.003

Pasupuleti, V., Thuraka, B., Kodete, C. S., & Malisetty, S. (2024). Enhancing Supply Chain Agility and Sustainability through Machine Learning: Optimization Techniques for Logistics and Inventory Management. *Logistics (Basel)*, *8*(3), 73. Advance online publication. DOI: 10.3390/logistics8030073

Perumalsamy, R., & Natarajan, J. (2010). Predictive analytics using genetic algorithm for efficient supply chain inventory optimization. *2010 2nd International Conference on Computing, Communication and Networking Technologies, ICCCNT 2010*. DOI: 10.1109/ICCCNT.2010.5591607

Punia, S., Singh, S. P., & Madaan, J. K. (2020). From predictive to prescriptive analytics: A data-driven multi-item newsvendor model. *Decision Support Systems, 136*, 113340. Advance online publication. DOI: 10.1016/j.dss.2020.113340

Radhika, K., Mohammed, S., Tadepalli, S. K., & Puvvula, K. (2024). Food supply chain management by leveraging AI, IoT, and blockchain technologies. In *The Future of Agriculture: IoT, AI and Blockchain Technology for Sustainable Farming*. Bentham Science Publishers., DOI: 10.2174/9789815274349124010010

Reddy, A. D., Gupta, S., Kumar, D., & Kashyap, S. (2024). AI-enabled food quality and safety assurance. In *Agriculture 4.0: Smart Farming with IoT and Artificial Intelligence*. CRC Press., https://www.scopus.com/inward/record.uri?eid=2-s2.0-85205659966&partnerID=40&md5=b3221c30a7f765e037a14d9eb5e96432 DOI: 10.1201/9781003570219-14

Román, A. R., & Menéndez, M. R. (2024). The Future of Retail: Harnessing Generative AI for Disruptive Innovation and Sector Transformation. In *Reshaping Marketing Science in Wholesaling and Retailing*. IGI Global., DOI: 10.4018/979-8-3693-6145-0.ch013

Senanayake, T. E., Mahanama, T. V., & Prasadika, J. (2024). Strategizing Inventory Write-Off Process in Sri Lankan Apparel Sector: A Systematic Literature Review. *ICARC 2024 - 4th International Conference on Advanced Research in Computing: Smart and Innovative Trends in Next Generation Computing Technologies*, 334 – 339. DOI: 10.1109/ICARC61713.2024.10499743

Sengar, S. S., Hasan, A., Kumar, S., & Carroll, F. (2024, August 14). Bin, Kumar, S., & Carroll, F. (2024). Generative Artificial Intelligence: A Systematic Review and Applications. *Multimedia Tools and Applications*. Advance online publication. DOI: 10.1007/s11042-024-20016-1

Shahin, M., Chen, F. F., Maghanaki, M., Firouzranjbar, S., & Hosseinzadeh, A. (2024). Evaluating the fidelity of statistical forecasting and predictive intelligence by utilizing a stochastic dataset. *International Journal of Advanced Manufacturing Technology*. Advance online publication. DOI: 10.1007/s00170-024-14505-8

Shrivastava, S., & Pal, S. N. (2019). A framework for next generation agricultural marketing system in Indian context. *2019 5th IEEE International WIE Conference on Electrical and Computer Engineering, WIECON-ECE 2019 - Proceedings*. DOI: 10.1109/WIECON-ECE48653.2019.9019983

Vegesna, V. V., Prabhakaran, S. M., & Whig, P. (2024). AI for food safety: Leveraging artificial intelligence to ensure a safe and reliable food supply. In *Nutrition Controversies and Advances in Autoimmune Disease*. IGI Global., DOI: 10.4018/979-8-3693-5528-2.ch019

Chapter 7
Reducing Carbon Footprint With AI-Driven Logistics Planning

K. Sasikala
R.P. Sarathy Institute of Technology, India

Ritu Dahiya
https://orcid.org/0000-0001-5788-006X
Chhotu Ram Arya College, India

P. Selvakumar
https://orcid.org/0000-0002-3650-4548
Department of Science and Humanities, Nehru Institute of Technology, Coimbatore, India

P. Sudheer
https://orcid.org/0009-0005-1912-6636
CVR College of Engineering, India

Kamal Kumar Rajagopalan
https://orcid.org/0000-0001-8833-0554
Uttaranchal Institute of Management, Uttaranchal University, India

T. C. Manjunath
https://orcid.org/0000-0003-2545-9160
Rajarajeswari College of Engineering, India

Mohit Sharma
https://orcid.org/0009-0007-2280-8077
Maharshi Dayanand University, Rohtak, India

ABSTRACT

Sustainable logistics pivotal focus in modern, driven by the urgent need to address environmental, social, and economic challenges. As global trade expands the sector faces significant pressure to reduce its carbon footprint while maintaining efficiency and competitiveness. Traditional logistics practices have often prioritized speed and cost over sustainability, leading to increased Consequently, the logistics industry is at a crossroads, compelled to rethink its operational frameworks and embrace sustainable practices.One of the most pressing challenges in sustainable logistics

DOI: 10.4018/979-8-3693-9856-2.ch007

is the reliance on fossil fuels, climate change. Transitioning to alternative fuels and electric vehicles is essential but fraught with obstacles, including high upfront costs, limited infrastructure, and technological readiness. Additionally, among various stakeholders, including suppliers, manufacturers, and retailers, each of whom may have differing priorities and capabilities.Another critical challenge is the efficient management of resources.

INTRODUCTION TO SUSTAINABLE LOGISTICS: CHALLENGES, OPPORTUNITIES, AND AI-DRIVEN SOLUTIONS

Companies that adopt sustainable practices can enhance their brand reputation, meet regulatory requirements, and tap into a growing market of enhance decision-making processes. For instance, AI can analyze vast amounts of data to identify inefficiencies in logistics networks, enabling companies to streamline operations, reduce fuel consumption, and minimize emissions Moreover, AI can facilitate the transition to alternative energy sources by improving the management of electric vehicle fleets and optimizing charging schedules based on demand patterns. Through route optimization, AI can also reduce empty miles, ensuring that vehicles operate at maximum capacity and efficiency. This not only contributes to sustainability goals but also enhances overall profitability. Collaboration among stakeholders is critical sharing, enabling companies to collectively address shared challenges. Public-private collaborations can also play a vital role in developing infrastructure for sustainable transportation, such as electric vehicle charging stations and alternative fuel sources. Moreover, engaging customers and communities in sustainability initiatives can foster greater awareness and drive collective action toward more sustainable practices. In conclusion, sustainable logistics presents a complex yet rewarding landscape filled with a commitment to sustainable practices, the logistics industry can play a crucial role in addressing global challenges while creating long-term value for all stakeholders. The future of logistics is not only about moving goods efficiently but also about doing so in a way that respects our planet and its resources.

CARBON FOOTPRINT IN LOGISTICS: MEASUREMENT, REPORTING, AND VERIFICATION

The logistics industry is at a critical juncture in addressing its carbon footprint, as environmental sustainability becomes a top priority for businesses and consumers alike. Understanding the carbon footprint in logistics involves assessing the total greenhouse gas emissions produced during the transportation and storage of goods, which encompasses direct emissions from . A comprehensive assessment requires data collection from multiple sources, including fuel consumption, vehicle types, load factors, and distance traveled. Advanced technologies, such as telematics and IoT devices, can enhance data accuracy and granularity by tracking vehicle movements and fuel usage in real-time. Additionally, companies must consider emissions associated with upstream and downstream activities, including production, distribution, and end-of-life disposal of products, to gain a holistic view of their carbon footprint.Once the carbon footprint is measured, the next step is effective reporting. Transparent and consistent reporting is crucial for, which provide standardized metrics for disclosing emissions data. Reporting not only highlights(Ahmed QW et al., 2022a) a company's progress toward its sustainability goals but also fosters accountability and encourages continuous improvement. However, the complexity of logistics operations can lead to challenges in data aggregation and reporting, necessitating robust data management systems that can handle diverse sources of information and produce reliable reports. Verification of emissions data is the final component in the MRV process and is vital for ensuring credibility and accuracy.

Independent third-party verification enhances the reliability of reported emissions and provides stakeholders with confidence in the data. Verification processes typically involve a systematic review of the data collection and reporting methodologies employed by a company, assessing whether they align with established standards and best practices. This external validation not only bolsters a company's reputation but also identifies areas for improvement in its sustainability initiatives. Furthermore, an increasing number of regulatory frameworks require verified emissions data for compliance, making verification a crucial step for businesses operating in jurisdictions with strict environmental laws.In addition to compliance and reputation management, measuring, reporting, and verifying carbon emissions offers substantial business advantages.. Moreover, the demand for sustainable logistics solutions is rising among consumers and businesses alike. Companies that actively manage their carbon footprint and demonstrate While the logistics industry faces challenges in measuring, reporting, and verifying carbon emissions, it also has significant opportunities for growth and innovation(Ahmed S et al., 2022b). As sustainability In conclusion, addressing the carbon footprint in logistics through effective measurement, reporting, and verification identify areas for improvement, enhance operational efficiency, and strengthen their market position. The logistics industry stands at a pivotal moment where proactive engagement in businesses, communities, and the planet. As stakeholders increasingly demand accountability and transparency, companies that embrace these practices will not only reduce their carbon footprint

AI IN LOGISTICS: FOUNDATIONS, APPLICATIONS, AND BENEFITS

At its core, AI leverages data to optimize logistics processes. This begins with data collection from various sources, including transportation management systems, warehouse management systems, IoT devices, and customer feedback. behavior patterns. This predictive capability enables logistics providers to align inventory levels with anticipated demand, reducing excess stock and minimizing stockouts. Accurate demand forecasting also facilitates better resource allocation and improves customer satisfaction through timely deliveries.Another significant application By optimizing routes, logistics companies can respond more effectively to real-time changes, such as traffic disruptions or last-minute customer requests. The incorporation of AI-powered navigation systems allows drivers to receive dynamic updates, ensuring they are always following the best possible route.Warehouse operations also benefit substantially from AI integration. Automated systems powered by AI can streamline processes such as inventory tracking, checks(Bagheri M et al., 2017), reducing errors and ensuring that products meet specified standards before shipment. By automating routine tasks, companies can free up human workers to focus on more complex problem-solving and customer service roles, thereby enhancing overall productivity., AI can analyze historical data and external factors to predict potential disruptions, allowing companies to develop contingency plans and enhance resilience.

The benefits of AI in logistics are manifold. First and foremost, AI enhances operational efficiency, leading to cost reductions and improved profitability. By optimizing processes, reducing waste, and enhancing decision-making, logistics companies can operate with greater agility and responsiveness. This increased efficiency translates to faster turnaround times and improved customer service, which are critical in a highly competitive market. Moreover, AI can contribute to sustainability efforts within the logistics sector. By optimizing routes and improving load management, AI reduces fuel consumption and associated greenhouse gas emissions. Enhanced inventory management ensures that resources are utilized effectively, minimizing waste. As consumers and businesses alike place greater emphasis on sustainability, the ability to demonstrate a commitment to environmentally responsible practices can enhance brand reputation and attract new customers.The integration of AI in logistics also fosters innovation and drives competitive advantage. Companies that adopt AI technologies can differentiate themselves by offering enhanced services, such as real-time tracking and personalized delivery options. This technological edge can attract new business and build customer loyalty. Furthermore, AI facilitates the development of new business models, such as on-demand logistics and autonomous delivery systems, which can reshape the industry landscape.. The successful integration of AI in logistics will not only reshape operational processes but will also redefine the standards of service and sustainability in the logistics sector.

AI-DRIVEN ROUTE OPTIMIZATION: ALGORITHMS, MODELS, AND CASE STUDIES

AI-driven route. As businesses strive to reduce costs, enhance delivery speeds, and minimize their carbon footprint, leveraging advanced algorithms and models powered by artificial intelligence has become essential. This approach not only streamlines operations but also adapts to real-time conditions, providing logistics managers with the tools necessary to make informed decisions that optimize routing processes. At the heart of AI-driven route optimization are sophisticated algorithms(Bahaloo S et al., 2022), making it ideal for static routing scenarios. However, in dynamic environments where real-time data plays a crucial role, more complex algorithms like the Ant Colony Optimization and Reinforcement Learning methods become beneficial. These algorithms simulate natural behaviors, enabling them to learn and adapt as conditions change, thus optimizing routes on-the-fly.Machine learning models, particularly those utilizing supervised and unsupervised learning, also play a critical role in enhancing route optimization. Supervised learning can predict traffic patterns based on historical data, while unsupervised learning identifies underlying patterns without predefined labels, allowing for the detection of anomalies or new trends. By integrating these models with real-time traffic data, weather conditions, and even social media reports, logistics companies can dynamically adjust routes to avoid delays, reduce fuel consumption, and enhance customer satisfaction.AI algorithms to manage fleets of delivery vehicles, optimizing routes for thousands of packages daily. For instance, UPS employs a proprietary algorithm called ORION, which analyzes package delivery points and traffic conditions to minimize both the distance traveled and the number of left turns, significantly reducing fuel consumption and emissions. By optimizing delivery routes, UPS has reported annual savings of millions of gallons of fuel and reduced carbon emissions by hundreds of thousands of tons.Another notable case study is that of DHL, which has incorporated AI into its logistics operations through advanced route optimization tools. DHL's "Resilience360" platform leverages AI to analyze potential disruptions in supply chain operations and recommends alternative routes in real time. By combining historical data with predictive analytics, DHL can preemptively address potential delays caused by weather events, road closures, or other unforeseen circumstances, ensuring timely deliveries and improving overall service reliability.In the food delivery sector, companies like Uber Eats have implemented AI-driven route optimization to enhance delivery efficiency. By using machine learning algorithms that analyze user behavior, order patterns, and traffic conditions, Uber Eats can optimize delivery routes in real time, minimizing Looking forward, continues to grow. The advent of autonomous vehicles and drone delivery systems presents new opportunities for applying advanced routing algorithms in innovative ways. As these technologies evolve, they will require even more sophisticated optimization models that can account for the unique challenges and operational parameters associated with autonomous systems. In conclusion, models, logistics companies can dynamically adapt to real-time conditions, ensuring optimal routing that enhances service levels while minimizing environmental impact. The successful implementation of these technologies is exemplified (Cai W et al., 2019)by leading logistics firms such as UPS, DHL, and Uber

AI-DRIVEN VEHICLE ROUTING AND SCHEDULING: REAL-TIME OPTIMIZATION AND DYNAMIC ROUTING

AI-driven vehicle routing and scheduling represent a significant evolution in logistics and transportation, fundamentally altering how companies approach the complexities of fleet management. With increasing customer demands for faster deliveries, coupled with the necessity to minimize operational costs and environmental impact, leveraging artificial intelligence for real-time optimization and dynamic routing has become essential. This innovative approach utilizes.At the core of AI-driven routing and scheduling logistics managers with a comprehensive view of current conditions, enabling them to optimize routes and schedules effectively. The real-time nature of this optimization is crucial, as traffic patterns can change rapidly due to accidents, road work, or weather events, requiring immediate adjustments to planned routes.Another significant approach involves machine learning models that predict various factors affecting routing decisions. Unsupervised learning methods can identify patterns in delivery operations, highlighting inefficiencies or opportunities for improvement. By harnessing these predictive capabilities, logistics providers can enhance their operational planning, making data-driven decisions that lead to more efficient use of resources.Dynamic routing, enabled by AI, provides a critical advantage over traditional static routing methods. In static routing, vehicles are assigned fixed routes based on predetermined schedules, which can become inefficient as real-time conditions change.

In contrast, dynamic routing allows for continuous adjustments throughout the delivery process. Case studies highlight the effectiveness of AI-driven vehicle routing and scheduling in various sectors. For instance, FedEx employs advanced routing algorithms to optimize its delivery network. By integrating, such as DoorDash and Uber Eats, leverage AI to manage their fleets effectively, ensuring that drivers are assigned the most efficient routes based on real-time demand and traffic conditions. These companies have reported significant improvements in delivery times and customer satisfaction due to their ability to respond quickly to changing conditions.In the public transportation sector, AI-driven vehicle routing has also shown promise. For instance, some cities have implemented AI systems to optimize bus routes in real-time based on passenger demand and traffic patterns. This not only enhances use of public transportation, contributing to reduced urban congestion and lower emissions. The benefits of AI-driven vehicle routing and scheduling extend beyond operational efficiencies., where consumers increasingly prefer companies that demonstrate a commitment to environmental responsibility. Additionally, improved efficiency in routing leads to cost savings, allowing companies to pass on benefits to customers through competitive pricing and enhanced service offerings.Looking expansion of 5G networks, enhance data connectivity and processing capabilities, the opportunities for real-time optimization will expand further. Moreover(Carranza-García M et al., 2019), the advent of autonomous vehicles presents new frontiers for AI-driven routing, requiring even more sophisticated algorithms that can navigate complex environments safely and efficiently.In conclusion, AI-driven vehicle routing and scheduling mark a significant advancement in logistics and real-time optimization and dynamic routing, companies can respond more effectively to changing conditions, enhance operational efficiency, and improve customer satisfaction. With successful case studies demonstrating the tangible benefits of these technologies, the logistics industry stands poised to embrace AI-driven solutions as a core component of its operational strategy.(Debrah C et al., 2022) As organizations vehicle routing and scheduling, driving sustainable growth and innovation across the industry.

AI-DRIVEN LOAD OPTIMIZATION: CARGO PLANNING, CONSOLIDATION, AND CAPACITY UTILIZATION

AI-driven load optimization has become an indispensable aspect of modern logistics, significantly enhancing cargo planning, consolidation, and capacity utilization. As global trade continues to expand and customer expectations for rapid delivery grow processes, and ultimately streamline logistics operations, leading to improved efficiency and sustainability.At the foundation of AI-driven load optimization are advanced algorithms and machine learning models that analyze complex data sets related to cargo dimensions, weight, destination, and transportation modes. Traditional load optimization methods often rely on manual calculations and heuristic approachescontributing to lower carbon emissions. This is particularly critical for intermodal transport, where cargo must transition seamlessly between different modes of transportation, such as trucks, trains, and ships. Optimized cargo planning not only enhances efficiency but also minimizes the risk of damage to goods during transit, ensuring that shipments arrive safely and intact.AI algorithms can analyze order data and delivery schedules to identify opportunities for consolidation, thereby maximizing the use of available capacity. For example, an AI system might evaluate various orders destined for the same geographic area and determine the optimal way to group these shipments into a single truckload. This approach not only reduces transportation costs but also minimizes the environmental impact by decreasing the number of vehicles on the road, contributing to more sustainable logistics practices. Capacity utilization is a critical metric for logistics companies, reflecting how effectively available space and resources are being used. AI-driven load optimization helps organizations achieve higher capacity utilization rates by providing insights into underutilized assets and identifying patterns that can inform more efficient load planning(Ebrahimi M et al., 2019). For instance, advanced analytics can highlight trends in shipping volumes, enabling companies to adjust their operations proactively based on expected demand. By optimizing capacity utilization, logistics providers can improve profitability while also reducing waste and inefficiencies associated with overcapacity or underutilization of resources.he integration of AI-driven load optimization also aligns with sustainability initiatives. By maximizing cargo space and improving consolidation, logistics companies can reduce their Despite the numerous advantages, implementing AI-driven load optimization poses challenges. Companies must invest in technology infrastructure, data management systems, and training to effectively leveragcapabilities and improved computational power. The proliferation of IoT devices will provide logistics companies with even more granular data, enabling more precise load planning and real-time adjustments. Furthermore, the potential for AI to integrate with emerging technologies, such as blockchain, could revolutionize load optimization by enhancing transparency and traceability in the supply chain.

AI-DRIVEN SUPPLY CHAIN NETWORK DESIGN: LOCATION-ALLOCATION, INVENTORY MANAGEMENT, AND TRANSPORTATION PLANNING

AI-driven supply chain network design has revolutionized how companies approach the complexities of logistics and operations management. In today's fast-paced, globalized economy, businesses face increasing pressure to optimize their supply chains for efficiency, cost-effectiveness, and resilience. Leveraging artificial intelligence enables organizations to enhance their location-allocation decisions, improve inventory management, and refine transportation planning location-allocation modeling involves determining the optimal placement of facilities, such as warehouses and distribution centers, based on proximity to customer demand, transportation infrastructure, and operational that minimize transportation costs while maximizing service levels.Location-allocation problems often utilize mixed-integer linear programming (MILP) and other optimization techniques. By incorporating AI, organizations can enhance these models with predictive analytics, enabling them to anticipate shifts in demand, AI-driven models can quickly reassess existing facility locations and recommend adjustments, such as increasing warehouse capacity in strategic areas or opening new facilities to meet customer needs effectively.(Elahi E et al., 2019b)In addition to location-allocation, inventory management is a critical component of supply chain network design that benefits significantly from AI integration. Traditional inventory management methods often rely on historical data and fixed reorder points, which can lead to inefficiencies such as stockouts or overstock situations. AI algorithms can enhance inventory optimization by analyzing various factors, including seasonal trends, consumer behavior, and supply chain disruptions. Machine learning models can predict future demand with greater accuracy, enabling businesses to optimize their inventory levels dynamically.. Efficient transportation planning involves determining the best routes, modes of transport, and scheduling to ensure timely deliveries while minimizing costs. Traditional methods often rely on static models that do not account for real-time variables, leading to inefficiencies.

AI-driven transportation planning solutions leverage real-time data, such as traffic conditions, weather forecasts, and vehicle performance metrics, to optimize routing and scheduling continuously -based location-allocation model that analyzed customer demand patterns and logistics costs, enabling the company to optimize its network of fulfillment centers. This strategic redesign not only reduced shipping times but also lowered operational costs significantly, resulting in improved customer satisfaction.In the automotive industry, manufacturers have leveraged AI to enhance their inventory management processes. By utilizing predictive analytics, companies can accurately forecast demand for specific vehicle models and parts, optimizing production schedules and inventory levels. This approach minimizes excess inventory and associated holding costs, contributing to a more efficient supply chain.Furthermore, transportation companies have adopted AI-driven solutions to enhance their routing and scheduling processes. A prominent freight carrier implemented The advantages of AI-driven supply chain network design extend beyond operational efficiency and cost savings. By enhancing responsiveness and agility, organizations can better adapt to market fluctuations and customer In conclusion, AI-driven supply chain network design is transforming how companies approach logistics and operations management. Through enhanced location-allocation, inventory management, and transportation planning, businesses can optimize their supply chains for greater efficiency, cost-effectiveness, and resilience. Successful case studies from various industries highlight the tangible benefits of AI integration, demonstrating its critical role in meeting

AI-DRIVEN INVENTORY MANAGEMENT: PREDICTIVE ANALYTICS, DEMAND FORECASTING, AND STOCK OPTIMIZATION

AI-driven inventory management represents a paradigm shift in how businesses handle stock levels, forecast demand, and optimize their supply chains. As companies strive to meet increasing consumer expectations for rapid fulfillment while simultaneously controlling costs, For instance, a retail company previous years to identify patterns and seasonal trends. By incorporating variables such as promotional events, market trends, and even weather conditions, the AI model can provide a nuanced forecast that informs inventory purchasing decisions. This capability not only helps businesses maintain optimal stock levels but also reduces the risk of Demand forecasting is a critical element of effective inventory management. AI-powered forecasting models employ techniques such levels with anticipated demand, thereby enhancing their responsiveness and agility.A case study highlighting the benefits of AI-driven demand forecasting can be found in the grocery sector. A leading supermarket chain implemented an AI-based forecasting system that analyzed not only historical sales data but also external factors such as local events and seasonal promotions.

Stock crucial aspect of AI-driven inventory management that enhances operational efficiency. Effective stock optimization involves determining the right quantity of products to hold at any given time while minimizing carrying costs and ensuring product availability(Gautam K et al., 2023). Traditional approaches often result in either excessive inventory, which ties up capital, or insufficient stock, AI-driven stock optimization leverages algorithms that consider multiple factors, such as lead times, supplier reliability, and historical sales patterns. By analyzing these variables, AI systems can recommend optimal reorder points and quantities, ensuring that businesses maintain the right levels of inventory. For example, an electronics retailer that implemented AI-driven stock optimization reported a significant reduction in inventory carrying costs while improving service levels. The system analyzed sales trends and supplier lead .(Farghali M et al., 2022)Despite the numerous benefits, implementing AI-driven inventory management poses challenges. foster a culture of collaboration between IT and operational teams to integrate AI solutions seamlessly into existing processes. Employee training is also crucial to ensure that staff can effectively utilize new technologies and understand the insights generated by AI systems.In conclusion, AI-driven inventory management is transforming the way businesses approach stock levels, demand forecasting, and optimization strategies.

Successful case studies across various industries demonstrate the tangible benefits of AI integration, highlighting its critical role in meeting the demands of modern supply chains. making, ensuring that businesses can meet the challenges of an increasingly dynamic market landscape.

AI-DRIVEN TRANSPORTATION MODE SELECTION: MULTI-MODAL TRANSPORT, INTERMODAL OPTIMIZATION, AND CARBON FOOTPRINT REDUCTION

In an era marked by rapid globalization and increasing environmental awareness, AI-driven transportation mode selection has emerged as a pivotal strategy optimization not only enhances logistical flexibility but also plays a crucial role in reducing carbon footprints, aligning with the growing demand for sustainable practices across industries.At the heart of AI-driven transportation mode selection is the integration of advanced algorithms that analyze a myriad of variables, including costs, transit times, capacity constraints, and environmental impacts. Traditional transportation planning often relies on fixed parameters and historical data, which can limit adaptability and responsiveness to real-time conditions. In contrast, AI systems can process real-time data from various.Multi-modal transport involves utilizing two or more modes of transportation to move goods, thereby capitalizing on the unique advantages of each mode. For example, a shipment might travel by rail for long distances due to its cost-effectiveness and then switch to trucks for final delivery.intermodal optimization is a critical component of multi-modal transport that focuses on enhancing the connections between different transportation modes. AI-driven models can optimize the scheduling and routing of shipments to ensure smooth transitions between modes, reducing delays and improving overall efficiency. For instance(Ghavami SM et al., 2017), AI systems can analyze historical transit times and potential bottlenecks at transfer points to recommend the best times for mode switches. By optimizing these handoff points, businesses can minimize idle time and ensure that goods reach their destinations promptly.A compelling example of AI-driven intermodal optimization can be seen in the operations of major logistics providers like DHL and FedEx.

These companies employ sophisticated algorithms to analyze shipment data and predict the most efficient transport modes for their operations. By optimizing their multi-modal transport networks, they have successfully reduced transit times and operating costs while enhancing service reliability. rail transport, while slower, typically has a significantly lower carbon footprint per ton-mile.AI systems can incorporate carbon emission data into their decision-making processes, allowing companies to balance cost and efficiency with environmental responsibility. By recommending transport modes that minimize emissions, organizations can align their logistics strategs with sustainability goals. This approach is increasingly relevant as consumers and stakeholders prioritize eco-friendly practices and demand transparency in corporate environmental impact.(Hahn D et al., 2021)Moreover, AI can facilitate real-time monitoring of emissions throughout the supply chain, enabling companies to track their progress toward sustainability targets. By analyzing data on transportation routes and modes, organizations can identify opportunities for further emission reductions, such as consolidating shipments or switching to lower-impact transport options when feasible. This ongoing analysis fosters a culture of continuous improvement, where businesses can adapt their strategies in response to evolving environmental standards and consumer expectations.(Jahanger A et al., 2023)While the benefits of AI-driven transportation mode selection are substantial, implementing these systems also presents challenges. Organizations must invest in robust data infrastructure to gather and analyze the diverse data sets necessary for effective decision-making. Additionally, the integration of AI into existing logistics frameworks requires a cultural shift toward data-driven decision-making, necessitating training and upskilling of employees to harness the full potential of AI technologies.In conclusion, AI-driven transportation mode selection is transforming how organizations approach logistics and supply chain management. By leveraging advanced algorithms for multi-modal transport and intermodal optimization, companies can enhance operational efficiency, reduce costs, and minimize their carbon footprints. Successful case studies from leading logistics providers demonstrate the tangible benefits enabling businesses to balance efficiency, cost-effectiveness, and environmental responsibility.

AI-DRIVEN WAREHOUSE LAYOUT OPTIMIZATION: STORAGE, HANDLING, AND MATERIAL FLOW

AI-driven warehouse layout optimization is transforming the way businesses manage their storage facilities, significantly enhancing operational efficiency and productivity. As e-commerce continues to expand and customer artificial intelligence, organizations can analyze complex data related to storage capacity, handling processes, and material flow, resulting in layouts that maximize space utilization, improve order picking efficiency, and reduce operational bottlenecks. At the core of AI-driven warehouse layout optimization order volumes, and seasonal trends, to generate dynamic layouts that adapt to fluctuating conditions. This agility allows businesses to respond proactively to changes in demand, ultimately enhancing their operational responsiveness.on their size, weight, and handling characteristics. AI-driven systems can categorize products and suggest optimal storage locations that minimize travel time for pickers. For example, fast-moving items can be placed closer to shipping areas, while slower-moving items can be stored in less accessible locations. This strategic arrangement not only reduces the time spent retrieving items but also improves overall productivity.Moreover, AI can facilitate the evaluation of different storage methods, such as By analyzing factors such as product dimensions, turnover rates, and handling frequency, AI systems can recommend the most suitable storage solutions for a given warehouse environment. This tailored approach ensures that each warehouse maximizes its space while accommodating the specific needs of the inventory it holds.Handling processes within a warehouse are equally important to consider when optimizing layout.

AI can analyze the flow of materials and products through the warehouse, identifying potential bottlenecks. (Ise et al., 2019). For instance, by studying historical data on order fulfillment, AI can pinpoint areas where delays frequently occur, such as during loading or unloading. With this insight, organizations can redesign their layouts to streamline these processes, potentially incorporating dedicated staging areas or optimizing dock door assignments to facilitate smoother operations.Material flow analysis is another critical component of warehouse layout optimization. AI-driven systems can simulate different layout scenarios to evaluate how materials move through the warehouse, taking into account factors like worker movements, equipment usage, and product flows. By modeling these interactions, businesses can identify the most efficient layouts that minimize travel distances and enhance overall throughput. This is particularly beneficial in high-volume warehouses where every second counts, and inefficiencies can lead to significant time and cost losses.A notable case study in AI-driven warehouse layout optimization can be found in the operations of major e-commerce retailers. These companies have implemented AI algorithms that continuously analyze data from their fulfillment centers, allowing them to, drastically reducing picking times and improving customer satisfaction through faster order fulfillment.Additionally, the integration of automated technologies, such as robotics and conveyor systems, can further, AI algorithms can optimize the paths of autonomous mobile robots (AMRs) as they transport goods throughout the warehouse, minimizing collisions and ensuring efficient use of space. This synergy between AI, automation(Javaid M et al., 2022), and human labor can lead to significant improvements in warehouse efficiency.While the benefits of AI-driven warehouse layout optimization are substantial, implementing such systems does present challenges. Organizations must invest in technology infrastructure that supports advanced data analytics and machine learning capabilities. Additionally, successful implementation requires buy-in from all stakeholders, including warehouse staff, management, Moreover, the rise of digital twins—virtual replicas of physical warehouses—will allow organizations to simulate and test various layout scenarios before implementing changes in the real world. This capability will enable businesses to visualize the impacts of layout adjustments on material flow and productivity, ensuring that decisions are backed by data and insights.In conclusion, AI-driven warehouse layout optimization is reshaping how businesses approach their storage and handling operations., ensuring that businesses (Jha SK et al., 2017) can optimize their layouts and operations in an increasingly competitive and dynamic marketplace.

AI-DRIVEN ORDER FULFILLMENT: PICK-AND-PACK OPTIMIZATION, AUTOMATED STORAGE AND RETRIEVAL SYSTEMS

With the increasing demand, reshaping how businesses manage the complexities of order processing and delivery. for faster, more efficient service, companies are turning to artificial intelligence to optimize their pick-and-pack operations and leverage automated storage and retrieval systems (AS/RS)..At the heart of AI-driven order fulfillment is the optimization of pick-and-pack processes, which are critical in determining how quickly and accurately orders are processed warehouse layouts to develop optimized picking strategies tailored to specific operational contexts. By evaluating factors such as order frequency and item location, AI systems can create efficient picking routes that minimize travel time for warehouse workers.For instance, in a typical warehouse scenario, an AI-driven system may identify that frequently ordered items should be stored closer to shipping areas, thereby reducing the distance pickers need to travel. Furthermore, AI can facilitate batch picking, where multiple orders are fulfilled simultaneously, allowing workers to gather several items in one trip. This strategy not only speeds up the picking process but also reduces the total time required to fulfill multiple orders, enhancing overall productivity.(Kadow C et al., 2020)In addition to optimizing picking strategies, AI can significantly improve packing processes. The packing phase is crucial in ensuring that items are securely and efficiently packed for shipment. AI-driven systems can analyze the dimensions and weights of items to recommend the best packing configurations, minimizing the use of materials while maximizing protection during transit. By utilizing machine learning algorithms, these systems can learn from historical packing data to identify optimal packaging materials and methods, ultimately reducing waste and lowering shipping costs.A notable workers to automate the retrieval and storage of products. These systems can quickly navigate the warehouse, fetching items and delivering them to packing stations with remarkable speed.

This not only accelerates the fulfillment process but also reduces the physical strain on employees, allowing them to focus on higher-value tasks.AI-driven order fulfillment is also enhancing inventory accuracy, which is paramount for effective supply chain management. Traditional inventory management methods often result in discrepancies due to human error during picking and packing. By employing AI systems that continuously monitor inventory levels in real-time, businesses can ensure that stock levels are accurately reflected in their databases. demand trends, enabling companies to proactively manage their inventory and fulfillment strategies. For example, during peak seasons such as holidays or special promotions, AI can predict spikes in demand for specific products, allowing businesses to allocate resources effectively and prepare their warehouses accordingly. This proactive approach not only enhances operational efficiency but also ensures that customer expectations for timely delivery are met.AI and automation. Additionally, workforce training is crucial to ensure that employees are equipped to work alongside automated systems and understand the insights generated by AI analytics. Fostering a culture of collaboration between human workers and machines will be essential for maximizing the potential of AI in order fulfillment Furthermore, advancements in machine learning will enable AI systems to continuously improve their algorithms based on new data and operational experiences (Lee & Yoo, 2021). This capability will allow organizations to refine their picking and packing strategies over time, ensuring that they remain competitive in an ever-evolving market landscape.In conclusion, AI-driven order fulfillment is reshaping the logistics and supply chain landscape by enhancing pick-and-pack optimization and leveraging automated storage and retrieval systems. Through the use of advanced algorithms and robotics,The future of order fulfillment lies in the ability to harness AI for informed decision-making and operational excellence, ensuring that businesses can meet the challenges of an increasingly dynamic marketplace.

AI-DRIVEN RETURNS MANAGEMENT: REVERSE LOGISTICS, REPAIR, AND REFURBISHMENT

AI-driven returns management is revolutionizing how businesses handle the complexities of reverse logistics, offering innovative solutions for repair and refurbishment processes. As e-commerce continues to grow, so does the volume of product returns, presenting both challenges and opportunities for retailers and manufacturers., AI algorithms can evaluate historical return data, enabling businesses to identify trends and predict future returns. For example, if a particular product consistently receives high return rates due to sizing issues, AI can flag this pattern, prompting businesses to reassess their product descriptions, sizing charts, or even supplier quality.Moreover, AI can facilitate the classification and assessment of returned products, determining whether items can be repaired, refurbished, or recycled. This capability is crucial for maximizing the value of returned goods and minimizing waste. Advanced imaging and machine learning technologies can analyze the condition of returned items, helping organizations decide on the most appropriate course of action. For instance, an AI system may use visual recognition to assess whether a returned electronic device can be repaired or if it should be sent for recycling.(Mrówczyńska M et al., 2019)Repair and refurbishment processes are increasingly inventory for refurbished goods. By analyzing historical data on repair times, costs, and parts availability, AI can forecast the resources needed for repairs, ensuring that the right parts are on hand when needed and reducing downtime.For example, a company that sells refurbished electronics can employ AI to track the most commonly required replacement parts for various devices. By maintaining optimal stock levels of these parts, the company can accelerate the repair process AI-driven offer tailored return solutions, such as flexible return windows or convenient drop-off locations.

This level of customization can foster customer loyalty and increase the likelihood of repeat purchases, as satisfied customers are more inclined to return to a retailer that prioritizes a hassle-free returns process.Furthermore, AI can facilitate improved communication throughout the returns process. Automated notifications can keep customers informed about the status of their returns, from initiation to resolution. This transparency enhances the customer experience and builds trust, as customers feel more engaged and informed about the handling of their returned items.Looking ahead, the potential for AI-driven returns management continues to expand. As technologies evolve, organizations can expect even greater capabilities in analyzing and optimizing returns processes. Additionally, advancements in predictive analytics will enable businesses to anticipate return trends more accurately, allowing them to adjust inventory levels, streamline repair processes, and refine product offerings. By leveraging these capabilities, organizations can not only improve their returns management but also reduce the overall environmental impact associated with product returns.In conclusion, AI-driven returns management is transforming reverse logistics, repair, and refurbishment processes within the supply chain complexities of product returns in today's dynamic marketplace(Nawaz R et al., 2021). As businesses continue to adopt these technologies, they will be better equipped to meet the challenges of reverse logistics, drive sustainable practices, and foster lasting customer relationships. The future of returns management lies in leveraging AI for informed decision-making, ensuring that organizations can navigate the intricacies of product returns with agility and responsiveness.

IMPLEMENTING AI-DRIVEN LOGISTICS PLANNING: CHANGE MANAGEMENT, INTEGRATION, AND SCALABILITY

The implementation of AI-driven logistics planning represents a significant shift in how organizations manage their supply chains, requiring careful consideration of change management, integration strategies, and scalability. As businesses increasingly recognize the potential of. This transformation demands a comprehensive approach that encompasses not only the adoption of AI technologies but also the alignment of organizational processes and stakeholder engagement.Change management is a critical component of implementing AI-driven logistics planning. The introduction of AI technologies can disrupt established workflows and practices, leading to resistance among employees who may feel threatened by automation or uncertain about new systems. To mitigate these challenges, organizations must adopt a proactive change management strategy that emphasizes communication, training, and involvement. Clear communication attitude toward the changes. Additionally, involving employees in the transition process through training and collaborative workshops can empower them to embrace new tools and technologies, thereby enhancing their confidence and competence in utilizing AI-driven solutions.the role of data is paramount in the implementation of AI-driven logistics planning. Successful AI A significant advantage of AI-driven logistics planning is its ability to enhance forecasting and demand planning. AI algorithms can analyze vast amounts of data to identify patterns and trends that inform inventory management, production scheduling, and distribution strategies.(Olabi AG et al., 2023) By leveraging predictive analytics, organizations can optimize their supply chains to better align with customer demand, minimizing excess inventory and reducing stockouts. This proactive approach to logistics planning enhances overall efficiency and responsiveness, ultimately improving customer satisfaction.Implementing AI-driven logistics planning also presents opportunities for continuous improvement through machine learning. As AI systems process more Despite the numerous benefits, organizations must remain vigilant about potential pitfalls during the implementation process. Over-reliance on AI without adequate human oversight can lead to unintended consequences,.Additionally, ethical considerations surrounding AI adoption in logistics should not be overlooked. As organizations leverage AI for decision-making, issues related to bias, transparency, and accountability must be addressed. Implementing ethical guidelines and best practices can help organizations navigate these challenges while fostering trust among stakeholders, including employees, customers, and partners.In conclusion, implementing AI-driven logistics planning involves a multifaceted approach that encompasses change management, integration, and scalability(Yue & Gao, 2018) . By proactively addressing cultural and operational challenges, organizations can successfully adopt AI technologies to enhance their logistics operations. The benefits of AI-driven logistics planning—improved forecasting, optimized transportation, and continuous improvement—position organizations to thrive in an increasingly competitive and dynamic marketplace. As businesses continue to embrace

CASE STUDIES IN AI-DRIVEN LOGISTICS: SUCCESS STORIES, CHALLENGES, AND LESSONS LEARNED

across various industries, yielding significant benefits while also presenting unique challenges. This examination of case studies in AI-driven logistics highlights success stories, the obstacles encountered, and the lessons learned from organizations that have embraced this technological evolution(Saheb T et al., 2022) . Through these insights, businesses can better understand how to effectively implement AI solutions and maximize their potential in enhancing operational efficiency and customer satisfaction.One prominent success story is that of Amazon, a leader in logistics innovation. Amazon employs AI This predictive capability is complemented by advanced robotics in its fulfillment centers, which enhance picking efficiency by assisting human workers. For instance, Amazon's Kiva robots autonomously transport shelves of products to pick stations, drastically reducing the time employees spend searching for items. This combination of AI-driven forecasting and automation has enabled Amazon to achieve unparalleled order fulfillment speeds, often delivering packages within a day or even hours, thereby setting new standards for customer expectations in e-commerce.Another compelling case study is that of DHL, which has successfully integrated AI into its logistics operations to enhance delivery routes and optimize supply chain visibility. DHL has developed an AI-based tool called Resilience360, which analyzes global supply chain risks and disruptions. By leveraging real-time data and predictive analytics, this tool enables DHL to identify potential threats—such as natural disasters or geopolitical instability—before they impact operations. This (Partel V et al., 2019)AI solutions to optimize its supply chain management but faced significant resistance from employees concerned about job security and changes to established workflows. To address these challenges, the company initiated comprehensive training programs aimed at reskilling employees and demonstrating how AI could augment their roles rather than replace them. This emphasis on change management was critical in fostering a culture of acceptance and Similarly, a major food retailer sought to implement AI-driven inventory management to reduce waste and optimize stock levels. While the technology demonstrated great potential, the company encountered difficulties in integrating AI with its existing systems. The disparate nature of its legacy software and data silos hindered the seamless flow of information necessary for effective AI analytics.

This experience underscored the importance of ensuring that data infrastructure is robust and adaptable before embarking on an AI implementation journey. In response, the retailer committed to investing in an integrated data platform that would support real-time data sharing across its supply chain.Another key lesson from these case studies discrepancies found that its AI-driven forecasting tool generated inconsistent predictions, undermining the potential benefits of the technology. By establishing clear data management practices and regularly auditing data quality, organizations can enhance the effectiveness of their AI systems.Moreover, ethical considerations surrounding AI adoption are gaining prominence. Companies must be vigilant about potential biases in their algorithms, particularly when AI systems are used for decision-making related to hiring, promotions, or customer interactions. For instance, a transportation company that utilized AI for driver scheduling discovered that its algorithm inadvertently favored certain demographics based on historical data. This realization prompted the company to reassess its data sources and implement measures to mitigate bias, highlighting the necessity for ethical oversight (Yue W et al., 2023)in AI applications.Scalability is another aspect that organizations must consider when implementing AI-driven logistics solutions. A retail chain that initially deployed AI for demand forecasting found its solution ineffective during peak seasons, as the system could not handle the surge in data volume. To overcome this limitation, the retailer shifted to a cloud-based AI platform that offered greater scalability and flexibility, allowing it to accommodate fluctuating demand without compromising performance.The insights garnered from these case studies emphasize the importance of a holistic approach to AI-driven logistics. Organizations must not only focus on the technology itself but also consider the broader organizational culture, data management practices, and ethical implications of their AI initiatives. In conclusion, the journey of implementing AI-driven logistics is characterized by both success stories and challenges. Companies like Amazon and DHL showcase the transformative potential of AI in optimizing logistics operations and enhancing customer satisfaction. However, the lessons learned from these experiences highlight the need for robust change management, seamless integration, high-quality data, and ethical considerations in AI adoption. By understanding these

AI-DRIVEN LOGISTICS IN PRACTICE: INDUSTRY-SPECIFIC APPLICATIONS AND EXAMPLES

logistics is reshaping how various industries operate, optimizing processes from warehousing and transportation to inventory management and customer service. AI-driven logistics applications are being tailored to meet the unique challenges of different sectors, demonstrating versatility and the potential for significant efficiency gains. This exploration delves into industry-specific applications and examples, highlighting how organizations leverage AI technologies to enhance their logistics operations., seasonality, and promotional activities. By leveraging vast amounts of transactional data, Walmart can predict demand with remarkable accuracy, ensuring that stores are stocked appropriately while minimizing excess inventory. This capability not only reduces costs associated with overstocking but also enhances customer satisfaction by ensuring product availability. Moreover, AI-driven chatbots are increasingly being employed in customer service, providing instant support for order tracking and returns, thereby improving the overall shopping experience.The has also embraced AI to. Through AI, Tesla can forecast potential supply chain disruptions and implement corrective measures proactively, ensuring uninterrupted production flow. Additionally, AI algorithms analyze data from connected vehicles to optimize spare parts logistics. By predicting which components are likely to fail based on driving patterns and conditions, Tesla can enhance its inventory management, ensuring that necessary parts are available when and where they are needed.. AI systems analyze patient demand, historical usage patterns, and regional health data to forecast the need for medications and medical supplies. This predictive capability enables healthcare providers to maintain optimal inventory levels, reducing waste and ensuring that essential products are readily available. Moreover, AI is being used to optimize the logistics of medical devices and equipment, improving delivery times and ensuring that healthcare professionals have the tools they need to provide timely care.presents unique challenges that AI-driven logistics can address effectively.

Companies like Pfizer utilize AI to manage complex supply chains for sensitive products that require stringent temperature control. AI systems monitor environmental conditions throughout the logistics process, ensuring that temperature-sensitive medications are stored and transported within safe parameters. This capability minimizes the risk of spoilage and enhances compliance with regulatory standards, ultimately protecting patient safety.In logistics management is critical for project success. AI-driven solutions are being used to optimize the delivery of materials and equipment to job sites. Companies like Caterpillar leverage AI to analyze project timelines and material requirements, ensuring that necessary resources arrive on time and in the right quantities(Rustia et al., 2020). By predicting potential delays and optimizing delivery schedules, AI helps construction firms maintain project timelines and reduce costs associated with idle equipment and labor.Successful organizations recognize the importance of change management and employee engagement in the AI implementation process. Providing training and reskilling opportunities can help alleviate concerns and empower employees to leverage AI as a tool that enhances their work rather than replaces it. For example, a major shipping company that introduced AI-driven forecasting tools offered comprehensive training programs, enabling employees to effectively utilize these technologies and improve their decision-making capabilities.Furthermore, ethical considerations surrounding AI in logistics are gaining attention(Zahraee SM et al., 2016). Organizations must ensure that their AI systems operate transparently and without bias discriminatory practices based on historical data. By establishing ethical guidelines and promoting accountability, organizations can foster trust among stakeholders and maintain their reputation in the marketplace.processes and enhancing operational efficiency. Success stories from sectors such as retail, automotive, healthcare, food and beverage, transportation, pharmaceuticals, and construction illustrate the diverse ways AI is being utilized to address industry-specific challenges. However, organizations must navigate hurdles related to data quality, integration, cultural resistance, and ethical considerations to realize the full potential of AI-driven logistics.

FUTURE OF AI-DRIVEN LOGISTICS: EMERGING TRENDS, OPPORTUNITIES, AND CHALLENGES

As we look toward, and evolving market demands. AI-driven logistics is poised to transform supply chain operations, enhancing efficiency, transparency, and sustainability. However, with these opportunities come significant challenges of the future of AI-driven logistics highlights key trends, opportunities, and the hurdles that lie ahead Another pivotal trend is the rise of autonomous vehicles and drones in logistics. As AI technology advances, autonomous delivery vehicles are becoming more feasible for last-mile delivery solutions. Companies such as Waymo and Amazon Prime Air are experimenting with drone delivery systems that promise faster and more cost-effective shipping options. The integration of autonomous vehicles not only enhances efficiency but also addresses labor shortages in the logistics sector, particularly in regions struggling to recruit drivers. However, regulatory challenges and infrastructure readiness remain significant barriers that must be overcome to realize the full potential of autonomous logistics.solutions to monitor container conditions during transport, enhancing the reliability of their logistics operations(Sun Y et al., 2019b). As the IoT ecosystem continues to expand, the integration of AI will enable companies to derive actionable insights from the vast amounts of data generated, further optimizing logistics processes.Sustainability is becoming an essential consideration in logistics, and AI-driven solutions are well-positioned to support green initiatives. As environmental concerns grow, organizations are under increasing AI algorithms to calculate the most environmentally friendly routes and modes of transportation. Additionally, AI can facilitate better waste management and recycling practices within supply chains, promoting a circular economy(Yuan X et al., 2020). Companies that prioritize sustainability will not only comply with regulatory requirements but also appeal to environmentally conscious consumers. despite the promising trends and opportunities, several challenges loom over the future of AI-y threats pose risks to sensitive information, including customer data and proprietary algorithms. Organizations must implement robust cybersecurity measures and establish clear data Looking ahead, the collaboration between AI and human expertise will become increasingly important. While AI can analyze data and generate insights at remarkable speeds, human intuition, creativity, and ethical considerations remain crucial in decision-making processes. The most successful logistics operations will likely be those that strike a balance between leveraging AI's capabilities and harnessing human judgment. In conclusion, the future of AI-driven logistics is filled with potential, driven by emerging trends such as predictive analytics, autonomous vehicles, cultural resistance, integration complexities, and the need for collaborative decision-making. By embracing change management, investing in employee training, and modernizing infrastructure, businesses can position

SUSTAINABLE LOGISTICS 2030: VISION, ROADMAP, AND AI-DRIVEN INNOVATION FOR A LOW-CARBON FUTURE

This vision outlines a roadmap for the logistics industry, emphasizing the critical role of AI-driven innovations in achieving sustainability goals.The roadmap to sustainable logistics in 2030 begins with a commitment to decarbonization. Companies must establish clear targets for reducing their carbon footprint, aligning with global climate agreements and national regulations. This includes adopting ambitious goals for emissions reductions in transportation, warehousing, and overall supply chain operations. in logistics.Another critical aspect of the roadmap is the adoption of circular economy principles in logistics. This approach focuses on minimizing waste and maximizing resource efficiency throughout the supply chain. AI can facilitate this by enabling better demand forecasting, inventory management, and end-of-life product recovery Logistics companies that embrace these principles will not only reduce their environmental impact but also enhance their competitive advantage in a market increasingly driven by sustainability.In the quest for sustainable logistics, reputation as responsible corporate citizens.Furthermore, investment in sustainable technologies will be crucial for achieving the vision for sustainable logistics in 2030. Companies should the performance of electric vehicle fleets, identifying opportunities for improved charging infrastructure and fleet management Employee engagement and training will also be vital components of the sustainable logistics roadmap. As organizations adopt new technologies and practices, it is essential to equip employees with the skills and knowledge necessary to drive sustainability initiatives forward. Training programs focused on sustainability innovation and collaboration, ensuring that sustainability becomes a core value embedded in the company's operations.As we look toward 2030, consumer expectations for sustainability will continue to rise. (Tian Z et al., 2021) Customers are increasingly seeking products and services that align with their values, including environmental responsibility. Logistics companies must respond to this demand by transparently communicating their sustainability efforts and demonstrating a commitment to low-carbon.In conclusion, the vision for sustainable logistics in 2030 is centered around a low-carbon future driven by AI-driven innovation and a commitment to sustainability. The roadmap for achieving this vision includes ambitious decarbonization targets, the adoption of circular economy principles, enhanced collaboration across the supply chain, regulatory compliance, investment in sustainable technologies, employee engagement, and responsiveness to consumer expectations.

CONCLUSION

In the quest for sustainable logistics, a transformative opportunity to AI-driven logistics planning emerges as a vital tool in navigating these complexities. This conclusion synthesizes the key insights into how AI can facilitate substantial reductions in carbon emissions across various aspects of logistics, ultimately paving the way for a more sustainable future.AI technologies enable businesses to optimize their operations by analyzing conditions, weather patterns, and delivery windows, resulting in more efficient transportation logistics. This optimization leads to fewer miles driven, lower fuel consumption, and, consequently,AI-driven logistics also supports the transition to greener transportation modes. By analyzing data on transportation options electric vehicles or alternative fuels. The ability to evaluate the environmental impact of various transport modes allows companies to make informed decisions that align with their sustainability Collaboration across the efforts. AI technologies facilitate data sharing and communication between stakeholders, allowing for coordinated logistics planning. This collaboration can lead to shared transportation networks and co-loading opportunities, which reduce empty miles and enhance overall efficiency. For instance, companies can leverage AI to identify overlapping delivery routes with partner organizations, consolidating shipments and maximizing vehicle utilization. By fostering collaboration,.of granularity allows for targeted interventions that can lead to significant carbon reductions.As organizations strive to meet sustainability targets, the supply chain. Innovations such as blockchain integration can enhance traceability and accountability, providing a clearer picture of the environmental impact of logistics operations. Furthermore, advances in machine learning will allow for continuous improvement of AI algorithms, adapting to changing market conditions and consumer behaviors in real time. In summary, AI-driven logistics planning is a powerful catalyst for reducing carbon footprints across the logistics sector. Through optimization of transportation routes, improved inventory management, support for greener transportation modes, enhanced collaboration, and real-time monitoring, AI technologies empower organizations to significantly diminish their environmental impact. However, as we embrace these advancements, it is crucial to remain vigilant about the ethical considerations associated with AI deployment. By fostering collaboration, transparency, and responsibility,

REFERENCES

Ahmed, Q. W., Garg, S., Rai, A., Ramachandran, M., Jhanjhi, N. Z., Masud, M., & Baz, M. (2022a). AI-based resource allocation techniques in wireless sensor internet of things networks in energy efficiency with data optimization. *Electronics (Basel), 11*(13), 2071. DOI: 10.3390/electronics11132071

Ahmed, S., Alshater, M. M., Ammari, A. E., & Hammami, H. (2022b). Artificial intelligence and machine learning in finance: A bibliometric review. *Research in International Business and Finance, 61*, 101646. DOI: 10.1016/j.ribaf.2022.101646

Bagheri, M., Bazvand, A., & Ehteshami, M. (2017). Application of artificial intelligence for the management of landfill leachate penetration into groundwater, and assessment of its environmental impacts. *Journal of Cleaner Production, 149*, 784–796. DOI: 10.1016/j.jclepro.2017.02.157

Bahaloo, S., Mehrizadeh, M., & Najafi-Marghmaleki, A. (2022). *Review of application of artificial intelligence techniques in petroleum operations*. Petrol Res., DOI: 10.1016/j.ptlrs.2022.07.002

Cai, W., Lai, K., Liu, C., Wei, F., Ma, M., Jia, S., Jiang, Z., & Lv, L. (2019). Promoting sustainability of manufacturing industry through the lean energy-saving and emission-reduction strategy. *The Science of the Total Environment, 665*, 23–32. DOI: 10.1016/j.scitotenv.2019.02.069 PMID: 30772553

Carranza-García, M., García-Gutiérrez, J., & Riquelme, J. C. (2019). A framework for evaluating land use and land cover classification using convolutional neural networks. *Remote Sensing (Basel), 11*(3), 274. DOI: 10.3390/rs11030274

Debrah, C., Chan, A. P. C., & Darko, A. (2022). Artificial intelligence in green building. *Automation in Construction, 137*, 104192. DOI: 10.1016/j.autcon.2022.104192

Ebrahimi, M., Sarikhani, M. R., Safari Sinegani, A. A., Ahmadi, A., & Keesstra, S. (2019). Estimating the soil respiration under different land uses using artificial neural network and linear regression models. *Catena, 174*, 371–382. DOI: 10.1016/j.catena.2018.11.035

Elahi, E., Weijun, C., Zhang, H., & Nazeer, M. (2019b). Agricultural intensification and damages to human health in relation to agrochemicals: Application of artificial intelligence. *Land Use Policy, 83*, 461–474. DOI: 10.1016/j.landusepol.2019.02.023

Farghali, M., Osman, A. I., Umetsu, K., & Rooney, D. W. (2022). Integration of biogas systems into a carbon zero and hydrogen economy: A review. *Environmental Chemistry Letters, 20*(5), 2853–2927. DOI: 10.1007/s10311-022-01468-z

Gautam, K., Sharma, P., Dwivedi, S., Singh, A., Gaur, V. K., Varjani, S., Srivastava, J. K., Pandey, A., Chang, J.-S., & Ngo, H. H. (2023). A review on control and abatement of soil pollution by heavy metals: Emphasis on artificial intelligence in recovery of contaminated soil. *Environmental Research, 225*, 115592. DOI: 10.1016/j.envres.2023.115592 PMID: 36863654

Ghavami, S. M., Taleai, M., & Arentze, T. (2017). An intelligent spatial land use planning support system using socially rational agents. *International Journal of Geographical Information Science, 31*(5), 1022–1041. DOI: 10.1080/13658816.2016.1263306

Hahn, D., Munir, A., & Behzadan, V. (2021). Security and privacy issues in intelligent transportation systems: Classification and challenges. *IEEE Intelligent Transportation Systems Magazine, 13*(1), 181–196. DOI: 10.1109/MITS.2019.2898973

Ise, T., & Oba, Y. (2019). Forecasting climatic trends using neural networks: An experimental study using global historical data. *Frontiers in Robotics and AI, 6*, 32. DOI: 10.3389/frobt.2019.00032 PMID: 33501048

Jahanger, A., Ozturk, I., Chukwuma Onwe, J., Joseph, T. E., & Razib Hossain, M. (2023). Do technology and renewable energy contribute to energy efficiency and carbon neutrality? Evidence from top ten manufacturing countries. *Sustainable Energy Technologies and Assessments, 56*, 103084. DOI: 10.1016/j.seta.2023.103084

Javaid, M., Haleem, A., Singh, R. P., & Suman, R. (2022). Artificial intelligence applications for industry 4.0: A literature-based study. *J Ind Integr Manag, 7*(1), 83–111. DOI: 10.1142/S2424862221300040

Jha, S. K., Bilalovic, J., Jha, A., Patel, N., & Zhang, H. (2017). Renewable energy: Present research and future scope of artificial intelligence. *Renewable & Sustainable Energy Reviews, 77*, 297–317. DOI: 10.1016/j.rser.2017.04.018

Kadow, C., Hall, D. M., & Ulbrich, U. (2020). Artificial intelligence reconstructs missing climate information. *Nature Geoscience, 13*(6), 408–413. DOI: 10.1038/s41561-020-0582-5

Lee, J., & Yoo, H. J. (2021). An overview of energy-efficient hardware accelerators for on-device deep-neural-network training. *IEEE Open Journal of the Solid-State Circuits Society, 1*, 115–128. DOI: 10.1109/OJSSCS.2021.3119554

Mrówczyńska, M., Sztubecka, M., Skiba, M., Bazan-Krzywoszańska, A., & Bejga, P. (2019). The use of artificial intelligence as a tool supporting sustainable development local policy. *Sustainability (Basel), 11*(15), 4199. DOI: 10.3390/su11154199

Nawaz, R., Akhtar, R., Shahid, M. A., Qureshi, I. M., & Mahmood, M. H. (2021). Machine learning based false data injection in smart grid. *International Journal of Electrical Power & Energy Systems*, *130*, 106819. DOI: 10.1016/j.ijepes.2021.106819

Olabi, A. G., Abdelghafar, A. A., Maghrabie, H. M., Sayed, E. T., Rezk, H., Radi, M. A., Obaideen, K., & Abdelkareem, M. A. (2023). Application of artificial intelligence for prediction, optimization, and control of thermal energy storage systems. *Thermal Science and Engineering Progress*, *39*, 101730. DOI: 10.1016/j.tsep.2023.101730

Partel, V., Charan Kakarla, S., & Ampatzidis, Y. (2019). Development and evaluation of a low-cost and smart technology for precision weed management utilizing artificial intelligence. *Computers and Electronics in Agriculture*, *157*, 339–350. DOI: 10.1016/j.compag.2018.12.048

Rustia, D. J. A., Lin, C. E., Chung, J.-Y., Zhuang, Y.-J., Hsu, J.-C., & Lin, T.-T. (2020). Application of an image and environmental sensor network for automated greenhouse insect pest monitoring. *Journal of Asia-Pacific Entomology*, *23*(1), 17–28. DOI: 10.1016/j.aspen.2019.11.006

Saheb, T., Dehghani, M., & Saheb, T. (2022). Artificial intelligence for sustainable energy: A contextual topic modeling and content analysis. *Sustainable Computing : Informatics and Systems*, *35*, 100699. DOI: 10.1016/j.suscom.2022.100699

Sun, Y., Gao, C., Li, J., Wang, R., & Liu, J. (2019b). Quantifying the effects of urban form on land surface temperature in subtropical high-density urban areas using machine learning. *Remote Sensing (Basel)*, *11*(8), 959. DOI: 10.3390/rs11080959

Tian, Z., Shi, X., & Hong, S.-M. (2021). Exploring data-driven building energy-efficient design of envelopes based on their quantified impacts. *Journal of Building Engineering*, *42*, 103018. DOI: 10.1016/j.jobe.2021.103018

Yuan, X., Wang, Y., Yang, C., & Gui, W. (2020). Stacked isomorphic autoencoder based soft analyzer and its application to sulfur recovery unit. *Information Sciences*, *534*, 72–84. DOI: 10.1016/j.ins.2020.03.018

Yue, W., Yao, Y., Su, M., Rong, Q., & Xu, C. (2023). Identifying distributions of urban ecosystem health based on Latin-hypercube sampling and multi-criteria decision analysis framework. *Ecological Indicators*, *147*, 109957. DOI: 10.1016/j.ecolind.2023.109957

Yue, X.-L., & Gao, Q.-X. (2018). Contributions of natural systems and human activity to greenhouse gas emissions. *Advances in Climate Change Research*, *9*(4), 243–252. DOI: 10.1016/j.accre.2018.12.003

Zahraee, S. M., Khalaji Assadi, M., & Saidur, R. (2016). Application of artificial intelligence methods for hybrid energy system optimization. *Renewable & Sustainable Energy Reviews*, *66*, 617–630. DOI: 10.1016/j.rser.2016.08.028

Chapter 8
Greening the Supply Chain:
AI's Role In Reducing Carbon Emissions In Logistics

Latha Manickam
https://orcid.org/0009-0001-6627-7227
MAHSA University, Malaysia

Manickam Ramasamy
https://orcid.org/0000-0002-8729-9337
UCSI University, Malaysia

Sofia Jonathan
https://orcid.org/0000-0002-3535-2634
Lady Doak College, India

Jonathan Samuel
Independent Researcher, India

ABSTRACT

The modern logistics sector, crucial to worldwide trade, is also a major source of carbon emissions. The environmental impact is significant, from transporting goods over long distances to the energy-intensive activities in warehouses. With the pressing necessity to cut down greenhouse gas emissions and address climate change, the logistics industry is at a crucial point. Introducing artificial intelligence (AI), an impressive technology that has the ability to transform logistics and greatly reduce its environmental impact.

DOI: 10.4018/979-8-3693-9856-2.ch008

INTRODUCTION

Industries all throughout the world are being urged to reconsider their operations and lessen their environmental impact in light of the fast changing climate and growing environmental concerns. Supply chain management, an essential part of international trade and logistics that has historically contributed significantly to carbon emissions, is one of the most urgent areas for innovation. There is increasing pressure on businesses to decarbonize their operations and lessen the environmental impact of their logistics networks as they attempt to meet sustainability goals. (McKinsey & Company, 2022)

Then comes artificial intelligence (AI), a game-changing technology that has the power to completely change supply chains by increasing productivity, making the best use of available resources, and promoting environmentally friendly behavior. (Wamba-T et al., 2020). Beyond simple automation, artificial intelligence (AI) helps the supply chain become greener by facilitating better decision-making, predictive analytics, and real-time monitoring, all of which provide innovative ways to lower carbon emissions. AI is assisting logistics organizations in identifying inefficiencies, reducing waste, and minimizing fuel use through predictive maintenance and route optimization, all of which improve the bottom line.

The global carbon emissions are greatly impacted by the logistics industry, as road freight alone accounts for 7% of greenhouse gas emissions. Major factors include ineffective transport routes and unused capacity. Energy consumption in warehouses, especially from fossil fuels, also contributes to emissions. The final stage of delivery, known as the "last mile," is inefficient and emits a high level of emissions. Establishing objectives and putting in place AI-powered logistics planning is essential in decreasing the industry's environmental impact.

The logistics and transportation sector is crucial to the global economy but significantly contributes to greenhouse gas emissions. Sustainable logistics aims to reduce environmental damage through efficient supply chain management, including optimized routes, reduced fuel use, and energy-saving practices. Companies are adopting AI to improve planning, cut emissions, and lower costs, enhancing sustainability, customer retention, and their environmental impact despite challenges.

This chapter examines how AI-driven innovations are helping businesses create greener, more efficient supply chains by studying the relationship between AI and sustainability in logistics. The main uses of AI in lowering carbon emissions will be discussed, along with the potential and difficulties associated with putting these technologies into practice. It becomes evident when we examine supply chain management's future that artificial intelligence (AI) is not only a tool for enhancing operations but also a vital component of a more carbon-conscious and sustainable global economy.

1. AI IN LOGISTICS: A REVOLUTION IN FIELD

AI is revolutionizing logistics through the use of data and algorithms to improve decision-making, streamline processes, and save expenses. AI is a crucial tool in addressing logistical difficulties with complicated supply chains and high delivery demands.

Predictive Analytics: By examining data and patterns, AI predicts demand, spots interruptions, and maximizes inventory levels.

Route Optimization: AI systems minimize fuel use and CO2 emissions by designing effective delivery routes depending on traffic, weather, and delivery windows.

Warehouse Automation: Artificial Intelligence lowers costs and errors by optimizing operations, automating processes, and improving inventory accuracy.

Drones and Autonomous Vehicles: Artificial Intelligence (AI) in drones and cars increases delivery efficiency and lowers the need for human intervention.

Supply Chain Visibility: AI systems provide in-the-moment data to trace shipments, keep an eye on efficiency and anticipate disruptions.

Smart Demand Forecasting: AI reduces waste and overproduction by analyzing market patterns and behaviors to deliver precise demand projections.

Energy Management: AI optimizes logistical operations' use of energy, which lowers costs and lowers greenhouse gas emissions.

Artificial intelligence is developing logistics by improving efficiency, lowering expenses, and, crucially, facilitating sustainability efforts. Through the adoption of AI-driven solutions, organizations may enhance operational management, minimize environmental impact, and sustain competitiveness in a progressively eco-aware market.

2. OPTIMIZING ROUTES AND REDUCING FUEL CONSUMPTION

AI helps to minimize the environmental impact of logistics by optimizing routes and reducing fuel consumption, ultimately finding the most efficient delivery paths to save fuel and lower emissions.

AI Algorithms for Route Optimization

1. **Shortest Path Algorithms**: Logistics planning focuses on determining the most efficient route among delivery locations, using algorithms like Dijkstra's Algorithm and A (A-star) Algorithm to minimize fuel use and reduce emissions. (Delling et al., 2009).

2. **The Traveling Salesman Problem (TSP)** is a fundamental optimization challenge in logistics, aiming to identify the optimal path for a salesperson or delivery truck to traverse a series of destinations and return to the origin. Artificial intelligence algorithms like Genetic Algorithms and Ant Colony Optimization are employed to address this problem, emulating evolutionary or natural processes to rapidly identify optimal solutions.
3. **Machine learning models** are essential for AI-based route optimization, analyzing extensive historical data to forecast optimal routes. Reinforcement learning algorithms enable logistics organizations to modify routes in real-time according to fluctuating conditions, such as traffic congestion, accidents, or weather disturbances. This ensures cars circumvent superfluous halts or delays, optimizing fuel use.
4. **Multi-objective optimization methods**, such as Particle Swarm Optimization (PSO) and Simulated Annealing, are employed for simultaneous optimization of multiple objectives, including fuel usage and delivery time minimization. These algorithms examine multiple alternative routes and negotiate trade-offs between competing objectives, ensuring the selected route is optimal in distance, fuel efficiency, and time.
5. **The Vehicle Routing Problem (VRP)** is a more intricate variant of the TSP, considering several vehicles and numerous delivery locations. Algorithms like Swarm Intelligence and Tabu Search optimize the assignment of cars to delivery sites, minimizing overall distance, fuel consumption, and delivery time.
6. **Real-time data integration** is another aspect of AI route optimization systems, assimilation of real-time data from GPS, traffic reports, and weather forecasts to deliver current route modifications.
7. **Geofencing technology** combined with AI algorithms can enhance route optimization by establishing designated zones for delivery vehicles, ensuring adherence to environmentally sustainable routes or avoiding congested areas. AI-powered predictive maintenance ensures proper vehicle maintenance, preventing failures and ensuring optimal fuel efficiency.
8. **Collaborative Logistics** driven by AI enables companies to minimize fuel consumption by sharing delivery resources and organizing joint deliveries through platforms such as Convoy and Uber Freight.

Benefits of Route Optimization Powered by Artificial Intelligence

- Reduced fuel consumption by minimizing driving distance and avoiding traffic.
- Decrease the amount of carbon emissions

- Reducing costs and saving money.
- Enhanced delivery speed - resulting in effective processes and notable economic and ecological advantages

AI-driven logistics planning optimizes routes and minimizes fuel use, significantly reducing the carbon footprint of transportation operations and aligning logistics organizations with global sustainability objectives.

3. EFFECTIVE INVENTORY CONTROL

Effective inventory control is vital in logistics to maintain equilibrium in stock levels. The use of AI in predictive analytics improves demand forecasting to optimize performance.

How Predictive Analytics Improves Demand Forecasting

1. **Data-Driven Forecasting**: Predictive analytics depends on extensive datasets to produce forecasts. AI models may accurately forecast future demand by examining past sales data, customer purchasing history, market trends, and external factors such as weather, holidays, and economic situations. This data-centric methodology enables enterprises to foresee shifts in consumer demands and modify their inventory levels accordingly. (Choi et al., 2023; Waller & Fawcett (2013))
2. **Seasonal and Trend-Based Forecasting**: A primary advantage of predictive analytics is its capacity to discern demand patterns caused by seasonal variations, market trends, or product life cycles. AI algorithms can identify heightened demand for particular products throughout specific seasons (e.g., holiday season or back-to-school time) and predict the necessary inventory levels. This assists enterprises in preventing stockouts during high-demand phases while minimizing surplus inventory during slower intervals.
3. **Real-Time Demand Monitoring:** Predictive analytics systems can monitor real-time sales and market data to dynamically change projections. This enables enterprises to swiftly react to abrupt fluctuations in demand, including those prompted by new product introductions, competitive maneuvers, or unforeseen occurrences such as natural disasters. AI can evaluate this data instantaneously and suggest inventory modifications, mitigating the risk of overstocking or understocking.

4. **Minimizing Inventory Holding Costs**: Overestimating demand may result in surplus inventory, leading to increased expenses like storage fees, insurance, and spoiling for perishable items. Predictive analytics assists organizations in mitigating expenses by delivering precise demand estimates that avert overstocking. By sustaining ideal inventory levels, enterprises can minimize waste, decrease operational costs, and enhance cash flow.
5. **Reducing Stockouts and Lost Sales**: Misjudging demand may cause stockouts, resulting in lost sales opportunities and consumer dissatisfaction. Predictive analytics enables organizations to precisely anticipate product demand, assuring adequate inventory to satisfy client requirements. This not only reduces stockouts but also enhances customer satisfaction and brand loyalty, as customers are more inclined to return to a business that regularly offers the things they require.
6. **Enhancing Supply Chain Efficiency**: Predictive analytics facilitates the optimization of the supply chain by improving coordination among suppliers, manufacturers, and retailers. By predicting demand with more precision, firms may convey their requirements to suppliers proactively, enabling adjustments to production timelines and delivery dates. This minimizes lead times, prevents urgent orders, and facilitates a more seamless movement of items along the supply chain, hence decreasing transportation expenses and lowering emissions.
7. **Administering Product Lifecycles**: AI-driven predictive analytics assists enterprises in overseeing the complete lifecycle of items. Through the analysis of temporal sales data, AI can forecast the anticipated fall in product demand, enabling enterprises to systematically eliminate outdated inventory and efficiently launch new items. This guarantees that enterprises avoid squandering resources on products that have beyond their peak demand, thereby facilitating a streamlined and economical inventory.
8. **Personalized Demand Forecasting:** Predictive analytics can be utilized to generate more tailored demand projections informed by client segmentation. Through the analysis of data regarding client preferences, purchasing behavior, and geographic location, firms can customize their inventory levels for particular customer segments or areas. This guarantees the availability of appropriate products in optimal amounts at designated places, resulting in enhanced customer satisfaction and less inventory waste.
9. **Demand Sensing:** Demand sensing is a sophisticated use of predictive analytics that utilizes real-time data from several sources, including social media, point-of-sale systems, and market reports, to identify demand signals more promptly. This enables organizations to modify inventory levels more rapidly than conventional forecasting techniques. AI-driven demand sensing may effectively identify the instant influence of external elements such as trending products,

competitor promotions, or abrupt changes in customer behavior, leading to enhanced responsiveness and agility in inventory management.

10. **Scenario Planning and What-If Analysis**: Predictive analytics facilitates scenario planning, wherein AI models emulate diverse demand scenarios predicated on varying assumptions (e.g., economic recessions or supply chain interruptions). This allows enterprises to anticipate various outcomes and modify their inventory strategies accordingly. "What-if" assessments assist organizations in evaluating the effects of potential modifications and formulating more resilient inventory strategies, thereby preparing them to address demand variations under diverse circumstances. (Waller & Fawcett, 2013).

Advantages of Predictive Analytics in Inventory Management

- **Enhanced Forecast Precision**: AI-driven predictive analytics yields more precise demand projections, diminishing uncertainty in inventory planning and decreasing inaccuracies.
- **Cost Minimization:** By sustaining ideal inventory levels and preventing surplus stock, enterprises can diminish holding expenses, spoilage, and write-offs, resulting in reduced total operational expenditures.
- **Improved Customer Satisfaction**: Predictive analytics guarantees that organizations possess the appropriate merchandise at the optimal time, averting stockouts and enhancing the consumer experience.
- **Sustainability**: Precise demand forecasting mitigates overproduction and surplus inventory, hence diminishing waste and the ecological footprint of logistical operations.
- **Agility and Flexibility**: Predictive analytics empowers firms to swiftly adapt to fluctuations in demand, facilitating real-time adjustments to inventory strategy and maintaining competitiveness in dynamic marketplaces.

AI-driven demand forecasting significantly improves inventory management by minimizing waste, decreasing expenses, and augmenting customer pleasure, so fostering efficiency and sustainability. Organizations that implement predictive analytics can anticipate market trends, enhance their supply chain efficiency, and promote a more sustainable logistics framework.

Minimizing Waste Through Enhanced Inventory Management

Effective inventory management is essential for minimizing waste in supply chains, and AI-driven solutions are empowering organizations to adopt a more proactive strategy. Utilizing modern analytics and automation, firms may optimize their inventory management procedures, guaranteeing product availability in appropriate quantities at the correct time, while reducing waste from surplus stock, spoiling, and obsolescence.

Principal Methods AI-Enhanced Inventory Management Minimizes Waste

1. **Preventing Overstock and Surplus Inventory**: Overstocking constitutes a significant source of waste, especially for perishable goods, as surplus items may expire or deteriorate over time. Predictive analytics enables organizations to enhance inventory management by precisely estimating demand, so averting over-ordering or excessive production. AI algorithms evaluate previous sales data, seasonal variations, and market trends to optimize inventory levels, ensuring businesses maintain the appropriate quantity of stock. By minimizing overstock, organizations save storage expenses, lessen the necessity for markdowns or liquidation, and mitigate the environmental consequences of discarding unsold merchandise.
2. **Reducing spoiling and Expiration**: In sectors handling perishable goods, including food, pharmaceuticals, or cosmetics, spoiling presents a considerable difficulty. AI-driven inventory management systems can monitor expiration dates and product shelf lives to prioritize the sale or distribution of things approaching the end of their viability. This guarantees that products are utilized or sold prior to spoilage, hence reducing waste. Furthermore, AI can be incorporated into first-expired, first-out (FEFO) systems, guaranteeing that the oldest inventory is sold first, hence minimizing the risk of waste due to expiration.
3. **Mitigating Obsolescence for Non-Perishable Goods**: Non-perishable goods, including electronics, fashion items, and various consumer products, are susceptible to obsolescence due to shifting client tastes or the introduction of new versions. AI-powered demand forecasting enables enterprises to predict trends and inventory levels using real-time data, thereby averting surplus stock of potentially obsolete products. By mitigating the risk of retaining outmoded inventory, organizations might diminish the necessity to discount or liquidate unsalable products.
4. **Enhanced Inventory Replenishment**: AI-driven replenishment solutions assist enterprises in sustaining appropriate stock levels by autonomously ordering products just when required. These systems utilize real-time data from sales, inventory levels, and supply chain circumstances to initiate reorders, hence preventing overstock and understock situations. By ordering only the requisite quantity of goods, firms mitigate waste linked to surplus stockpiling.

5. **Demand-Driven Production Planning**: AI-enhanced demand forecasting is included into production planning to guarantee that producers create goods in amounts that correspond with projected demand. This mitigates the overproduction of commodities, which may result in waste if items remain unsold. Just-in-time (JIT) manufacturing, augmented by AI, aligns output with real-time demand, hence decreasing inventory accumulation and minimizing material waste in the manufacturing process. (Guo, & Mantravadi, 2024)
6. **Enhancing Product Lifecycle Management**: AI may evaluate sales trends and consumer behavior to forecast the conclusion of a product's lifecycle. This enables organizations to systematically discontinue products and timely launch new ones, hence avoiding surplus inventory of outdated things. Accurate demand data enables organizations to optimize product launches, promotional cycles, and phase-outs, thereby minimizing waste linked to unsold or obsolete inventory.
7. **Efficient Returns and Reverse Logistics**: Inventory waste frequently arises from ineffective management of product returns. Artificial intelligence can enhance reverse logistics by forecasting return rates and enabling expedited, more efficient product recovery, resale, or recycling. AI systems can spot patterns in return data, allowing organizations to modify their inventory rules and prevent overstocking items that are often returned.
8. **Sustainable Packaging and Transportation**: Artificial intelligence in inventory management can enhance the optimization of packaging and transportation of commodities. AI algorithms can optimize shipping techniques to minimize excessive packaging and decrease transportation waste. Effective route design and load optimization minimize resource wastage during transportation, hence decreasing fuel consumption and emissions.
9. **Minimizing Dead Stock**: AI assists enterprises in detecting dead stock—inventory that remains unsold for an extended duration—prior to it becoming an issue. Through the analysis of sales patterns, AI can identify slow-moving inventory promptly, enabling managers to execute strategies such as discounts or marketing campaigns to liquidate the stock. Mitigating dead stock diminishes the waste linked to the eventual disposal of unsold merchandise.
10. **Improved Supplier Collaboration**: AI-driven inventory solutions enhance collaboration between suppliers and retailers, allowing organizations to sustain lean inventory models. Through the dissemination of real-time data, enterprises and suppliers can modify inventory orders in accordance with prevailing demand, so preventing overproduction or surplus accumulation. This partnership minimizes waste in manufacturing and distribution, enhancing the sustainability of the supply chain.

Advantages of AI-Enhanced Inventory Management for Minimizing Waste

1. **Diminished Environmental Impact**: Minimizing surplus inventory, waste, and obsolescence mitigates the environmental footprint of logistics and supply chain activities. By reducing landfill waste, companies can substantially decrease their carbon footprint and promote sustainable practices.
2. **Cost Savings**: Minimizing waste results in reduced operational expenses, as enterprises conserve on storage, transportation, and disposal costs. By sustaining ideal inventory levels, organizations may alleviate the financial strain of surplus inventory, unsold merchandise, and returned items.
3. **Enhanced Resource Efficiency**: Effective inventory management guarantees optimal utilization of raw resources and finished goods, hence reducing waste across the supply chain. This enhances resource allocation and optimizes overall efficiency in manufacturing and logistical operations.
4. **Advancing Sustainability Objectives**: Numerous corporations under pressure to fulfill corporate social responsibility (CSR) and sustainability objectives. AI-driven inventory management supports these goals by encouraging sustainable consumption, minimizing waste, and decreasing emissions across the supply chain.

AI-driven inventory management serves as an effective instrument for minimizing waste in the logistics and supply chain sector. Through the optimization of inventory levels, the prevention of overproduction, and the reduction of spoiling, AI enables enterprises to attain enhanced sustainability and cost-effectiveness. As organizations endeavour to achieve environmental objectives, AI technology will assume a progressively significant role in mitigating waste produced by ineffective inventory management systems.

4. ENERGY-EFFICIENT WAREHOUSING

Automation and Smart Warehousing

As enterprises endeavor to diminish their carbon impact and enhance operational efficiency, energy-efficient warehousing has emerged as a primary priority. Intelligent warehousing systems, driven by artificial intelligence and automation, provide substantial enhancements in energy efficiency by streamlining processes, minimizing waste, and increasing productivity. These technologies not only reduce energy consumption but also advance sustainability objectives by fostering more environmentally friendly, adaptive, and economical operations.

What Constitutes Smart Warehousing?

Smart warehousing denotes the amalgamation of sophisticated technologies such as artificial intelligence (AI), machine learning, robotics, the Internet of Things (IoT), and big data analytics to automate and enhance warehouse operations. It facilitates real-time data processing, monitoring, and decision-making, hence enhancing operational efficiency and diminishing energy consumption.

How Automated and Smart Warehouse Increase Energy Efficiency

1. Automated Storage and Retrieval Systems (AS/RS):
    - o Overview: Automated Storage and Retrieval Systems (AS/RS) employ robots, conveyors, and cranes to facilitate the precise movement of goods in and out of storage. AI algorithms govern these systems, optimizing the most rapid and energy-efficient pathways for item retrieval or storage.
    - o Energy Efficiency: By reducing the energy required for physical labor, AS/RS systems decrease overall energy consumption in the warehouse. These systems are engineered for minimal power use, and by optimizing storage, they diminish the need for lighting and climate regulation in expansive warehouse environments. Furthermore, AS/RS systems can operate during off-peak hours, capitalizing on reduced energy costs.
2. AI-Enhanced Warehouse Management Systems (WMS):
    - o Overview: AI-enhanced Warehouse Management Systems (WMS) streamline multiple facets of warehousing, including inventory management and picking and packaging processes. The WMS can automate processes such as stock replenishment, item organization for expedited retrieval, and minimization of journey distances inside the warehouse through real-time data analysis. (**Mariam et al.,2022**)
    - o Energy Efficiency: WMS enhances space usage, minimizing the physical footprint necessary for storage and facilitating improved climate control. It also diminishes the necessity for energy-intensive operations such as lighting and heating/cooling by improving warehouse configuration and minimizing the number of trips needed for pickers or robots to fulfil their responsibilities.
3. Energy-Efficient Lighting with IoT Integration:

- Overview: Intelligent lighting systems, together with IoT sensors, provide adaptive management of warehouse illumination. These systems can reduce or deactivate lighting in unoccupied areas and modify illumination levels according to real-time data from occupancy sensors and daylight availability.
- Energy Efficiency: Smart lighting solutions minimize superfluous energy use by ensuring that lights are utilized solely when necessary. Moreover, the implementation of energy-efficient LED lighting with automated controls can substantially decrease electricity use, especially in expansive warehouses where lighting constitutes a significant energy expenditure.

4. Robotics and Automation:
    - Overview: Robots are essential in automating operations such picking, packing, sorting, and transferring goods inside the warehouse. AI-driven robots employ machine learning algorithms to enhance their movements, minimize idle periods, and collaborate with human workers more effectively.
    - Energy Efficiency: Robots can function more effectively than human laborers, especially in repeated jobs, thereby diminishing the necessity for lighting, heating, and air conditioning in specific areas of the warehouse. Robots typically utilize renewable energy sources or energy-efficient batteries, resulting in decreased overall energy consumption compared to conventional manual processes.
5. Climate Control Optimization:
    - Overview: Heating, ventilation, and air conditioning (HVAC) systems substantially contribute to energy usage in warehouses. AI-driven climate control systems evaluate real-time data from temperature, humidity, and air quality sensors to enhance heating and cooling efficiency across the facility.
    - Energy Efficiency: Intelligent climate control systems modulate HVAC output according to real-time conditions, guaranteeing that only the essential quantity of energy is utilized to sustain optimal working environments. For example, they can reduce temperatures in unoccupied regions or deactivate climate control systems in non-temperature-sensitive parts, resulting in significant energy savings.
6. Predictive Maintenance for Equipment:
    - Overview: AI-powered predictive maintenance solutions assess the condition and functionality of warehouse machinery such as conveyors, cranes, forklifts, and HVAC systems. Through the analysis of sensor

data, AI models may forecast maintenance requirements and avert unforeseen equipment malfunctions.

- o Energy Efficiency: Predictive maintenance enhances equipment performance, hence decreasing the energy consumption necessary for mechanical operation. Properly maintained machinery consumes less energy, functions more effectively, and diminishes the probability of energy-intensive failures that could impede warehouse operations.

7. **Smart Energy Management Systems (EMS)**:
   - o Overview: Smart Energy Management Systems (EMS) utilize artificial intelligence to monitor energy consumption within the warehouse, pinpointing inefficiencies and opportunities for energy reduction. These systems can autonomously deactivate equipment during non-operational hours or modulate energy use during peak demand periods to circumvent elevated costs.
   - o Energy Efficiency: An Energy Management System (EMS) assists enterprises in diminishing total energy consumption by regulating and optimizing energy utilization across several warehouse systems. This may encompass the automatic modulation of lighting, climate regulation, and machinery functioning in accordance with real-time demand and energy costs.
8. **Dynamic Shelving and Space Utilization:**
   - o Overview: Intelligent warehouses frequently employ dynamic shelving systems that autonomously adapt according to the dimensions and classification of the material kept. AI algorithms may identify optimal storage sites for products by analyzing demand patterns, dimensions, and weight, hence reducing the area necessary for storage and retrieval.
   - o Energy Efficiency: Optimal space usage minimizes the warehouse's overall footprint, hence reducing the energy required for climate control and illumination. By arranging products in a more compact and systematic fashion, enterprises can markedly decrease their energy usage.
9. AI-Enhanced Forklifts and Autonomous Vehicles:
   - o Overview: AI-enhanced autonomous forklifts and vehicles are progressively utilized in intelligent warehouses for the transportation of items between storage areas and loading ports. These cars employ artificial intelligence to explore the warehouse effectively, circumventing barriers and optimizing routes to minimize energy consumption.
   - o Energy Efficiency: Autonomous trucks function with greater precision and reduced energy consumption compared to conventional forklifts, hence minimizing fuel or electricity usage. They are frequently fueled

by energy-efficient batteries or renewable energy sources, thereby minimizing their carbon footprint.

10. Renewable Energy Integration:
    - o Overview: Numerous smart warehouses are incorporating renewable energy sources, including solar panels, wind turbines, or geothermal energy, to sustain their operations. AI algorithms oversee the allocation and storage of renewable energy in the warehouse to guarantee optimal utilization.
    - o Energy Efficiency: AI facilitates the integration of renewable energy into warehouse operations, diminishing dependence on fossil fuels and decreasing overall energy usage. Through the optimization of energy storage and consumption, AI systems guarantee the efficient utilization of renewable energy and the reduction of waste.

Advantages of Intelligent Warehousing and Automation for Energy Efficiency

1. **Reduced Operational Expenses**: Automation and intelligent energy management diminish the necessity for manual work, decrease power expenses, and curtail energy waste. This results in substantial cost reductions for enterprises, particularly for extensive operations.
2. **Diminished Carbon Footprint**: Through the optimization of energy consumption, smart warehouses can markedly decrease their carbon emissions. Employing AI-driven automation, energy-efficient technology, and renewable energy sources enables organizations to achieve sustainability objectives and enhance a greener supply chain.
3. **Augmented Productivity**: Automation enhances productivity by accelerating operations and minimizing errors, resulting in expedited order fulfillment and decreased downtime. This enhancement in productivity diminishes the energy consumed per unit of output, hence augmenting energy efficiency.
4. **Enhanced Sustainability Compliance**: Numerous enterprises face pressure to adhere to environmental legislation and sustainability requirements. Intelligent warehousing systems assist organizations in minimizing their ecological footprint and attaining adherence to green certifications like LEED (Leadership in Energy and Environmental Design).

AI-driven robots, automated storage systems, real-time climate control, and energy management technologies empower organizations to function more sustainably, minimize waste, and save operational expenses. Implementing intelligent warehousing solutions enables firms to establish more resilient, sustainable, and economical supply chains.

5. ARTIFICIAL INTELLIGENCE IN THE MANAGEMENT OF ENERGY USE IN WAREHOUSES

Energy consumption is a significant operating expense in warehousing, influenced by lighting, heating, ventilation, air conditioning (HVAC), equipment operation, and automated systems. Artificial intelligence is progressively utilized to optimize energy consumption, enabling enterprises to decrease expenses, minimize carbon emissions, and enhance overall energy efficiency.

Key Ways AI Manages Energy Use in Warehouses

1. **Smart Energy Management Systems (EMS):**

AI-driven Energy Management Systems (EMS) are warehouse systems that provide real-time oversight and regulation of energy usage. These systems monitor energy trends, detect inefficiencies, and implement modifications to enhance efficiency. AI algorithms analyze energy consumption data from lighting, HVAC, and equipment, adjusting settings based on real-time demand. EMS can deactivate idle equipment, reduce illumination, and regulate HVAC systems to maintain optimal temperatures with low energy use. This reduces total energy consumption, reduces costs, and minimizes the warehouse's environmental impact.

2. Predictive Maintenance of Energy-Consuming Machinery:

AI-powered predictive maintenance systems are used to manage energy-intensive machinery, such as conveyors, forklifts, HVAC systems, and automated storage solutions. These systems can predict when machinery needs repair, preventing malfunctions and ensuring optimal energy efficiency. Real-time data from sensors on machinery is transmitted to AI systems, which evaluate the information to identify wear, malfunctions, or inefficiencies. This helps prevent energy waste due to malfunctioning equipment. Predictive maintenance reduces the likelihood of sudden equipment failures, prolongs machinery's lifespan, and reduces the need for costly replacements.

3. **Optimization of HVAC Systems Utilizing Artificial Intelligence:**

Artificial intelligence can improve HVAC systems in warehouses by analyzing data from temperature, humidity, and occupancy sensors. AI systems use machine learning algorithms to assess factors like weather conditions, warehouse usage, and equipment thermal output. They calibrate the HVAC system to maintain ideal temperature and humidity levels while minimizing energy use. This can be done in low-activity zones or off-peak periods.

4. **Intelligent Lighting Systems:**

Intelligent lighting systems use real-time data from occupancy and daylight sensors to enhance lighting levels in warehouses. AI algorithms regulate lighting based on personnel or machinery presence, deactivating or lowering lights when vacant zones are present. These systems also modulate lighting levels based on natural light availability, reducing artificial illumination during daylight hours. By minimizing unnecessary lighting, AI-driven systems reduce electricity usage and prolong the lifespan of fixtures, reducing energy expenditures and maintenance costs.

5. **Load Shifting and Energy Consumption Scheduling:**

AI can help warehouses optimize energy consumption by scheduling energy-intensive operations during off-peak hours when energy costs are lower. AI systems analyze past usage data and utility pricing schedules to identify optimal times for executing these processes. Load shifting allows warehouses to capitalize on reduced energy rates, decreasing total expenses and mitigating excessive consumption during peak periods. This also enhances power system stability by ensuring a more even distribution of energy usage throughout the day.

6. **AI-Driven Renewable Energy Integration:**

Warehouses are increasingly utilizing renewable energy sources like solar panels and wind turbines to reduce their reliance on grid electricity. AI systems are helping to integrate renewable energy into warehouse operations by optimizing its use. Additionally, AI can manage energy storage technologies like batteries to retain surplus renewable energy for future use. This approach reduces the warehouse's reliance on fossil fuels, reduces energy costs, and reduces its carbon footprint.

7. **Smart Battery Management for Forklifts and Equipment:**

Warehouses are increasingly utilizing renewable energy sources like solar panels and wind turbines to reduce their reliance on grid electricity. AI algorithms can monitor the generation of renewable energy and prioritize it for energy-intensive applications during high solar output. Additionally, AI can manage energy storage technologies like batteries to retain surplus renewable energy for future use. This approach reduces the warehouse's reliance on fossil fuels, reduces energy costs, and reduces its carbon footprint.

8. **Energy-Efficient Inventory and Workflow Management:**

AI improves warehouse efficiency by reducing energy-consuming tasks like equipment and personnel relocation. Warehouse Management Systems and AI-powered robotics facilitate efficient storage and retrieval, reducing travel distance for forklifts and conveyors. This reduces energy consumption in warehouse transportation.

Advantages of AI in Optimizing Warehouse Energy Consumption

- Enhanced Sustainability: AI reduces energy waste and increases renewable energy utilization, reducing warehouse carbon footprint.
- Improved Operational Efficiency: AI automates energy-intensive processes, increasing productivity and reducing energy consumption.
- Prolonged Financial Savings: AI eliminates energy waste, saves maintenance, and prolongs equipment durability, leading to substantial savings.

Artificial intelligence is revolutionizing energy management in warehouses by facilitating more intelligent and efficient resource utilization. AI utilizes real-time monitoring, predictive analytics, and automation to ensure energy is utilized exclusively when and where necessary, aiding warehouses in achieving energy efficiency, minimizing expenses, and fulfilling sustainability objectives. Integrating AI-driven solutions enables organizations to enhance energy efficiency, hence fostering a more sustainable and environmentally friendly logistics sector.

6. AI-ENHANCED FLEET MANAGEMENT

Real-Time Surveillance and Upkeep

Fleet management is an essential component of logistics operations, and the emergence of AI technology enables firms to enhance the efficiency, reliability, and sustainability of their vehicle fleets. AI-powered fleet management employs real-time monitoring and predictive maintenance to enhance fleet efficiency, minimize downtime, and save operating expenses. Utilizing sophisticated sensors, machine learning algorithms, and real-time data analysis,

Real-Time Surveillance in AI-Enhanced Fleet Administration:

- **Vehicle Health Monitoring**: AI systems continuously monitor vehicle health, identifying anomalies for prompt identification.
- **Fuel Consumption Optimization:** AI systems optimize fuel consumption, reduce fuel consumption, and enhance operational efficiency.
- **Route Optimization:** AI systems use real-time information to optimize routes, reducing fuel consumption and enhancing efficiency.
- **Real-Time Fleet Tracking:** AI systems monitor fleet vehicles using GPS and telematics data for operational visibility and security.

Artificial Intelligence-Driven Predictive Maintenance

1. **Predicting Vehicle defects**:
   - o **Overview**: Predictive maintenance use artificial intelligence to anticipate possible mechanical defects before to their occurrence. Through the analysis of past maintenance records, real-time sensor information, and performance trends, AI systems can forecast the likelihood of vehicle malfunctions.

AI algorithms examine trends in vehicle performance, like engine vibrations, fluid levels, and brake wear, to forecast potential component failures. Upon detection of a probable failure, the system notifies fleet managers, enabling them to arrange maintenance at a suitable moment.

Advantage: Predictive maintenance mitigates the likelihood of unforeseen failures, which can incur significant expenses related to repairs, operational interruptions, and diminished productivity. By proactively addressing issues prior to their escalation into failures, firms can guarantee prolonged operational functionality of vehicles, hence enhancing fleet availability.

2. **Optimized Maintenance Scheduling**:
   - **Overview**: AI-driven fleet management solutions enhance maintenance schedules by evaluating the real-time performance and condition of each vehicle. Rather than depending on predetermined maintenance schedules, AI systems arrange maintenance according to actual vehicle utilization and deterioration.

Advantage: The advantage of optimized maintenance scheduling is in its ability to minimize unnecessary downtime and maintenance expenses by ensuring that vehicles receive servicing solely when required. This aids organizations in sustaining fleet efficiency, enhancing vehicle longevity, and reducing total operational expenses.

3. **Parts and Inventory Management**:
   - **Overview**: AI-driven predictive maintenance systems can anticipate the necessity for replacement components and oversee spare parts inventory. This guarantees the availability of essential components during maintenance, preventing delays caused by parts shortages.

Advantage: By optimizing parts inventory management, AI diminishes downtime caused by awaiting components and decreases inventory holding expenses. This guarantees that cars are promptly repaired and reinstated into operation without postponements.

4. **Energy-Efficient Maintenance Practices**:
   - **Overview**: AI systems enhance maintenance protocols to minimize energy usage during service. AI optimizes energy use in maintenance facilities, ensuring optimum power usage and optimal machinery operation.

Advantage: Energy-efficient maintenance procedures decrease operational expenses by minimizing energy usage in repair facilities. This enhances overall sustainability initiatives and mitigates the environmental impact of fleet operations.

By means of ongoing data analysis, AI guarantees that fleets function at peak efficiency, resulting in diminished breakdowns, lower fuel consumption, and enhanced resource utilization. By implementing AI-driven solutions, organizations may establish more sustainable, economical, and dependable logistics operations.

Use of electric and autonomous vehicles

As logistics prioritize sustainability and efficiency, electric and autonomous vehicles with AI integration revolutionize fleet management by reducing emissions and costs.

1. Electric Vehicles in Fleet Management

Electric vehicles are increasingly favored over traditional internal combustion engine (ICE) vehicles owing to their environmental advantages, reduced operating expenses, and less dependence on fossil fuels. AI-powered fleet management solutions enhance the efficiency and advantages of electric vehicles by maximizing their utilization and incorporating them into the logistics network.

Advantages include less carbon emissions, regulatory compliance, and lower operational expenses.

AI-Driven Solutions for Electric Vehicle Management

1. Battery Management and Optimization:
    - AI forecasts recharging needs and enhances charging during off-peak periods.
    - AI can suggest appropriate driving velocities and pathways to conserve battery energy.
2. Charging Infrastructure Optimization:
    - AI manages and optimizes charging stations for electric car fleets.
    - Optimized charging schedules reduce the risk of vehicle inactivity.
3. Route Optimization for Electric Vehicles:
    - AI analyzes vehicle range, road conditions, weather, and traffic patterns to optimize routes.
    - AI dynamically adjusts routes to accommodate real-time traffic conditions and charging station availability.
4. Predictive Maintenance for Electric Vehicles:
    - AI monitors the status of components like electric motors, battery integrity, and charging mechanisms.
    - Predictive maintenance enhances vehicle availability and reduces repair costs.

2. Autonomous Vehicles (AVs) in Fleet Administration

Autonomous vehicles (AVs) signify the future of fleet management, offering enhanced safety, efficiency, and cost-effectiveness. Artificial intelligence is fundamental to autonomous vehicle technology, facilitating self-driving automobiles and trucks in navigating roadways, optimizing routes, and executing real-time choices independently of human involvement.

Advantages of Autonomous Vehicles in Fleet Management:

- Enhanced Safety
- Optimized Efficiency
- Financial Savings

AI-Driven Solutions for Autonomous Vehicle Management

- **Advanced Route Optimization:** AI systems optimize routes based on real-time data from sensors, GPS, and road infrastructure. This improves delivery efficiency, reduces fuel consumption, and reduces operational costs.
- **Real-Time Decision Making:** AI systems analyze extensive data from sensors, cameras, radar, and LiDAR to make safe and efficient driving decisions. This enhances road safety, vehicle performance, and facilitates smoother driving, reducing fuel consumption and component wear.
- **Fleet Coordination and Platooning**: AI systems assist fleet managers in coordinating multiple autonomous vehicles, reducing air resistance and enhancing fuel economy. This can decrease fuel consumption by up to 10%, reducing emissions and operational costs.
- **Maintenance and Diagnostics of Autonomous Vehicles**: AI systems continuously monitor and maintain essential equipment, ensuring system functionality and reducing the likelihood of system failures.

AI-powered fleet management utilizing electric and autonomous cars is revolutionizing the logistics sector by improving efficiency, lowering expenses, and promoting environmental initiatives. Electric vehicles (EVs) and autonomous vehicles (AVs) signify the future of fleet operations, with artificial intelligence (AI) playing a pivotal role in performance management, operational efficiency, and benefit optimization. Implementing AI-driven solutions for electric and autonomous cars enables firms to enhance operational efficiency, minimize environmental impact, and establish more robust logistical networks.

7. REDUCING EMPTY MILES

Aligning Cargo and Reducing Empty Hauls

Vacant hauls lead to fuel wastage, elevated operational expenses, and augmented carbon emissions. AI-driven solutions provide a method to minimize empty miles through enhanced load matching, refined route planning, and the optimization of freight operations.

When trucks or delivery vehicles return to their base or go to their next destination unladen, they expend fuel, time, and resources without producing money. The U.S. Department of Transportation estimates that approximately 30% of truck miles are driven without cargo, underscoring the necessity for improved load matching and route optimization techniques.

Minimizing empty miles advantages logistics firms by:

- Decreasing operational expenses through diminished fuel consumption and vehicle depreciation.
- Enhancing profitability by optimizing the productive utilization of trucks.
- Reducing carbon emissions and promoting sustainability activities.

AI-Powered Solutions to Reduce Empty Hauls and Match Loads

1. **AI-Enhanced Dynamic Load Matching:**
    - Prioritizes loads based on proximity to the truck's current location.
    - Reduces the duration trucks travel without cargo, enhancing efficiency and revenue.
2. **Freight Aggregation and Consolidation**:
    - Aligns smaller shipments into complete truckloads.
    - Maximizes truck capacity, reducing unutilized mileage.
    - Reduces the required number of trucks for transportation, lowering empty miles and fuel expenses.
3. **AI-Enhanced Route Optimization:**
    - Determines the most efficient routes for trucks.
    - Streamlines routes, reducing unutilized mileage and fuel usage.
4. **Collaborative Freight Networks**:
    - Aligns available loads among various logistics companies.
    - Reduces empty miles and carbon emissions.
5. **Real-Time Freight Visibility and Tracking:**

- Provides instantaneous visibility into freight locations and vehicle availability.
- Facilitates expedited load matching and reduces the likelihood of vehicles operating without cargo.

6. **Predictive Freight Demand Forecasting**:
   - Uses past transportation data, market patterns, and seasonality to forecast increased freight demand.
   - Maximizes fleet efficiency and reduces trips made by unoccupied vehicles.
7. **Load Sharing and Platooning**:
   - Enables platooning, allowing trucks to travel in convoys.
   - Minimizes the presence of unoccupied trucks on the road.
8. **Digital Freight Marketplaces:**
   - Facilitates the connection between shippers and carriers.
   - Aligns freight with carriers based on parameters such as location, truck capacity, delivery schedules, and pricing.

AI-driven load matching, route optimization, and collaborative freight networks offer effective ways to reduce idle miles in logistics. These technologies facilitate the transportation of loads on both outward and return journeys, thereby diminishing fuel consumption, enhancing fleet utilization, and reducing carbon emissions. The outcome is a more efficient, economical, and environmentally friendly logistics operation.

8. DATA-DRIVEN DECISION MAKING

In the contemporary, intricate, and rapid logistics sector, big data is essential for decision-making and operational optimization. The incorporation of big data into logistics planning enables firms to examine extensive information from diverse sources, facilitating more informed, accurate, and data-driven decisions. Utilizing big data, logistics firms may optimize routes, enhance inventory management, predict demand, improve customer service, and ultimately augment efficiency while minimizing costs.

What Constitutes Big Data in Logistics?

Big data denotes the vast quantity of organized and unstructured data produced from diverse sources, such as IoT devices, GPS trackers, meteorological reports, client orders, supply chain management systems, social media, and transaction records. In logistics, big data includes real-time vehicle positions, warehouse inventory levels, historical shipment patterns, and customer feedback.

Big data is defined by its volume (substantial amounts of data), velocity (the rapidity of data generation and processing), diversity (many forms of data from multiple sources), and veracity (the dependability and precision of data). The function of big data in logistics is to get actionable insights from intricate data, facilitating better informed decision-making and enhanced operational efficiency. (Wang et al., 2016)

Key Roles of Big Data in Logistics Planning

1. **Route Optimization and Traffic Management:**
    - Real-time data from GPS trackers and traffic management systems provide current information on traffic congestion, accidents, road closures, and weather conditions.
    - Route optimization via big data reduces fuel usage, shortens delivery times, and cuts operational expenses.
2. **Demand Forecasting and Inventory Management:**
    - Big data facilitates enhanced demand forecasting, enabling logistics organizations to optimize inventory management and avert stockouts or overstocking.
    - Precise demand forecasting minimizes surplus inventory, averts stockouts, and enhances order fulfilment rates. (Choi et al., 2023)
3. **Supply Chain Visibility and Tracking:**
    - Big data improves visibility across the supply chain, enabling logistics managers to monitor shipments, oversee warehouse operations, and predict disruptions.
    - IoT sensors affixed to shipping containers, lorries, and warehouses gather data regarding the location, temperature, humidity, and movement of commodities.
4. **Predictive Maintenance for Fleet and Equipment**:
    - Big data analytics enables firms to conduct predictive maintenance on their vehicles, machinery, and equipment, minimizing downtime and averting expensive failures.
    - Predictive maintenance diminishes unforeseen vehicle and equipment malfunctions, enhances fleet dependability, and minimizes maintenance expenditures.
5. **Customer Service and Personalization:**
    - Big data offers insights into consumer preferences, purchase history, and delivery expectations, enabling organizations to customize their

services, enhance delivery efficiency, and pre-emptively resolve possible problems.
- Data-driven customisation results in increased customer happiness, recurrent patronage, and enhanced brand loyalty.

6. **Risk Management and Disruption Mitigation:**
    - Big data enables logistics organizations to identify and manage risks through the analysis of diverse internal and external data sources.
    - Proactive risk management diminishes the probability of expensive delays, product deficiencies, and customer discontent.
7. **Sustainability and Environmental Impact**:
    - Big data enables logistics businesses to monitor their environmental footprint and make informed decisions to minimize fuel usage, emissions, and waste.
    - Data-driven sustainability programs assist firms in minimizing expenses, enhancing fuel economy, and decreasing carbon emissions.
8. **Warehouse and Labor Optimization:**
    - Big data enhances warehouse management through the optimization of labour allocation, the reduction of picking times, and the increase of overall operational efficiency.

Big data revolutionizes logistics planning to optimize operations, reduce costs, enhance satisfaction.

AI Technologies in Logistics for Data Analysis and Decision Assistance

1. Machine Learning Frameworks

Machine learning (ML) is a branch of artificial intelligence (AI) that enables systems to acquire knowledge from data and generate predictions or judgments autonomously, without direct programming. In logistics, machine learning platforms examine previous data to discern patterns, forecast future trends, and facilitate decision-making.

- **Essential Attributes:** Predictive Analytics, anomaly detection, and demand forecasting.

2. Predictive Analytics Instruments

Upstream, by using real-time data processing and predictive analytics, AI can revolutionize inventory management and supplier relationships. To ensure the integrity and confidentiality of sensitive data, this transformation calls for a thorough analysis of data security and privacy issues. Furthermore, there is a fine balance to be struck because an over-reliance on AI algorithms could possibly overshadow human expertise.(Richey et al., 2023)

Predictive analytics use artificial intelligence and machine learning algorithms to examine historical and real-time data, allowing logistics firms to foresee future occurrences and enhance operational efficiency accordingly. Predictive analytics technologies are crucial for decision support in domains such as demand forecasting, inventory management, and risk mitigation.

- Principal Attributes: Demand Forecasting and Predictive Maintenance.

3. Optimization Algorithms

Optimization algorithms in AI assist logistics organizations in resolving intricate issues, including identifying optimal delivery routes, maximizing vehicle loads, and reducing transportation expenses. These algorithms are crucial for enhancing operational efficiency and minimizing costs.

- Essential Attributes: Route Optimization and Load Optimization.

4. AI-Driven Business Intelligence Platforms

AI-driven business intelligence (BI) technologies assist logistics managers in visualizing and analyzing data from many sources, offering insights into operational performance, customer behavior, and supply chain efficacy. These technologies facilitate data-driven decision-making by delivering actionable insights through dashboards, reports, and visualizations.

- Essential Characteristics: Data Integration and Automated Reporting.

5. Natural Language Processing (NLP) Tools

Natural language processing (NLP) techniques facilitate AI systems in analyzing and comprehending human language, hence simplifying data interaction and insight extraction for logistics professionals. These instruments are frequently employed in chatbots, voice assistants, and text analytics, assisting organizations in enhancing customer service, automating reporting, and optimizing decision-making.

- Essential Attributes: Chatbots and Virtual Assistants.
- Text Analytics.

6. Prescriptive Analytics Tools

Prescriptive analytics solutions employ AI to forecast future events and suggest particular measures to attain optimal results. These technologies are especially beneficial for decision support in logistics, since they assist organizations in determining the most effective course of action to achieve corporate objectives.

- Essential Attributes: Practical Suggestions and Scenario Planning.

7. Robotic Process Automation (RPA) Tools

Robotic process automation (RPA) systems enable logistics organizations to automate repetitive processes and optimize workflows, enhancing productivity and minimizing human error. RPA tools can collaborate with AI systems to improve decision-making through the automation of data gathering, processing, and analysis.

- Essential Attributes: Process Automation and Workflow Optimization.

Artificial intelligence technologies for data analysis and decision support are revolutionizing logistics by delivering the insights required to enhance operations, minimize expenses, and increase efficiency. AI empowers logistics managers to utilize predictive analytics, optimization algorithms, business intelligence, and NLP tools for data-driven decision-making that improves performance, enhances agility, and provides superior customer experiences. By implementing these AI solutions, logistics organizations may maintain competitiveness, foster innovation, and attain more sustainable, cost-efficient operations.

Optimal strategies that organizations can implement to improve sustainability and operational efficiency:

1. Establish Explicit Sustainability Objectives
    - Establish Quantifiable Goals: Formulate explicit and quantifiable sustainability objectives prior to the incorporation of AI. Establish objectives for decreasing fuel consumption by a specific percentage or reducing CO2 emissions by a predetermined quantity of metric tons within a designated timeframe.
    - Integrate AI Initiatives with Sustainability Strategy: Guarantee that AI solutions are congruent with overarching business sustainability ob-

jectives. This establishes a compelling rationale for AI investment and guarantees that the environmental advantages are accurately assessed and conveyed.

2. Utilize Data for Informed Decision-Making
    - Invest in Superior Data: Guarantee access to dependable and extensive data from IoT devices, GPS systems, telematics, and additional sources. The accuracy of AI forecasts and recommendations is directly contingent upon the quality and quantity of the data utilized.

      Utilize Real-Time Data for Optimization: Employ AI solutions that assess real-time data to enhance routing, oversee inventories, and adapt to fluctuating circumstances. This can aid in minimizing fuel usage and mitigating energy waste in dynamic settings.
3. Initiate with Minor, Scalable Pilot Programs
    - Implement Pilot AI Solutions in Specific Domains: Initiate a pilot program concentrating on a designated area, such as route optimization, minimizing fleet idle time, or enhancing energy efficiency in a warehouse. Evaluate AI technologies inside this regulated setting and analyze their effects on sustainability.
    - Incrementally Expand: Following the successful pilot, progressively extend AI projects to additional routes, fleets, or facilities. This enables organizations to mitigate risks and enhance AI performance prior to complete implementation.
4. Prioritize Optimization Over Mere Automation
    - Emphasize Optimization Algorithms: Deploy AI tools that concentrate on enhancing resource utilization, like optimizing vehicle capacities, identifying the most fuel-efficient routes, or forecasting optimal maintenance schedules to prevent energy waste.
    - Integrate Automation with Sustainability: Although automation enhances productivity, it is imperative that AI systems are developed to mitigate environmental concerns. An AI-driven warehouse management system should prioritize not just expedited picking and packing but also the reduction of energy consumption within the facility.
5. Employ Predictive Maintenance to Prolong Asset Longevity
    - Initiate Predictive Maintenance Promptly: Employ AI to oversee the condition of vehicles, machinery, and equipment to identify possible problems prior to their escalation into breakdowns or inefficient fuel consumption. This can prolong asset longevity and diminish energy usage and emissions from inefficient equipment.
6. Collaborate and Disseminate Data for Enhanced Impact

- Promote Collaboration Throughout the Supply Chain: Utilize AI-driven collaboration tools to disseminate data with partners and clients. Collaborative systems facilitate enhanced load consolidation, efficient routing, and greater resource efficiency, hence resulting in reduced fuel consumption and emissions.
- Utilize Open Data and Collaborations: Formulate data-sharing alliances with critical stakeholders in your supply chain to utilize shared insights. Disseminating real-time information regarding traffic, routes, demand projections, and shipment status can enhance overall sustainability.

7. Implement AI-Driven Solutions for Inventory and Energy Management
   - Implement AI within Inventory Systems: Utilize AI to enhance inventory management, hence decreasing the necessity for surplus stock and mitigating the energy consumption and emissions linked to superfluous storage and transportation. AI-enabled demand forecasting can assist organizations in preventing overproduction.
   - Enhance Energy Efficiency in Warehouses and Facilities: AI can be employed to monitor and optimize energy consumption in warehouses, thereby decreasing electricity usage for lighting, heating, cooling, and equipment. AI-driven energy management systems can autonomously modify operations according to real-time conditions.
8. Utilize AI for Ongoing Enhancement
   - Oversee and Enhance AI Models: Persistently gather new data to improve AI algorithms. Regularly update AI models with the most current data on traffic, weather, fuel prices, and consumer demand to enhance predictive accuracy and improve operations.
   - Monitor Environmental KPIs: Consistently assess environmental KPIs, including fuel consumption, emissions, and energy utilization, to verify that AI solutions are promoting sustainable results. Utilize this data to implement modifications as necessary to further diminish your carbon impact.

Organizations that effectively employ AI to diminish their carbon footprint prioritize explicit sustainability objectives, data-informed optimization, and cooperative efforts. Initiating small pilot projects, enhancing maintenance, energy, and inventory management, and utilizing ongoing AI refinement are essential best practices. By using these measures, firms can diminish their environmental footprint while attaining substantial cost reductions and operational efficiencies, so establishing themselves as frontrunners in the sustainable logistics sector.

Future Directions for AI in Logistics

The future of AI in logistics is promising, with various new trends poised to influence corporate adoption of AI in the forthcoming years:

1. AI and Autonomous Vehicles: The use of autonomous vehicles (AVs) in logistics is anticipated to rise as AI-driven self-operating trucks and drones gain greater reliability. These technologies can markedly diminish emissions, fuel consumption, and labor expenses.
2. AI-Enhanced Collaborative Platforms: As logistics networks get increasingly interconnected, AI-enhanced collaborative platforms will facilitate more efficient data and resource sharing among enterprises. These platforms will enhance load distribution, minimize unutilized mileage, and augment overall supply chain efficacy.
3. Sustainability and AI: Artificial Intelligence will assume a progressively significant role in assisting logistics firms in achieving sustainability objectives. AI-driven sustainability initiatives will prioritize route optimization and energy consumption reduction in warehouses, becoming essential for firms aiming to mitigate their environmental effect.
4. AI-Driven Robotics in Warehousing: The utilization of AI-powered robots in warehousing will persist in its expansion. These robots can independently navigate, categorize, and package items, enhancing warehouse efficiency and diminishing energy usage.
5. AI for Real-Time Supply Chain Optimization: Real-time AI analytics will allow organizations to perpetually refine their supply chain operations, making instantaneous adjustments in response to variations in demand, traffic circumstances, and market swings. (Joel et al.,2024)

Addressing the obstacles to AI adoption in logistics necessitates tackling issues such as data quality, financial constraints, insufficient skills, and opposition to change. By investing in appropriate infrastructure, cultivating an innovative culture, and utilizing scalable AI technologies, logistics organizations can realize substantial advantages. The future of logistics will be characterized by intelligent, efficient, and sustainable operations, propelled by AI-driven decision-making and optimization as AI technologies progress.

Future Developments in Artificial Intelligence and Sustainable Logistics

As enterprises worldwide pursue enhanced sustainability, AI increasingly assumes a pivotal role in modernizing logistics operations to mitigate environmental effect. The future of AI in logistics will be characterized by technologies that promote sustainability, enhance resource efficiency, and diminish carbon emissions. The following are the principal future trends in artificial intelligence and sustainable logistics:

1. AI-Driven Autonomous Vehicles and Drones

Trend:

AI will enable self-driving trucks, electric AVs, and drones to autonomously navigate routes, deliver goods, and optimize last-mile delivery operations, reducing reliance on fossil fuels and human drivers.

Benefits:

1. Decreased fuel consumption due to enhanced fuel efficiency.
2. Efficient last-mile delivery through AI-driven drones and autonomous vehicles.
3. Uninterrupted functionality of autonomous logistics vehicles.

2. AI-Powered Supply Chain Visibility and Transparency

Trend: AI will enable logistics companies to monitor emissions, track sustainability metrics, and ensure compliance with environmental regulations.

Benefits:

1. Real-time emission monitoring.
2. Improved reporting and compliance.
3. Sustainability-oriented decision-making.

3. Circular Supply Chains with AI

Trend: AI will optimize processes like reverse logistics, waste management, and product lifecycle management.

Benefits:

1. Enhances reverse logistics, waste management, and product lifecycle management.
2. Reduces waste and environmental impact.
3. Facilitates efficient reutilization of materials
4. Sustainable Route Optimization and AI-Powered Mobility

Trend:

AI-powered **Mobility-as-a-service (MaaS)** will enable the adoption of shared, low-emission transportation solutions.

Benefits:

1. Minimizes emissions and fuel usage.
2. Integrates with electric and hybrid vehicles for charging station accessibility.
3. Facilitates widespread adoption of shared electric vehicles and autonomous trucks.

5. AI-Enhanced Smart Warehousing and Energy Management

Trend:

Smart warehousing solutions powered by AI will manage energy-efficient lighting, heating, cooling, and equipment operations based on real-time occupancy and demand.

Benefits:

1. Optimizes energy use in warehouses and distribution centers.
2. Minimizes packaging waste.
3. Facilitates carbon-neutral warehousing.

6. AI for Sustainable Packaging Optimization

Trend:

AI will revolutionize sustainable packaging by optimizing material usage, reducing waste, and designing more eco-friendly packaging solutions.

Benefits:

1. Minimizes packaging waste.
2. Optimizes sustainability.
3. Facilitates smart packaging design.

7. AI and Green Freight Corridors

Trend:

AI will help develop **green freight corridors**—routes optimized for low-emission transportation through the use of electric vehicles, renewable energy, and collaborative logistics.

Benefits:

1. Defines pathways for low-emission vehicles and sustainable transportation modalities.

2. Promotes collaborative logistics and renewable energy integration

AI will continue to play a central role in the future of sustainable logistics by optimizing operations, reducing waste, and minimizing the carbon footprint of transportation and supply chains. As autonomous vehicles, smart warehousing, sustainable packaging, and green freight corridors gain traction, AI-driven technologies will empower businesses to meet ambitious sustainability goals while remaining competitive in a rapidly evolving logistics landscape. The future of logistics will be smarter, greener, and more efficient, with AI driving the way forward.

Current Advances in AI Technology in Logistics

The logistics sector is experiencing a substantial upheaval due to advancements in AI technology. Here are few critical domains in which AI is effecting change:

1. **Predictive Analytics for Demand Projection**

AI systems can evaluate extensive past data to forecast future demand with considerable precision. This assists logistics firms in optimizing inventory levels, minimizing waste, and assuring effective resource utilization.

2. **Route Optimization and Planning**

Algorithms for route optimization powered by AI can determine the most effective paths for delivery vehicles, considering variables such as traffic, weather conditions, and delivery timeframes. This decreases fuel usage and emissions by limiting superfluous travel.

3. **Autonomous Delivery Vehicles**

Autonomous vehicles and drones are being engineered to manage deliveries, particularly for the "last mile" section. These cars may function continuously and adhere to optimum routes, so minimizing emissions.

4. **Intelligent Warehousing and Inventory Control**

AI-driven systems can oversee warehouse operations, encompassing inventory management and automated picking and packing. This enhances efficiency and decreases energy usage in warehouses.

5. **Real-Time Monitoring and Transparency**

The integration of AI with IoT devices facilitates real-time shipment tracking, enhancing coordination and minimizing the necessity for surplus inventory and faster dispatch. This results in more efficient resource utilization and reduced emissions.

6. **Improved Customer Service using AI Chatbots.**

AI chatbots may manage consumer inquiries and deliver real-time shipment updates, minimizing reliance on paper communication and enhancing overall productivity.

7. **Energy-Efficient Fleet Management**

Artificial intelligence can oversee and enhance the operational efficiency of delivery vehicles, guaranteeing their correct maintenance and optimal performance. This encompasses the utilization of electric and driverless vehicles, which possess a reduced carbon impact.

Targeting Emission Sources with AI

By leveraging these AI technologies, logistics companies can target specific emission sources:

- **Fuel Consumption**: Route optimization and autonomous vehicles diminish fuel consumption by guaranteeing that vehicles traverse the most efficient routes and function continuously without superfluous halts.
- **Warehouse Energy use**: Intelligent warehousing systems can optimize energy use, thereby diminishing the carbon impact of storage activities.
- **Inventory Waste**: Predictive analytics assist in sustaining ideal inventory levels, minimizing waste resulting from overstocking and stockouts.
- **Last Mile Delivery**: Autonomous delivery trucks and optimized routing enhance the efficiency of the last mile, reducing emissions from tiny, frequent deliveries.

These advancements not only diminish the carbon footprint but also improve the overall efficiency and cost-effectiveness of logistical operations. (Xiaonan et al., 2023)

CONCLUSION

Integrating AI-driven logistics planning enables firms to minimize their carbon impact and foster a more sustainable supply chain. AI facilitates logistics by utilizing real-time data, predictive analytics, and automated decision-making, ensuring efficiency and environmental sustainability.

The integration of AI with sustainability facilitates regulatory compliance while simultaneously bolstering a company's reputation as an environmentally responsible brand. Consequently, integrating AI in logistics is a strategic necessity for firms aiming to reconcile profitability with environmental responsibility. (Divya & Santhanakrishnan,2024)

As the globe increasingly faces the difficulties of climate change, the logistics and supply chain sectors must adapt to more sustainable practices. The incorporation of Artificial Intelligence (AI) has become a pivotal element in this transition, offering novel methods to diminish carbon emissions and enhance resource efficiency. AI facilitates enhanced operations through route optimization and predictive maintenance, so minimizing waste, decreasing fuel consumption, and reducing overall environmental impact.

By leveraging AI, firms may adopt more intelligent, data-informed strategies that improve profitability and substantially advance global sustainability objectives. The capacity of AI to transform logistics into more environmentally friendly and sustainable operations is substantial, facilitated by real-time data analytics, automation, and AI-driven forecasting. This change necessitates a comprehensive approach that integrates technical innovation, strategic investments, and collaboration among industries and stakeholders.

In the future, the ongoing advancement of AI technology, together with an increasing dedication from enterprises, governments, and consumers, will be crucial in facilitating the decrease of carbon emissions in logistics. As these technologies advance, the supply chain sector has a unique chance to harmonize economic prosperity with environmental responsibility, thereby securing a more sustainable and resilient future for future generations.

REFERENCES

Choi, T. M., Wallace, S. W., & Wang, Y. (2023). Big data analytics in operations management. *Production and Operations Management, 27*(10), 1868–1883. DOI: 10.1111/poms.12838

Delling, D., Goldberg, A. V., Pajor, T., & Werneck, R. F. (2009). Engineering fast route planning algorithms. In Vol. 5868, pp. 52–66). Lecture Notes in Computer Science. Springer., DOI: 10.1007/978-3-642-05465-5_5

Divya, D., & Santhanakrishnan, D. (2024). Artificial intelligence in sustainable supply chain management: A comprehensive review.

Guo, D., & Mantravadi, S. (2024). The role of digital twins in lean supply chain management: Review and research directions. *International Journal of Production Research, 63*(5), 1851–1872. DOI: 10.1080/00207543.2024.2372655

Joel, O., Oyewole, A., Odunaiya, O., & Soyombo, O. (2024). Leveraging Artificial Intelligence for Enhanced Supply Chain Optimization: A Comprehensive Review of Current Practices and Future Potentials. *International Journal of Management & Entrepreneurship Research., 6*(3), 707–721. DOI: 10.51594/ijmer.v6i3.882

Richey, R.G, Glenn, R., Chowdhury, S., Davis-Sramek, B., Giannakis, M., & Dwivedi, Y. K. (2023). Artificial Intelligence in Logistics and Supply Chain Management: A Primer and Roadmap for Research. *Journal of Business Logistics, 44*(4), 532–549. DOI: 10.1111/jbl.12364

McKinsey & Company. (2022, June 22). *Decarbonizing logistics: Charting the path ahead.* https://www.mckinsey.com/capabilities/operations/our-insights/decarbonizing-logistics-charting-the-path-ahead

Waller, M. A., & Fawcett, S. E. (2013). Data science, predictive analytics, and big data: A revolution that will transform supply chain design and management. *Journal of Business Logistics, 34*(2), 77–84. DOI: 10.1111/jbl.12010

Wamba-Taguimdje, S.-L., Fosso Wamba, S., Kala Kamdjoug, J. R., & Tchatchouang Wanko, C. E. (2020). Influence of artificial intelligence (AI) on firm performance: The business value of AI-based transformation projects. *Business Process Management Journal, 26*(7), 1893–1924. DOI: 10.1108/BPMJ-10-2019-0411

Wang, G., Gunasekaran, A., Ngai, E. W. T., & Papadopoulos, T. (2016). Big data analytics in logistics and supply chain management: Certain investigations for research and applications. *International Journal of Production Economics, 176*, 98–110. DOI: 10.1016/j.ijpe.2016.03.014

Xiaonan, Z., Lai, P.-L., Wang, X., & Ma, F. (2023). Evolution, Challenges, and Opportunities of Transportation Methods in the Last-Mile Delivery Process. *Systems*, *11*(10), 509. DOI: 10.3390/systems11100509

Chapter 9
Smart Warehousing: Automation and Real-Time Monitoring

Harleen Aggarwal
https://orcid.org/0009-0001-4504-5259
Western Community College, Canada

Parinita Malhotra
https://orcid.org/0000-0002-3626-6885
Western Community College, Canada

ABSTRACT

Warehousing is an integral component of supply chain management, serving as a link between procurement and distribution. After witnessing a downfall during the COVID-19 pandemic, the 3PL market has experienced a rebound due to the adoption of digital solutions and technology. Gone are the days when organisations were using traditional warehouse management systems to manage and track the activities of a warehouse. Modern warehouse managing systems are software-based and are designed to be cost-effective, improve order fulfillment and delivery times, optimise inventory levels, improve lead times, reduce logistics costs and increase competitive advantage. This chapter focuses on introducing and analyzing the concept of smart warehousing. This chapter aims to enable the students to understand how the supply chain industry is embracing digital transformation by using Artificial Intelligence, IoT and analytics, advanced robotics and automation, RFID, automated guided vehicles (AGV) and Blockchain Technology.

DOI: 10.4018/979-8-3693-9856-2.ch009

LEARNING OBJECTIVES

After reading this chapter the learners will be able to:

1) Understand the importance of automation and integration of AI in supply chain management.
2) Describe the processes and technologies used in smart warehousing.
3) Identify the future trends and challenges related to AI-driven warehouse automation systems.
4) Analyse the importance of regulatory provisions regarding the use of AI in warehousing.

OPENING STORY – AMAZON 'THE AUTOMATION JOURNEY'

Amazon, an American MNC is the world's largest online retailer. Amazon has a variety of distribution facilities including- CDC (cross-dock centers), fulfillment centers, sortation centers, receive centers, delivery stations, and hubs. Amazon expanded its distribution network from two fulfillment centers in 1997 to over 185 fulfillment centers globally. These fulfillment centers are large warehouses also known as "Amazon FBA -Fulfillment by Amazon" where Amazon receives, stores, packs, and ships customer orders and handles customer service. These fulfillment centers are equipped with advanced technology and software systems for quick and efficient order processing. Amazon has been relying heavily on automation and robotics for managing its warehousing and logistics system. As a commitment to invest in and encourage innovation, Amazon has set up a $ 1 billion corporate venture capital fund known as 'Amazon Industrial Innovation Fund' which provides direct investment to emerging technology companies.

Amazon's next-generation fulfillment center in Shreveport, Louisiana is highly powered by AI and robotics where the employees work hand in hand with a fleet of robotic systems. Following are a few landmarks of Amazon's automation journey:

- 2010 -AWS collaborated with NVIDIA to improve product search capability.
- 2012 - Amazon acquired Kiva Systems now known as Amazon Robotics.
- 2014 - Amazon launched Alexa which used natural language processing and machine learning algorithms to fulfill customer's requests.
- 2019 -Amazon purchased Canvas Technology thus entering into building robotic vehicles with guided vision.
- 2019 - Amazon collaborated with other vendors like CMC, Smart Pac, etc.
- 2024 -Amazon partnered with AI-RAN Alliance for wireless communications.

Amazon has deployed almost '750000 robots by 2023' to support its operations in warehousing which are mentioned below (Dresser,2023):

- Kiva
- Palletizer
- Robo-stow
- Pegasus AGV
- Hercules
- Xanthus
- Canvas Cart
- Xanthus
- Sparrow
- Proteus
- Robin
- Titan
- CMC Carton Wrap Machines
- Kermit
- Digit- a humanoid robot for supporting warehouse operations
- Drone delivery

Figure 1. Growth of amazon's robot workforce (2013-2023)

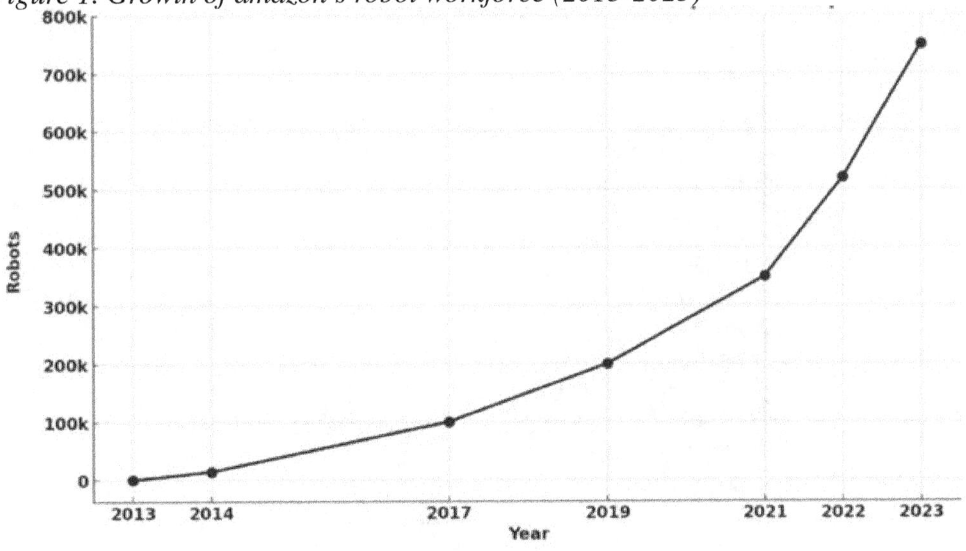

Source: Amazon Grows to More Than 750,000 Robots- January 25, 2024, by Ben Thompson

Figure 1 illustrates the growth of robots in Amazon's workforce from 2013-2023 starting from scratch and rapidly rising in a decade. This growth indicates Amazon's commitment to improving the processes through the use of AI and robotics (Thompson, 2024).

However, with an increase in investment in automation, the injury rates in Amazon's warehouses have also significantly risen as compared to the industry average. Humans co-working with robots also poses great safety risks.

To quote Brady, the Amazon Robotics chief technologist, "It's not humans against machines, it's humans and machines working together to do that task, and we're pioneering that movement (Bishop,2023)."

Amazon is trying to build a productive collaboration between their employees and robots where robotics can be used to complement human capabilities and reduce injuries from repetitive jobs by creating safer ergonomic workstations in warehouses. Amazon plans to ensure employee safety by deploying robotic systems like Sequoia and Digit. Amazon has been able to create '700 categories of new job types in skilled roles' by setting up new systems like Sequoia and Digit thereby generating new employment opportunities for skilled labour (Dresser,2023).

Figure 2. Amazon's safety data numbers for 2022

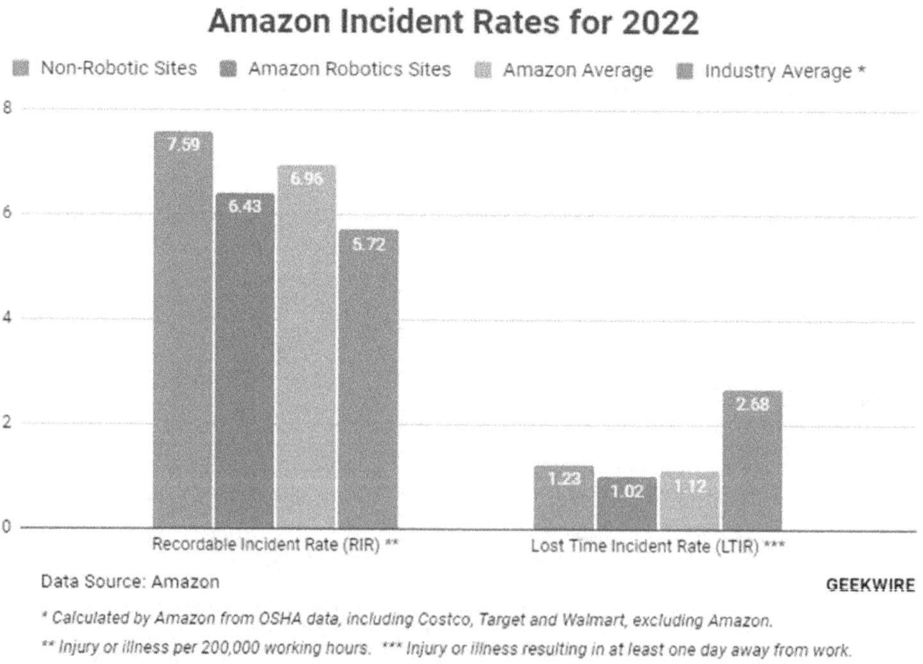

Source: Amazon Data Safety Numbers for 2022- September 7, 2023, by Bishop

Figure 2 reflects Amazon's 2022 safety data numbers which "show that recordable incident rates and lost-time incident rates were 15% and 18% lower, respectively, at Amazon Robotics sites than they were at its non-robotics sites in 2022''(Bishop,2023).

Companies all over the world have realised the importance of training their existing workforce to handle the technology being used in various processes. Similarly, Amazon is also investing heavily in training their existing employees, ''Amazon is investing more than $1.2 billion to provide free skills training opportunities to more than 300,000 U.S. employees by 2025, helping them further their careers in high-paying, in-demand roles."(https://www.aboutamazon.com/workplace/upskilling-commitments)

Amazon has started various programs to provide skills training to their employees which are listed as follows: -

- Career Choice
- Amazon Technical Apprenticeship Program
- AWS Grow Our Own Talent
- AWS Training & Certification
- Mechatronics and Robotics Apprenticeship
- AWS Intelligence Initiative
- Amazon's Upskilling 2025
- Machine Learning University
- Surge2IT
- The UXDR Apprenticeship Program

This case study summarises Amazon's automation and AI journey along with the challenges faced by the company along this route. Amazon has devised its own solutions over a period of time to handle the technological and digital issues which has become a benchmark for other organisations to follow. While embracing technology and automation, Amazon has neither neglected the welfare of its customers nor the safety of its employees.

INTRODUCTION

Warehousing Concept and Definition

Warehousing is an integral component in supply chain management which involves storing raw materials, work-in-process, and finished goods for short as well as long term before distribution or sale. Warehouses add a twofold value to the supply chain by providing storage on one hand and ensuring transport economies on the other hand. The traditional view of a warehouse as a storage facility only has undergone a drastic change with modern-day warehouses performing value-added functions like packaging, order-fulfillment, labeling, customization and assembly.

Definition

According to the Association for Supply Chain Management, "Supply chain warehousing encompasses the activities related to receiving, storing, and shipping materials to and from production or distribution locations."

(https://www.ascm.org/topics/supply-chain-management)

Smart Warehousing Concept

Gone are the days when organisations were using traditional warehouse management systems to manage and track the activities of a warehouse. Modern warehouse managing systems are software-based and are designed to be cost-effective, improve order fulfillment and delivery times, optimise inventory levels, improve lead times, reduce logistics costs and increase competitive advantage. The concept of smart warehousing is gaining popularity driven by e-commerce expansion, rising logistics complexity, and the need for increased efficiency.

The fourth industrial revolution known as Industry 4.0 has led to the emergence of smart warehouses. In their research, Geest et al., (2021) pointed out that, "Smart warehouses aim to increase the overall service quality, productivity, and efficiency of the warehouse while minimizing costs and failures."The key to developing smarter, more agile and more efficient warehouses in the future lies in the integration of AI-driven systems such as autonomous m.obile robots, robotic arms and automated guided vehicles (Geest,2021). Oluwademilade et al. (2024) examine the importance of "AI in warehouse automation encompassing autonomous mobile robots (AMRs), robotic arms, and automated guided vehicles (AGVs)."

Definition

According to Anita (2023), "Smart warehousing, powered by the integration of technologies such as AI, IoT, and robotics, is transforming the logistics industry by providing unprecedented efficiency, accuracy, and cost-effectiveness."

Many authors have described the impact of integrating AI in warehouse management systems (WMS). In their research, Javaid et al., (2022) pointed out that "From inventory management to order fulfillment, AI-driven warehouse automation systems have become indispensable tools for streamlining operations, increasing efficiency, and improving overall productivity".

The **advantages** of smart warehouses over traditional warehouses are as follows:

- Smart warehouses reduce human error by maximising the automation of tasks.
- Smart warehouses provide real-time information on managing inventory.
- Improve operational scalability.
- Have better predictive analysis capabilities as they use prediction models to make automated decisions.
- Can easily be modified to adapt to changes in customer demands or process changes
- Minimise downtime.

Disadvantages of Smart Warehouses

- Unemployment - Rendering the employees unemployed especially those who cannot be retrained.
- High cost – High initial investment in acquiring robots, hiring skilled employees to operate technology, retraining existing employees and maintenance costs.
- Data protection and privacy- Automated systems rely on generating and storing huge amounts of data and any breach in this respect is a critical concern.

Smart Warehousing Market size, trends and examples of companies

After witnessing a downfall during the Covid-19 pandemic the 3PL market has experienced a rebound due to the adoption of digital solutions and technology. According to Statista the 3PL (Third Party Logistics) market in Canada is projected 'to reach US$25.38 bn in 2024' with an expected 'annual growth rate (CAGR 2024-2029) of 2.78%, resulting in a projected market volume of US$29.11bn by 2029'. Sandy Gosling et al., in their article, "Digital logistics: Technology race gathers momentum" for McKinsey and Company (McKinsey Logistics Survey, 2023) analyse the impact of growing investment in technology on logistics. A recent survey report of 250 global shippers and logistics providers by McKinsey reveals that currently is the right time to invest in logistics technology and investment in technology is bound to increase in the future (Gosling et al.,2023).

Figure 3. Trends of investment in digital logistics by shippers & providers for diverse sectors

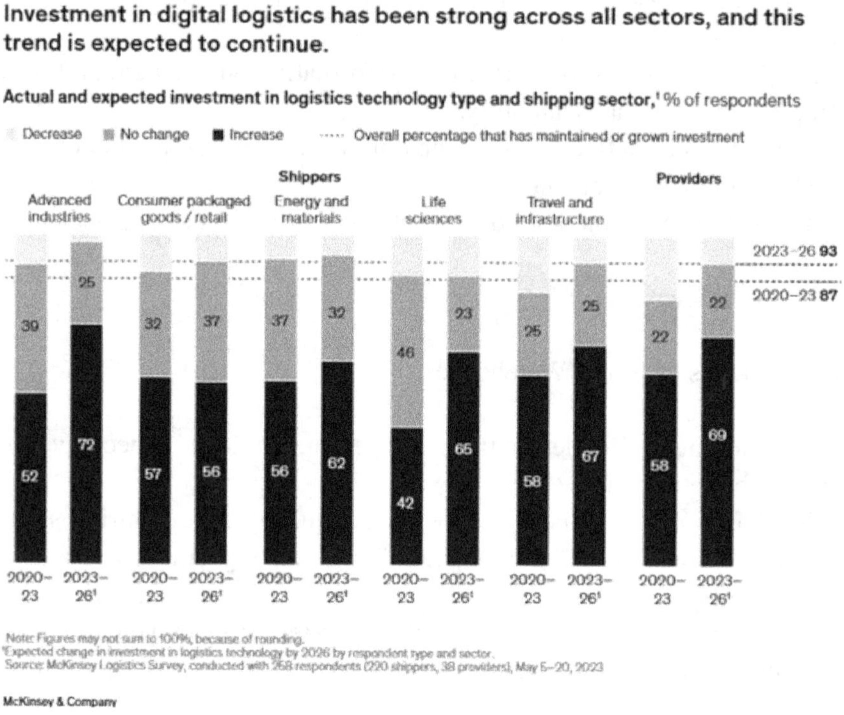

Source: Trends of Investment in Digital Logistics by Shippers & Providers for Diverse Sectiors- 2023, by Gosling

Figure 3 illustrates the McKinsey survey which indicates that the investment in digital logistics by both shippers and providers has increased significantly since 2020. According to Figure 3, the technology investments by shippers have grown by 87% since 2020 and 93% of providers reported to increase or grow their investment in the coming years. The survey also points out that while foundational technologies like warehouse management systems and transportation management systems used by both shippers and providers are now mainstream, they need to focus on 'addressing advanced use cases like robotics, network digital twins and real-time insights'(Gosling,2023).

Figure 4. Types and development stage of technology used in warehouses

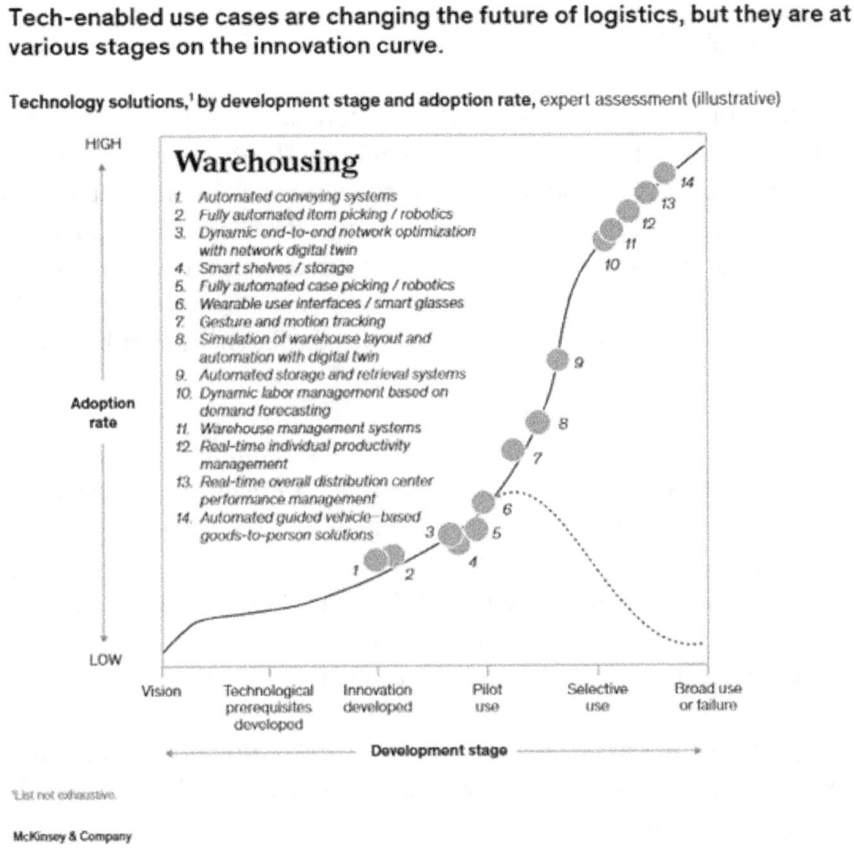

The diagram given above illustrates the widely used technology in warehouses surveyed by McKinsey based on the development stage of the technology and adoption rate in warehousing. It indicates that warehousing technologies that are in wider use and have a high adoption rate in warehousing are - dynamic labor management (10), warehouse management systems (11), real-time individual productivity management (12), real-time distribution center performance management (13), AGV-based goods-to-person solutions (14). Other technologies like gesture motion tracking (7), simulation of warehouse layout and digital twin (8) and automated storage and retrieval system (9) are selectively being used in warehouses. Other technologies are still in pilot use and their adoption rate is also low.

WAREHOUSING AUTOMATION

Meaning and Types

Warehouse automation is the process of using software and technology to boost operational effectiveness. It is achieved by combining a flexible network of sensors, robotics and advanced software solutions that are coordinated by smart automation. Warehouse automation helps businesses improve workflow by reducing repetitive physical activity and tasks, tracking inventory and reducing human errors. Abby Jenkins in her article 'Warehouse Automation Explained: Trends, Types & Best Practices' describes that warehouse automation can take the shape of either physical or digital automation:

- Physical automation is done in large warehouses to minimize the employee's physical activity and enhance safety within the warehouse so that workflow can be improved through the use of robotics e.g. AS/AR system, AGVs / ARMs, smart shelves and pallets, etc.
- Digital automation involves the use of data and software to reduce manual processes and errors by processing real-time data for managing inventory tracking e.g. RFID, resource planning, etc. (Jenkins,2024)

Yuliia explains the difference between physical and digital process automation by elaborating in the diagram given below about the types of digital technology used in each of the automation processes.

Figure 5. Types of warehouse automation

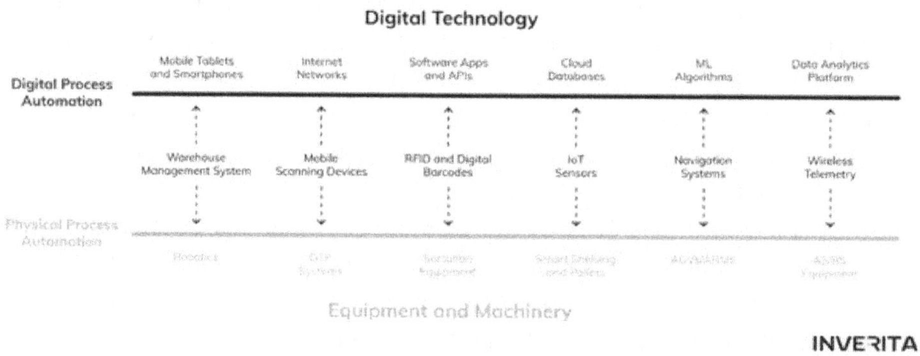

Source:https://inveritasoft.com/article-warehouse-automation-explained:-types-benefits-&-best-practices

Processes in warehousing which can be automated

Advanced warehouse automation can be carried out at various stages in warehouse operations:

A. Receiving
B. Storage
C. Picking
D. Packing
E. Staging items in designated loading areas
F. Loading on delivery vehicles
G. Cross-docking
H. Shipping
I. Processing returns
J. Inventory management
K. Quality checks

Technology used for warehouse automation

Smart warehouses use various types of tools to improve their operational efficiency:

- 5G Technology for monitoring and controlling the warehouse operations.
- Mobile Robots to assist human labour in picking, packing and moving goods.
- Cobotics - a collaboration between humans and robots e.g.Amazon is using over 200000 cobots in its warehouses.
- Automated Guided Vehicles (AGV)
- RFID – Radio Frequency Identification technology used in the form of warehouse tags.
- WMS – Warehouse Management Systems is a software equipped to manage and control warehouse operations.
- Automated Storage and Retrieval Systems (AS/AR)
- Wearable technology like Smart glasses and AR-enabled devices
- Warehouse Drones
- IoT Devices
- Big Data and Analytics
- Cloud warehouse servers
- Gamification
- Same day shipping
- Optical recognition

Case example of IKEA using automated warehousing systems

IKEA is the largest furniture retailer with an operation supply chain to effectively support its stores in different parts of the world and its clients. Due to increasing customer needs and production costs, IKEA integrated smart supply technologies such as automation, IoT, and monitoring devices into its warehouse systems.

Integration of automation in the IKEA Warehousing System:

1. IKEA uses Automated Storage and Retrieval Systems (AS/RS) to IKEA to manage stored places and coordinate the circulation of products within storage zones. Forklifts and automatic robots require only a few minutes to complete a pickup operation and handle delicate objects accurately. Real-time recording has also become a rarity as piping robots scan inventories and robot-type entry data. For example, robots that are fitted with sensors and cameras can therefore be used to scan for barcodes in stores and can immediately update stock as opposed to through human intervention. This precision eliminates cases in which the wrong product is picked, or incorrect orders are packed (Mecalux, n.d.).
2. Real-time monitoring with IoT Sensors like Temperature and humidity sensors, sensors to detect inventory status, etc. are placed in the warehouses which enables the organization to monitor product quality, especially for sensitive products such as textiles and food.
3. IKEA has collaborated with Verity to enforce a fully autonomous inventory control system based on drones that can scan pallets and update IKEA's stock status (Netland, 2023).
4. IKEA uses a demand forecasting tool known as 'Demand Sensing' which applies AI for supply chain management with a remarkable success rate. According to Peter Grimvall, Supply Chain Development Area Manager for IKEA, "The new tool can forecast demand from a day-to-day basis to up to 4 months. With us trying out new store formats and new selling capabilities, and fast-changing customer behaviors, the tool is agile enough to capture the unique needs" (Using AI for smarter demand forecasting, 2021).
5. Collaborative Robots (Cobots): Cobots perform some of the activities that human workers carry out, for example, pick-and-place, packaging, and sorting. These robots make operations better and ease human effort by reducing physically exhausting tasks (Netland, 2023).
6. IKEA uses a 'flat pack design' for storage which saves warehouse space by minimizing wasted storage space (Damotech,2024).
7. Sustainability Initiatives: IKEA's distribution centers have also been constructed power effectively, incorporating computerized technologies such as LED lighting as well as renewable power. The status of energy use is also monitored

through the application of real-time monitoring of energy use to help in making sustainable decisions for the operations.

Real-World Application - IKEA's warehouse located in Älmhult Sweden has integrated advanced-level robots and an Artificial intelligence system to manage their stocks. These systems can predict future trends in demand and the company can therefore organize its production and distribution systems well. Also, optimizations and some processes are tested on digital twins of their warehouses, and the presence of bottlenecks is revealed.

IKEA's practices of smart warehousing technology integration are the best demonstration of how the organization upholds innovativeness and sustainability. IKEA is using automation and real-time monitoring not only to refine its operations but also to become a role model for other retailers concerning supply chain management.IKEA serves as a model not only to reinforce sustainability and automation in its warehouses but also to ensure that safety protocols and employee training are well taken care of.

REAL-TIME MONITORING

Real-time tracking is one of the core elements of designing the new generation of intelligent warehouses, which provides businesses with instant command of operations, stocks, and tools. Through IoT-connected technologies, AI, and cloud technologies such as smart glasses, warehouse activities are updated in real time.

Real-time operations refer to the continuous tracking of activities in smart warehousing. As opposed to habitual set-check or interval updates, the current technique creates regular data flow, yielding prompt information on warehouse conditions.

Technologies Involved

- **IoT Sensors and Devices:** Integrated to check stocks and other physical environmental conditions such as temperature and humidity levels.
- **RFID and Barcode Scanning Tools:** Ease the process of tracking inventory and automatic data collection.
- **Warehouse Management Systems (WMS):** Applications for real-time inventory and operation planning (Shriyam, Palkar, & Srivastava, 2023).
- **Cloud Computing:** Allows for the application of centralized databases and remote administrative control.

Implementation of Real-Time Monitoring in Smart Supply Chain Systems

1. **Enhanced Inventory Visibility:** Real-time monitoring allows businesses to keep track of their inventory to avoid overstocking or running out of stock. For instance, Amazon has adopted RFID tags and smart shelves in its inventory management to increase the accuracy of its records (Smart Warehousing, 2024).
2. **Improved Operational Productivity:** Real-time monitoring eliminates unnecessary time gaps. For example, Walmart utilizes IoT sensors and Avnet technologies to successfully enhance supply chain management (Avnet, n.d.).
3. **Superior Customer Service:** Transponder technologies enhance delivery transparency, increasing customer confidence. DHL implements RFID and GPS technologies to communicate real-time shipment information (Tong, 2023).
4. **Reduced Costs and Waste:** Product inspection through flaw identification and imperfection tracking reduces waste. IoT sensors in cold rooms help avoid food wastage, benefiting small-scale businesses (Light Guide Systems, 2024).
5. **Risk Mitigation:** Predictive maintenance systems monitor equipment conditions to prevent failures. For example, Siemens uses IoT to track conveyor belts and replace them before failures occur (Rimel, Grungier, Gamache, Gendreau, & Rousseau, 2021).

Technologies Supporting Real-Time Monitoring

- **IoT Devices:** Operational sensors integrated into inventory and/or physical environments to continually collect data (Ship Bob, n.d.).
- **RFID Technology:** Tracks the location and status of inventories throughout stocking and transportation, such as in Zappos' case.
- **WMS:** Provides data and monitoring for order handling and distribution coordination (Oracle's WMS example).
- **GPS and RTLS:** Enable stock and vehicle tracking, improving the efficiency of fleets and forklifts (Zebra Technologies example).
- **Cloud Platforms:** Support data processing and storage through real-time analytics, e.g., Microsoft Azure.

Challenges in Real-Time Monitoring

1. **High Implementation Costs:** Technologies such as IoT and WMS may be costly for small businesses.
    a. *Solution:* Staged implementation of cost-effective solutions.
2. **Large Data Volumes:** Processing large volumes of data can be challenging.

   a. *Solution:* AI tools can screen and sort alert signals and relevant data (Confluent, 2024).
3. **Compatibility with Legacy Systems:** Implementing new systems in the presence of existing infrastructures can be difficult.
   a. *Solution:* Use modular integration solutions.
4. **Cybersecurity Risks:** Real-time systems are vulnerable to breaches.
   a. *Solution:* End-to-end security using encryption and multi-factor authentication (AWS Marketplace, n.d.).

Future Trends in Real-Time Monitoring

- **AI and Machine Learning:** Increase efficiency during peak hours by forecasting inventory demands (Shriyam, Palkar, & Srivastava, 2023).
- **Blockchain:** Enhances transparency and helps identify inventory issues (e.g., Walmart's implementation).
- **Autonomous Systems:** Use of drones and robots for stock checks with minimal human intervention (Tong, 2023).

Real-time monitoring in smart warehousing aims to make systems more effective by enhancing operational visibility, reducing costs, and improving customer satisfaction. Emerging technologies like AI, IoT, and blockchain continue to shape the future of smart warehousing, driving efficiency, reducing expenses, and fostering customer trust.

FUTURE TRENDS AND CHALLENGES

Smart warehousing means a revolutionized method of logistics by having a combination of an advanced system like automation and monitoring. These innovations bring to warehousing operations the possibility of having higher efficiency, more accurate work, and improved flexibility. However, the implementation and deployment of such technologies are not without some hitches. In as much as this chapter aims at addressing the prospects of smart warehousing, it also seeks to engage on the future perspective as well as the main challenges likely to face smart warehousing on aspects of data management, transparency of processes, partnership with technologies, sustainability of systems, regulatory realms, and skills development needs.

Data Management Issues - The management of data is central to smart warehousing. Warehouses have also adopted the use of many data inputs derived from various sources such as inventory operations, IoT devices, and predictable analytic tools. Integrated and well-structured data makes work more efficient, projections precise, and decisions sustainable.

However, it becomes quite complex to handle and assimilate data sets. This means that while warehouses exist in isolation, they employ different systems that may not be integrated well (Abidi et al., 2021). The coordination of these systems to foster real-time data sharing is a challenging and costly process. Another consideration is the practical issue of data protection and security, with more organizations storing data in the cloud and connecting more devices to IoT systems. This can put at risk business critical and customer data and erode trust compounded by the failure to satisfy the General Data Protection Regulation (GDPR) requirements.

Real-time data accuracy is another key consideration in this framework. Data processing can cause significant disconnections or delays in an organization's functioning or decision-making. In turn, the solution is emerging in the form of edge computing, which implies data processing closer to the source to increase speed and reliability (Ghosh et al., 2022).

Nevertheless, proper application of advanced data management systems can be very useful, but it implies certain costs and qualified staff. Customers and stakeholders want to have full visibility of products in transit to monitor delivery schedules and disruptiveness.

This visibility is made possible using IoT devices and sensors to track details of inventory and shipment. Hiccups are however experienced when it comes to the **issue of transparency** integration. One is the management of the visibility of the customer, although this is important it remains a challenge when the visibility must be done while keeping some data privileged. Huge data sharing could be risky to the warehouses for instance, it brings in issues to do with compromise in information security, competition, and affiliation to wrong channels or wrong people. Also, introducing easy-to-use control panels and visualization tools for tracking supply chain actions necessitates the use of complex software and knowledge (Kumar et al., 2023).

Real-time visibility also means real-time issue-solving. People expect to be informed of issues instantly and similarly, expect them to be solved the same way. For instance, where there is a delay in logistics or a sensor is off, something needs to be done to ensure continuity of operations. These challenges can only be countered by providing investments in monitor tools and in efforts to optimize work.

Technology Partnership Issues - Outsourcing is central to smart warehousing as companies cannot implement and sustain the solutions independently. Strategic partnering with technology players allows the warehouse to adopt high-end tools including robotics, AI, and automation systems. The above-stated partnerships however come with their pulling factors. Another main challenge is between the requirements set for the selected warehouse and the partners' competencies. The choice of a competent partner depends on their experience in a certain field, their credibility, and sustainability (Bose et al., 2021).

Additionally, communications between systems from different vendors can be integrated into the systems resulting in integration difficulties. If materials and processes do not complement each other, there are increases in costs and ineffective customization. The other issue is the cost of implementing such superior technologies and their expensive upkeep. The obvious challenges are to manage the direct and indirect operating costs of warehouses while at the same time, technology expenses must equate to tangible benefits. That is why knowing favorable terms and developing lasting partnerships constitute how to handle these concerns.

Sustainability Issues In recent years, sustainability has become one of the most important factors for running a smart warehouse since companies are focused on the negative impact on the environment. Smart warehousing solutions could also serve as inclusive improvements to sustainability by applying more efficient technologies, and resources and incorporating environmental measures. The problem integrates with the high initial costs that are associated with the integration of renewable energy technologies and energy-efficient facilities. Although these investments seem promising in the long run, the initial cost may be prohibitive, and possibly off-putting particularly to small warehouses (Zhou & Li, 2022).

Another challenge that organizations face is the good management of the flow and transportation to and from to reduce greenhouse gases. This means that there ought to be improved routing algorithms as well as accord with green logistics suppliers. Moreover, warehouses must seek environmental concerns in packaging and waste disposal. Purchasing must understand the switching costs of using recyclable and biodegradable materials and affirm that it entails a change in the purchasing section. To overcome these challenges, there is a need to enhance a strategic commitment toward sustainability and improvement of environmental conformity.

Changing Global Policies - and regulations have a great impact on warehousing operations. For example, WTO trade policies and labor and environmental laws change frequently posing a huge challenge to warehouses.

One difficulty is an intricate and constantly changing legal environment concerning international trade. Fluctuations in tariffs and customs, changes in customs regulation, and trade agreements affect supply chain management and cost. These shifts must be managed adequately to ensure that the warehouses continue being responsive, (Liu et al., 2023). Another important issue is to meet labor and safety requirements regarding the use of automation and AI, it is also important for the warehouses to respect the regulations regarding the local workforce.

REGULATION OF AI SUPPLY CHAIN IN CANADA

"Artificial intelligence is one of the greatest technological transformations of our age, and our government is committed to providing Canadians and Canadian businesses with the tools they need to thrive in this digital era. In undertaking the Voluntary Code of Conduct on the Responsible Development and Management of Advanced Generative AI Systems, leading Canadian organizations continue to adopt responsible measures for advanced generative AI systems that will help build safety and trust well into the future as the technology grows."

— The Honourable François-Philippe Champagne, Minister of Innovation, Science and Industry (Innovation, Science and Economic Development Canada, Government of Canada. 2024, May 27).

Canada is one of the leading countries to support and promote responsible use of AI across various sectors. On one hand, the government of Canada is promoting the use of AI by launching new programs and providing funding while on the other hand, it is also encouraging the adoption of safe and responsible measures while using AI systems. In October 2024, Canada's Minister of Finance 'announced the launch of two programs to grow Canada's AI ecosystem by supporting the development of new generative AI applications and enabling AI adoption among SMEs to increase productivity'(Innovation, Science and Economic Development Canada, Government of Canada. 2024, October 22).To support these programs the federal budget 2024 has allocated a $2.4 billion package to these AI-focused initiatives.

The Government of Canada is taking manifold measures to secure Canada's AI advantage which include the following:

- Creating Innovation Funds and clusters for providing industrial support and gearing investment towards AI.
- Supporting research excellence by giving grants and launching skill development programs.
- Ensuring the development of strategies and legal approaches by creating a Pan Canada AI strategy and enacting the Artificial Intelligence and Data Act.
- Globally promoting responsible development and use of AI by being the founder member of Global Partnership in AI.

Figure 6. Canadian government strategy to support and manage the use of ai

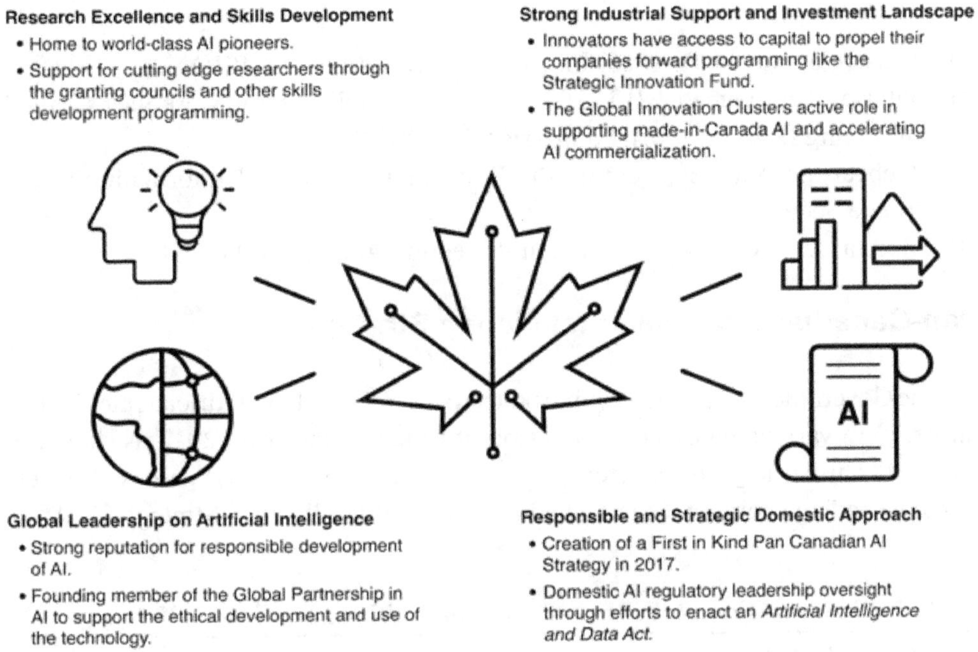

Source: Statistics Canada, Figure: 4.1 (https://budget.canada.ca/2024/report-rapport/chap4-en.html#a4)

Levels at which AI is Regulated in Canada:

At Federal Level

A) Artificial Intelligence and Data Act (AIDA)
B) Guide on the Use of Generative AI- issued in 2023 as a guideline for careful use of AI tools in the public sector entities.
C) Voluntary Code of Conduct on the Responsible Development and Management of Advanced Generative AI Systems – issued in 2023 for the private sector.
D) Privacy Act
E) Personal Information Protection and Electronic Documents Act (PIPEDA)
F) Pan-Canadian Artificial Intelligence Strategy

At Provincial Level (Stobbe, 2024)

A) British Columbia - Personal Information Protection Act (PIPA)
B) Ontario – introduced BILL-194 to regulate public sector workplaces called Enhancing Digital Security and Trust Act, 2024
C) Quebec- the Act Representing the Protection of Personal Information in the Private Sector
D) Alberta - the Personal Information Protection Act, SA 2003

Pan-Canadian Artificial Intelligence Strategy

The Government of Canada has introduced a 'Pan-Canadian Artificial Intelligence Strategy' in various phases over a period of time starting from 2017 as shown in Table. 4. This strategy is an effort to promote the adoption of AI across all facets of the Canadian economy (Pan-Canadian Artificial Intelligence Strategy,2024).

Table 1. Phases of Canada's Pan – Canadian Artificial Intelligence Strategy

A snapshot of the strategy	
Phase I (2017–2022) **Securing Canada's advantage** **$125 million through Budget 2017**	**• Canada CIFAR AI Chairs program** **Attracts, retains and supports top AI researchers at Canadian universities** **• National artificial intelligence institutes** **Amii, Mila and the Vector Institute** **• Research, training and collaboration** **CIFAR leads national programs and helps increase understanding of the effects of advances in AI**
Phase II (2021–2031) Driving adoption and use $443.8 million through Budget 2021	**• Pillar 1: Commercialization** Helping Canadian businesses develop and use AI and promoting the adoption of made-in-Canada AI solutions in key industries **• Pillar 2: Standards** Standards Council of Canada advancing the development and adoption of AI-related standards **• Pillar 3: Talent and research** Expanding the Canada CIFAR AI Chairs program, continuing national programs and ensuring AI researchers have access to advanced computing resources

Source: (*Artificial intelligence ecosystem*,2024)

Artificial Intelligence and Data Act (AIDA)

Canada is one of the countries that have taken the lead in developing a regulatory framework by introducing the Artificial Intelligence and Data Act (AIDA) as part of Bill C-27, the Digital Charter Implementation Act, 2022. This bill is currently under consideration in committee in the House of Commons and if passed will look into several concerns encompassing responsible development and ethical use of AI technologies, safety, transparency and accountability.

Purposes

The purposes of this Act (*Government Bill (House of Commons) C-27 (44-1) - first reading - digital charter implementation act, 2022 - Parliament of Canada*) are:

> "(a) to regulate international and interprovincial trade and commerce in artificial intelligence systems by establishing common requirements, applicable across Canada, for the design, development and use of those systems; and
>
> (b) to prohibit certain conduct in relation to artificial intelligence systems that may result in serious harm to individuals or harm to their interests".

Main implications of AIDA

- Violation or offense of AIDA regulations will result in administrative monetary penalties.
- Criminal liability and penalties will be imposed for the use of AI for intentional physical harm and fraud.
- Overseeing compliance of AI by establishing an AI and Data Commissioner who will liaise between the government and private sector.
- Establishment of advisory councils and committees to monitor compliance with AIDA.

Type of measures to regulate the use of AI worldwide include

Governments all over the world are developing and implementing new AI laws, policies, and strategies. The Organization for Economic Co-operation and Development (OECD) has put together a comprehensive 'repository of more than 1,000 AI policy initiatives from 69 countries, territories, and the European Union' (AI Laws Around the World,2024). This document identifies the four types of AI policy instruments being used by the various countries which are as follows:

1) Governance policies – which include developing national strategies & plans, conducting public consultations, setting up monitoring bodies, etc.
2) Guidance and regulation bodies– which includes developing AI regulations, standards, certifications and ethical advisory bodies.
3) AI Enablers and other incentive policies – which include developing and monitoring networking and collaborative platforms, promoting AI research infrastructure and skills, conducting competitions and awarding prizes and awards, developing labour market policies and providing knowledge transfers and advisory services.
4) Financial support policies – which include providing grants for projects, fellowships, loans, scholarships, funding and equity financing.

According to the OECD report (AI Laws Around the World,2024) given below are the AI policy initiatives from 69 countries, the top 3 countries:

- United States-
- U.S.A. leads in the development of AI policies and strategies as it has developed 82 AI initiatives.
- The U.S.A. also has a National AI Strategy named 'Executive Order on the Safe, Secure, and Trustworthy Development and Use of Artificial Intelligence'.
- European Union -
- The EU ranks second after the U.S.A. as it has 63 AI policies and strategies.
- The EU also has a national AI policy called the 'Coordinated Plan on Artificial Intelligence' which has been allocated a budget exceeding $500 million.
- The EU adopted the world's first comprehensive horizontal law governing artificial intelligence on 13 March 2024 known as the 'EU Artificial Intelligence Act'.
- U.K. – ranks 3rd by claiming 61 AI policies and strategies and also has a National Artificial Intelligence Strategy.
- Besides the 3 countries mentioned above many other countries have undertaken initiatives in regulating AI like – Australia, Germany, etc.

Employers' strategies to handle legal obligations of the use of AI

The use of AI in organisations has raised legal considerations and obligations on the part of employers whereby they have to redefine and redevelop their strategies throughout the employment cycle.

- Employers should make a provision to get employees' **consent** before making any change in employees' duties due to the adoption of AI. Additionally, valid consent should be taken from the employees willing to share information with an AI tool for training or work-related purposes.
- Employers should provide adequate advance **notice** to the employees before implementing technological changes.
- Also, employers need to **revise the employment contract** if there is a significant change in employees' duties due to the adoption of AI. Wherever the employment contract cannot be revised or becomes unenforceable in light of AI-induced changes, a termination provision has to be introduced in the contract.
- Review and revise the **language of collective agreements,** especially before implementing AI-related changes for unionised employees.
- Disclose the use of AI in employment practices such as hiring, recruitment, performance appraisal, etc. so that the risk of discriminatory assessments and recommendations is reduced and the grounds protected by **human rights legislation** are handled effectively.
- Develop **privacy policies** within the organisation that ensure that the AI systems being used are collecting, using and disclosing personal information in compliance with privacy and data protection laws.
- **Data sharing with a third party** while using an AI tool has to be monitored carefully so that confidential information of the company or a third party is not mistakenly inputted into the AI tool being used.
- Conduct regular **impact audits and assessments** to evaluate the impact of AI tools and technology on various organisational processes.
- Securing the **copyrights** for AI-generated content by humans by creating new rights.
- Electronic employee **surveillance should be consented** to by the employees. In Canada, PIPEDA governs the workplace surveillance regulations.
- Provide employees with a method to **dispute** inaccurate information and raise their concerns regarding the use and impact of AI.
- Employers planning to use AI in machinery or robotic systems should provide adequate training to the employees and **comply with the Occupational Safety and Health guidelines** to ensure a safe and hazard-free workplace.

How Organisations can Introduce AI at workplaces without creating a major conflict in the existing workforce

Gilbert noted that AI has three key impacts on employment namely 'job displacement, augmentation and creation' (Gilbert, 2023). As mentioned above employers will face three challenges with the workforce while adopting AI at different levels in the supply chain.

The first challenge will be handling the 'displacement' (Gilbert, 2023) of employees at a massive scale. Alexa Lardieri in her article 'Robots Will Replace 20 million Jobs by 2030, Oxford Report Finds' states that by 2030 '8.5% of the global workforce could be displaced by robots' (Alexa,2019).

The second challenge posed by adopting AI for employers would be 'augmentation' (Gilbert, 2023) of the existing employees which can also be translated as a reskilling of existing employees so that they can use the technology and software to assist in routine work. It is a challenge as the organisation adopting the new technology will also have to take up the responsibility for spending their time and money on reskilling and upskilling their employees. This augmentation trend of training the existing employees will boost employee morale on one hand and save the organisation from incurring layoff-related costs. The organisations will have to develop extensive training and support programs to promote digital literacy and AI skills among employees.

The third challenge will come in the shape of creating new jobs according to the new requirements after adopting automation and AI. The creation of a resilient AI supply chain will certainly transform the role of professionals working in the field by downsizing the workforce involved in clerical and data entry jobs on one hand and creating jobs in the field of software developers, chatbot developers, supply chain analysts bias analysts, etc. on the other hand. With the increase in integration of technology and AI in supply chains, there will be less demand for employees for routine and repetitive jobs as these jobs can be easily automated. AI automation will also require organisations to create a career development plan for the employees so that new opportunities can be created for upskilled and trained employees within the organisation.

Also, organisations will have to develop a plan for introducing new systems in the organisations and communicate the need for change to the employees. The change management process involved in introducing employees to AI will include removing their apprehensions about integrating AI with their routine work on one hand and describing the benefits in the long run about how AI systems can be used to monitor the safety of the employees working in the warehouses. Already organisations like Amazon are using smart robots with cameras and sensors to determine the most efficient, time-saving and safe route through a warehouse. Other AI-powered tools like wearable sensor-enabled safety vests, AI-based root cause analysis, using computer vision to ensure compliance with safety standards, etc. could also enhance worker and material safety standards.

Case Study of Canadian Tire - The AI-based planogram optimization project

Canada's No.1 retail business Canadian Tire was founded in 1922 to provide Canadians with products for life in Canada. Over the last 100 years, the business has grown to become Canadian Tire Corporation (CTC) Limited which currently is a group of companies that includes a Retail segment, a Financial Services division and CT Real Estate Investment Trust. CTC currently has 1700 retail locations selling varied products and services to customers from coast to coast. CTC invests heavily in innovation in order to survive a business environment that has become highly competitive (McQuarrie,2022).

In March 2022 CTC announced an investment strategy to the amount of $3.4 billion for enhancing omnichannel customer experience spread over four years. A major part of this investment initiative to the tune of $675 million is geared towards the development of supply chain fulfillment infrastructure and automation. This investment will include:

- Adding warehouse space across Canada –
- Expanding Montreal distribution centre by 322000 sq. feet.
- New state-of-the-art e-commerce fulfillment facility in the Greater Toronto Area expanding to 1.3 million sq. feet of space.
- All CTC banners will use a single platform for order fulfillment.
- Introducing robotic automated systems in distribution centers across Calgary, Brampton and Montreal.

In June 2023, CTC and Microsoft entered into a seven-year partnership to introduce Azure AI tools for enhancing customer service.

Even before announcing this huge investment strategy CTC had introduced new digital AI-based platforms across its stores and warehouses. Some of these measures have been outlined as given below:

- **TETRIS** – Canadian Tire has developed one of the largest AI-driven projects known as 'Tetris' to solve retail problems regarding optimal store layouts and customisable store solutions (Himmelsbach & Billot). The concept of this project is to use store-specific planograms to solve shelf stocking concerns the size of which is dependent on the needs of the customers in a given location. Tetris gives the managers the freedom to customise aisle dimensions as well as the products they want to include according to the requirements of the customers located in their region. For this project, Canadian Tire received financial support from SCALE AI – the federal government's AI supercluster and was also technically supported by Montreal-based Ivado Labs. Tetris which started as a pilot project in Canadian Tire's six stores has now expanded successfully to its 244 locations and is projected to capture its remaining locations as well (Himmelsbach & Billot).

- **CHATCTC** – is a chatbot used by CTC's 4300 employees which saves 30-60 minutes a day for each employee in performing routine tasks and assisting in administrative tasks (Toneguzzi, 2024).

Figure 7.

Image: Canadian Tire (Covent, 2023)

- **CEETEE** – According to Mario Toneguzzi, "Canadian Tire's new shopping assistant, CeeTee, is available on the Canadian Tire app under the Tires category and helps customers easily select the right tires for their vehicles (Toneguzzi, 2024)." Canadian Tire started using an AI-powered shopping assistant that can voice or text chat with customers helping them choose tires and also giving them personalised recommendations.

Figure 8.

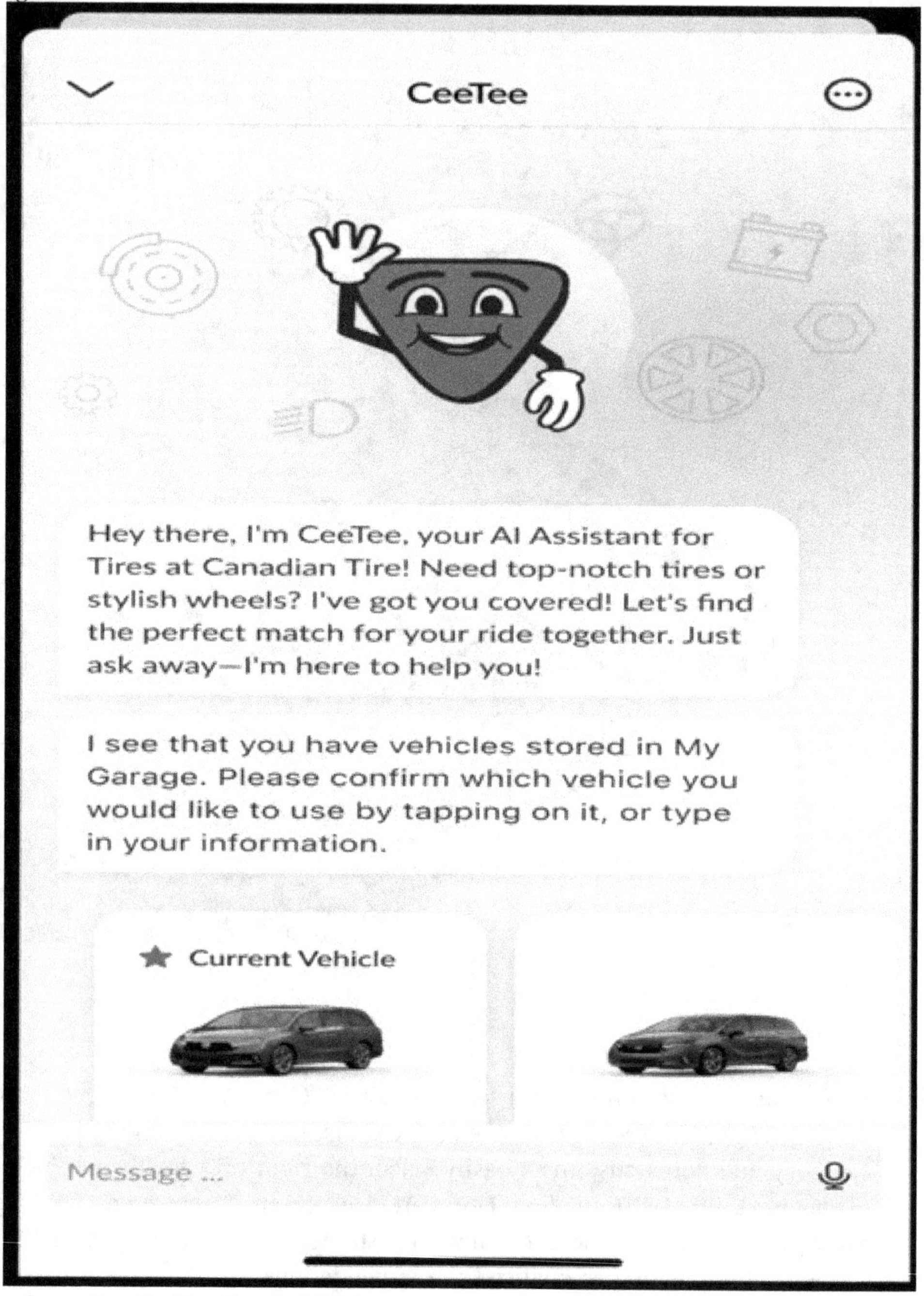

Image: Canadian Tire (Toneguzzi, 2024)

- **Humanoid in-store** - CTC partnered with Sanctuary to introduce a general-purpose robot in-store with human-like intelligence which completed 110 tasks successfully including packing, stacking, cleaning, labeling, etc. The picture below depicts a robot recently deployed at a CTC retail store on an experimental basis – Mark's in Langley, British Columbia to scan and package for shipment.

Figure 9.

Image: Sanctuary AI (Toneguzzi, 2024)

- **Autonomous trucking –** CTC and Nu Port collaborated to introduce AI in Canadian Tire's trucks so that the autopilot feature can be introduced for fixed shorter routes between distribution centers and warehouses (Scott, 2021).

Canadian Tire has journeyed over 100 years from its original tire store to a giant corporation spanning 1700 locations,34000 employees and 13 brand banners yet keeping its focus alive all through these years on innovative customer service. Mario Toneguzzi quotes Cari Covent, Head of AI and Emerging Technology, Canadian Tire Corporation on how the company is using AI to solve labour issues and develop employee skills, "We see AI, generative AI, as an opportunity to automate a lot of work that doesn't need to be done by humans and it really allows them to focus on higher value work(Toneguzzi, 2024)."

CONCLUSION

Smart warehousing, driven by automation and real-time monitoring, offers immense potential for transforming logistics and supply chain operations. However, its adoption involves addressing significant challenges related to data management, transparency, technology partnerships, sustainability, global policies, and workforce skills. To navigate these challenges, warehouses must adopt a strategic approach that includes investing in advanced technologies, fostering collaborative partnerships, and committing to sustainability and workforce development. By addressing these issues, smart warehousing can achieve its full potential and play a pivotal role in shaping the future of logistics.

Warehouses are the foundation of the logistics industry. The rapidly evolving supply chain industry has added complexities to the modern warehousing system, especially with the introduction of AI. Smart warehousing on the one hand is revolutionising the warehousing trends but on the other hand also posing challenges for organisations in terms of investment costs, risks associated with implementing technology, workplace safety, partnering with vendors, national and international restrictions, resistance from employees, etc. In the future smart warehouses will become a standard practice for every business and automation and technology will continue to evolve and create more innovative and advanced warehouse systems.

Discussion Questions:

1) What key features define smart warehousing, and how do automation and real-time monitoring contribute to operational efficiency?
2) What are organizations' primary challenges when implementing smart warehousing technologies?
3) How does IKEA leverage automation to streamline its supply chain and improve inventory management?
4) In what ways has real-time monitoring helped IKEA adapt to fluctuations in consumer demand?
5) Discuss how IKEA integrates sustainability into its smart warehousing practice.
6) What unique challenges does Canadian Tire face as a Canadian retailer, and how has smart warehousing helped address them?
7) How does Canadian Tire's use of automation and monitoring compare to global leaders like IKEA and Amazon?
8) Explore how Canadian Tire uses data analytics from real-time monitoring to enhance customer satisfaction.
9) Amazon is often seen as a leader in smart warehousing. What specific innovations set its warehousing practices apart?

10) How has Amazon's robotics and artificial intelligence use reshaped industry standards?
11) Compare and contrast the approaches by IKEA, Canadian Tire, and Amazon in their smart warehousing strategies.
12) What lessons can smaller companies learn from these case studies about adopting automation and real-time monitoring?
13) Which emerging technologies are more likely to transform smart warehousing in the next decade?
14) How might customer expectations for faster delivery influence future developments in warehouse automation?
15) What role should government policies play in regulating the implementation of smart warehousing to ensure fair competition and sustainability?

REFERENCES

Abidi, N., Battaıa, O., & Guinet, A. (2021). Data integration in logistics: Challenges and opportunities. *Journal of Logistics and Supply Chain Management, 12*(3), 45–58.

AI Laws Around the World. AIPRM. (2024, April 28). https://www.aiprm.com/ai-laws-around-the-world/

Amoo, O., Sodiya, E., Umoga, U., & Atadoga, A. (2024). AI-driven warehouse automation: A comprehensive review of systems. *GSC Advanced Research and Reviews., 18*(2), 272–282. DOI: 10.30574/gscarr.2024.18.2.0063

Avnet. (n.d.). *Smart warehouse monitoring solution.* Retrieved from https://www.iotconnect.io/smart-warehouse-monitoring.html

Bilicka, A., & Warehousing, S. Revolutionizing Connectivity and Efficiency-September 1, 2023.https://www.autostoresystem.com/insights/smart-warehousing-revolutionizing-connectivity-and-efficiency

Bishop, T. (2023, September 7). New robots are making Amazon's warehouses more efficient — can they also make them safer? https://www.geekwire.com/2023/new-robots-are-making-amazons-warehouses-more-efficient-can-they-also-make-them-safer/

Bose, I., Mahapatra, S., & Roy, S. (2021). Evaluating technology partnerships in smart logistics. *International Journal of Operations Management, 18*(2), 67–80.

Confluent. (March 12, 2024). *Real-time data streaming for smart warehouses.* Retrieved from https://www.confluent.io/blog/data-streaming-for-smart-warehouse/

Covent, C. (2023). *Exciting Announcement! Introducing ChatCTC - Empowering People to Work Smarter!* LinkedIn. Retrieved December 19, 2024, from https://www.linkedin.com/posts/caricovent_ctc-chatctc-aiinnovation-activity-7124762472997126144-hJXl/

Damotech (2024, August 15). *IKEA warehouses: Efficient Warehousing and distribution strategies.* Damotech. https://www.damotech.com/blog/ikea-efficient-warehousing-distribution-strategies

Department of Finance, Government of Canada. (2024, April 16). *Chapter 4: Economic Growth for Every Generation: Budget 2024.* https://budget.canada.ca/2024/report-rapport/chap4-en.html#a4

Dresser, S. (2023, October 18). *Amazon announces 2 new ways it's using robots to assist employees and deliver for customers*. Amazon news. https://www.aboutamazon.com/news/operations/amazon-introduces-new-robotics-solutions

Fedyk, Y. (n.d.). Warehouse Automation [A Complete Guide]: Types, Benefits & Best Examples. *INVERITA*. December 17, 2024, https://inveritasoft.com/article-warehouse-automation-explained:-types-benefits-&-best-practices

Ghosh, A., Singh, R., & Sharma, P. (2022). Edge computing in real-time warehousing systems. *IEEE Transactions on Industrial Informatics*, *19*(4), 1345–1355.

Gosling, S. Digital logistics: Technology race gathers momentum, *16Nov.2023, Accessed 22 October 2024.*https://www.mckinsey.com/capabilities/operations/our-insights/digital-logistics-technology-race-gathers-momentum

Government Bill (House of Commons) C-27 (44-1) - first reading - Digital Charter Implementation Act, 2022 - Parliament of Canada. Government Bill (House of Commons) C-27 (44-1) - First Reading - Digital Charter Implementation Act, 2022 - Parliament of Canada. (n.d.). https://www.parl.ca/DocumentViewer/en/44-1/bill/C-27/first-reading

Government of Canada. (2024, December 17). *Artificial intelligence ecosystem*. Innovation, Science and Economic Development Canada. https://ised-isde.canada.ca/site/ised/en/artificial-intelligence-ecosystem

Government of Canada. (2024, December 3). *Pan-Canadian Artificial Intelligence Strategy*. Innovation, Science and Economic Development Canada. https://ised-isde.canada.ca/site/ai-strategy/en#pillar1

Himmelsbach, V., & Billot, J. (n.d.). *AI in the Real World-Canadian Tire*. https://cdnscaleai.ca. https://www.google.com/url?sa=t&source=web&rct=j&opi=89978449&url=https://cdn.scaleai.ca/uploads/2022/05/scaleai-whitepaper-canadiantire-vf.pdf

Innovation, Science and Economic Development Canada, Government of Canada. (2024, May 27). *Eight organizations to join Canada's voluntary AI Code of Conduct*. Canada.ca. https://www.canada.ca/en/innovation-science-economic-development/news/2024/05/eight-organizations-to-join-canadas-voluntary-ai-code-of-conduct.html

Javaid, M., Haleem, A., Singh, R. P., & Suman, R. (2022). Artificial intelligence applications for industry 4.0: A literature-based study. *Journal of Industrial Integration and Management*, *7*(01), 83–111. DOI: 10.1142/S2424862221300040

Jenkins, A. (2024, November 3). Warehouse Automation Explained: Trends, Types & Best Practices. *Oracle Netsuite*. December 17, 2024, https://www.netsuite.com/portal/resource/articles/inventory-management/warehouse-automation.shtml

Kumar, V., Rao, P., & Shenoy, N. (2023). Transparency in smart supply chains: IoT-enabled solutions. *Supply Chain Review*, *9*(1), 12–18.

Tao, J., Li (2016). Design Optimization of Amazon Robotics. Automation. Control and Intelligent Systems, 4(2), 48-52.

Light Guide Systems. (June 2024). *Smart warehousing technologies: A comprehensive guide*. Retrieved from https://www.lightguidesys.com/resource-center/blog/smart-warehousing-technologies-a-comprehensive-guide/

Liu, Y., Zhang, J., & Wang, H. (2023). Adapting to global trade policy shifts: A logistics perspective. *Global Trade Journal*, *15*(2), 89–105.

McQuarrie. J. (2022). "Canadian Tire". *The Canadian Encyclopedia, Historica Canada*. www.thecanadianencyclopedia.ca/en/article/canadian-tire-corporation-limited. Accessed 19 December 2024.

Netland, T. (2023, August 11). *IKEA's Journey With Warehouse Drones*. Better Operations. https://better-operations.com/2023/08/11/ikeas-journey-with-warehouse-drones/

Rimel, A., Grungier, P., Gamache, M., Gendreau, M., & Rousseau, L.-M. (2021). Supervised learning and tree search for the real-time storage allocation in robotic mobile fulfillment systems. *arXiv preprint arXiv:2106.02450*. Retrieved from https://arxiv.org/abs/2106.02450

Scott, J. (2021, March 16). *NuPort Robotics receives $1 million from Ontario for an autonomous trucking project with Canadian Tire*. Canadian Startup News & Tech Innovation. https://betakit.com/nuport-robotics-receives-1-million-from-ontario-for-autonomous-trucking-project-with-canadian-tire/

Shriyam, S., Palkar, P., & Srivastava, A. (2023). On fulfilling the exigent need for automating and modernizing logistics infrastructure in India: Enabling AI-based integration, digitalization, and smart automation of industrial parks and robotic warehouses. *arXiv preprint arXiv:2310.01077*. Retrieved from https://arxiv.org/abs/2310.01077

Singh, A., Patel, M., & Verma, R. (2021). Upskilling logistics professionals for smart warehousing. *Human Resource Development Quarterly*, *32*(1), 78–95.

Stobbe, R. (2024, August). *The state of AI Regulation in Canada (2024)*. Field Law. https://www.fieldlaw.com/News-Views-Events/237102/The-State-of-AI-Regulation-in-Canada-2024)

Thompson, B. (2024, January 25). *Amazon Grows To More Than 750,000 Robots.* Global Trade News. https://incodocs.com/blog/amazon-robot-growth/

Toneguzzi, M. (2024, May 14). *Canadian Tire launches AI shopping assistants and humanoid robots to enhance customer service and operations [interview].* Retail Insider. https://retail-insider.com/retail-insider/2024/05/canadian-tire-launches-ai-shopping-assistant-and-humanoid-robots-to-enhance-customer-service-and-operations-interview/

12types of Amazon Warehouse Robots. Amazon Warehouse Robots. https://www.agvnetwork.com/robots-amazon

Using AI for smarter demand forecasting. IKEA global. (2021, May 27). https://www.ikea.com/global/en/stories/design/using-artificial-intelligence-for-smarter-demand-forecasting-210527/

Van Geest, M., Tekinerdogan, B., & Catal, C. (2021). Design of a reference architecture for developing smart warehouses in industry 4.0. *Computers in Industry, 124*, 103343. Advance online publication. DOI: 10.1016/j.compind.2020.103343

Warehousing, S. (August 2024). *The role of real-time data in enhancing inventory turnover rates.* Retrieved from https://www.smartwarehousing.com/blog/role-of-real-time-data-in-enhancing-inventory-turnover-rates

"What is Warehousing?" Ascm.Org, www.ascm.org/topics/warehousing/. Accessed 05 Dec. 2024.

Zhou, X., & Li, F. (2022). Sustainability challenges in automated warehousing. *Journal of Environmental Management, 14*(3), 89–97.

Chapter 10
Enhancing Inventory Control and Reducing Holding Costs Using Generative AI

R. Velmurugan
https://orcid.org/0000-0002-7925-9757
Karpagam Academy of Higher Education, India

R. Bhuvaneswari
https://orcid.org/0009-0003-9396-6129
Dr. Mahalingam College of Engineering and Technology, Pollachi, India

Maksud A. Madraswale
https://orcid.org/0000-0001-9565-0775
Presidency University, India

Ravi Thirumalaisamy
https://orcid.org/0000-0002-5954-5940
Modern College of Business and Science, Oman

ABSTRACT

The role of generative AI in enhancing inventory control and reducing holding costs in supply chain management. Traditional inventory management methods often struggle to balance stock levels, leading to overstocking or understocking, which incurs additional costs. Generative AI, through advanced predictive models, allows businesses to simulate various demand scenarios, improving forecasting accuracy and optimizing stock levels. By incorporating real-time data and automating decision-making, generative AI helps reduce overstocking, minimizes holding

DOI: 10.4018/979-8-3693-9856-2.ch010

costs, and improves operational efficiency. This paper discusses key applications of generative AI, including demand forecasting, stock optimization, and automated reordering processes, while also addressing challenges such as data quality and system integration. The successful implementation of AI in inventory control provides companies with a competitive edge in an increasingly complex and dynamic supply chain landscape

INTRODUCTION

Inventory control and the reduction of holding costs are pivotal aspects of supply chain management that significantly impact a company's operational efficiency and profitability. Effective inventory management ensures that a business maintains optimal stock levels to meet demand without incurring unnecessary costs. Generative AI, an advanced subset of artificial intelligence that focuses on generating data and creating models, offers new avenues for enhancing inventory control by improving demand forecasting, automating decision-making, and optimizing supply chains. The traditional challenges of inventory management have long plagued businesses across industries. Companies must carefully balance the risk of stockouts against the costs of excess inventory, while considering factors such as seasonal demand fluctuations, lead times, and storage capacity. The holding costs associated with inventory, including warehouse space, insurance, depreciation, and opportunity costs, can represent a significant portion of operational expenses. In some industries, these costs can amount to 20-30% of the inventory value annually (Christopher, 2016).

Generative AI is revolutionizing this landscape by introducing sophisticated predictive capabilities and automation tools. Unlike traditional forecasting methods that rely primarily on historical data and simple statistical models, generative AI can process vast amounts of diverse data sources to create more accurate and nuanced predictions. These systems can identify complex patterns and relationships that human analysts might miss, leading to more precise inventory management decisions. According to Brynjolfsson and McAfee (2014), AI systems are capable of learning from data and continuously improving their accuracy, making them increasingly reliable in dynamic environments such as inventory control.

One of the key advantages of generative AI in inventory control is its ability to create synthetic data scenarios. This capability allows businesses to simulate various supply chain conditions and test different inventory strategies without real-world risks. For example, a retailer can generate multiple demand scenarios based on factors such as weather patterns, economic indicators, and social media trends, then optimize their inventory levels accordingly. This approach is particularly valuable for new product launches or entering new markets where historical data may be limited.

Highsmith (2009) notes that such simulations enable organizations to respond with greater agility, a crucial aspect of modern supply chain operations.

The implementation of generative AI in inventory management typically focuses on several key areas. First, demand forecasting becomes more sophisticated through the analysis of multiple data streams, including point-of-sale data, social media sentiment, weather forecasts, and economic indicators. The AI systems can generate detailed demand predictions at various levels of granularity, from individual SKUs to entire product categories. This approach has been highlighted in research by Lee et al. (2021), who assert that the integration of AI-driven insights into forecasting processes significantly enhances accuracy, thereby reducing the likelihood of costly stockouts and overstock situations.

Second, generative AI enhances order optimization by automatically determining optimal order quantities and timing. These systems can account for various constraints and variables, such as storage capacity, transportation costs, bulk purchase discounts, and perishability. The AI can generate multiple ordering scenarios and recommend the most cost-effective approach while maintaining desired service levels. According to Tufano and Giustiniano (2020), this aspect of AI is key to reducing operational inefficiencies and improving service levels in a cost-effective manner, especially for industries with fluctuating demand patterns.

Third, generative AI contributes to risk management by identifying potential supply chain disruptions and generating contingency plans. By analyzing historical disruption patterns and current market conditions, these systems can predict potential issues and recommend preventive actions, such as adjusting safety stock levels or diversifying suppliers. Beck (2002) highlights that such predictive capabilities are crucial for mitigating risks in global supply chains, where unforeseen disruptions, such as natural disasters or geopolitical events, can have significant impacts on inventory management.

Real-world applications of generative AI in inventory control have demonstrated impressive results. For instance, a major automotive manufacturer implemented a generative AI system that reduced their inventory holding costs by 15% while maintaining service levels. The system achieved this by generating more accurate demand forecasts and optimizing order quantities across their global supply chain network. This case aligns with the findings of Sheffi (2005), who emphasizes the importance of integrating advanced technologies for improving resilience and efficiency in supply chains. Similarly, a large retail chain utilized generative AI to optimize their seasonal inventory management. The system analyzed historical sales data, weather patterns, and social media trends to generate detailed demand forecasts for different regions and stores. This implementation resulted in a 20% reduction in excess seasonal inventory and a 10% improvement in in-stock rates for high-demand items. Such results illustrate the broader potential for generative AI to improve inventory accuracy and reduce waste, aligning with the sustainability goals of modern supply chains (Westerman et al., 2014).

The benefits of generative AI in inventory control extend beyond cost reduction. These systems can also improve customer satisfaction by ensuring better product availability and reducing stockouts. Additionally, they can contribute to sustainability goals by minimizing waste from excess inventory and optimizing transportation routes to reduce carbon emissions. A study by Chesbrough (2003) supports the idea that integrating sustainability practices with AI-driven inventory management can provide long-term benefits both in terms of environmental impact and business profitability.

However, implementing generative AI in inventory control also presents challenges. Organizations must ensure they have quality data available to train the AI systems effectively. They also need to invest in appropriate infrastructure and develop new processes to integrate AI-generated insights into their decision-making workflows. Furthermore, staff training and change management are crucial for the successful adoption of these new technologies. Tidd and Bessant (2009) emphasize the importance of organizational readiness and support for innovation adoption, noting that companies must develop a culture that embraces technological advancements to realize their full potential.

Looking ahead, the role of generative AI in inventory control is likely to expand further. Emerging technologies such as digital twins and blockchain integration will enable even more sophisticated inventory management solutions. Digital twins, virtual representations of physical supply chains, can work in conjunction with generative AI to simulate and optimize inventory decisions in real time. Blockchain technology can provide enhanced visibility and traceability across the supply chain, feeding more accurate data into AI systems. This integration of technologies aligns with the views of Doz and Hamel (1998), who advocate for the adoption of collaborative technologies to enhance supply chain transparency and efficiency.

The future of inventory control will likely see increased automation of routine decisions, with human managers focusing on strategic planning and exception handling. Generative AI systems will continue to evolve, incorporating more advanced capabilities such as natural language processing for better interaction with users and autonomous decision-making capabilities for routine inventory adjustments. According to Gulati (1998), organizations that embrace these technologies will gain a competitive advantage by becoming more responsive to market demands and agile in adapting to changing conditions. Thus, generative AI represents a significant advancement in inventory control and holding cost reduction. Its ability to process complex data sets, generate accurate forecasts, and optimize inventory decisions offers substantial benefits for businesses across industries. While challenges exist in implementation, the potential returns in terms of cost savings, improved efficiency, and enhanced customer service make it an increasingly important tool in modern supply chain management. As technology continues to evolve, organizations that successfully leverage generative AI in their inventory control systems will likely gain significant competitive advantages in their respective markets.

THE IMPORTANCE OF INVENTORY CONTROL

Inventory control is a critical component of supply chain management, ensuring that businesses maintain the right amount of stock to meet customer demand while minimizing costs. Effective inventory management helps businesses avoid costly stockouts and reduce excess inventory, which ties up capital and increases holding costs. A key challenge in inventory management is balancing the risks of overstocking and understocking. Overstocking leads to higher holding costs, such as storage fees and depreciation, while understocking can result in lost sales and customer dissatisfaction (Silver et al., 2017). As Nahmias (2013) notes, the goal is to minimize costs while maintaining high service levels, ensuring that customers have access to the products they want when needed. One common tool for managing inventory is the Economic Order Quantity (EOQ) model, which helps businesses determine the optimal order quantity that minimizes both ordering and holding costs (Hill, 2000). While EOQ is effective in many traditional inventory management scenarios, it has limitations in industries with highly volatile or seasonal demand. In such cases, companies may adopt strategies like just-in-time (JIT) inventory management. JIT, introduced by Ohno (1988), reduces waste by ordering inventory only when needed, minimizing stockpiles and holding costs. However, JIT is highly dependent on the reliability of suppliers and is vulnerable to supply chain disruptions.

In recent years, advanced technologies like artificial intelligence (AI) and machine learning (ML) have transformed inventory control. AI-powered systems can analyze large datasets, including historical sales, customer demand patterns, and external factors like weather and economic conditions, to generate more accurate demand forecasts. This helps businesses reduce the risks of overstocking and understocking (Choi et al., 2018). AI-driven systems also optimize inventory replenishment by recommending the ideal order quantities and reorder points based on factors like lead time variability and supplier reliability, allowing businesses to respond more flexibly to changes in demand and supply conditions. Automation also plays a key role in modern inventory control systems. Technologies like robotic process automation (RPA) and autonomous vehicles (AVs) are being used to streamline inventory tracking and management, improving operational efficiency and reducing human error (Müller et al., 2020). These technologies enable real-time inventory tracking, improve stock accuracy, and ensure timely replenishment, contributing to more efficient inventory control.

Inventory control is especially important in industries with high inventory turnover or short product life cycles, such as fashion and technology. In these sectors, companies must continuously monitor inventory levels to meet demand without accumulating excess stock. Fisher (1997) emphasizes the importance of demand forecasting, supply chain flexibility, and rapid response times to stay competitive. With changing consumer preferences and significant holding costs for unsold products, businesses must manage inventory efficiently to remain competitive. Supply chain disruptions, such as those caused by natural disasters, political instability, or pandemics, can severely impact inventory management. During the COVID-19 pandemic, businesses struggled to manage inventory as global supply chains were disrupted and demand fluctuated unpredictably. Sheffi (2005) highlights that businesses with contingency plans, such as diversifying suppliers or maintaining safety stock, were better able to adapt to disruptions. AI and advanced analytics can further improve risk management by predicting potential disruptions and recommending actions like adjusting safety stock or finding alternative suppliers.

Sustainability is another key consideration in inventory control. By optimizing inventory levels and reducing waste, businesses can minimize their environmental impact. Excess inventory, especially in industries like food and fashion, can lead to waste if products become obsolete or perishable. Ageron et al. (2012) note that sustainable supply chain practices, including efficient inventory management, are essential for reducing waste and improving environmental outcomes. AI and data analytics can also help businesses optimize supply chain routes, reducing carbon emissions associated with transportation and warehousing. Thus, inventory control is essential for minimizing costs and ensuring that businesses can meet customer demand. Traditional models like EOQ, along with advanced technologies such as AI and automation, help businesses achieve more accurate demand forecasting, reduce holding costs, and improve overall supply chain efficiency. As the business landscape evolves, the integration of AI and machine learning will play an increasingly vital role in shaping the future of inventory control and supply chain management.

THE RISE OF GENERATIVE AI IN INVENTORY MANAGEMENT

Generative AI has emerged as a transformative tool in addressing complex challenges across industries, including inventory management. Unlike traditional AI models that rely heavily on historical data to identify patterns, generative AI can create synthetic data and simulate a wide range of scenarios, offering businesses enhanced decision-making and operational efficiency. This is particularly valuable in inventory control, where demand fluctuations and supply chain disruptions can significantly affect a company's performance. Generative AI, especially through techniques like Generative Adversarial Networks (GANs), enables businesses to forecast demand more accurately and optimize stock levels. Introduced by Goodfellow et al. (2014), GANs use two neural networks—the generator and the discriminator—that collaborate to create synthetic data. This technique has revolutionized many industries, from creating realistic images to generating synthetic financial data. In inventory management, GANs help forecast demand by simulating a variety of scenarios, taking into account external variables such as economic conditions, weather patterns, or social media trends.

In traditional inventory management, businesses rely on historical sales data and statistical models to predict future demand. However, these methods are limited when confronted with unpredictable events, like supply chain disruptions or shifts in consumer behavior. Generative AI offers a more adaptable approach by simulating different possible future scenarios. For example, GANs can generate synthetic sales data, allowing businesses to better understand how factors like weather or market shifts might impact demand. This flexibility is especially crucial in industries with short product lifecycles, such as fashion or electronics. Moreover, generative models can improve inventory optimization by considering a broader range of potential disruptions. Traditional inventory management models, such as Economic Order Quantity (EOQ) or Just-In-Time (JIT), are based on fixed assumptions and may not account for dynamic changes. In contrast, generative AI can simulate supply chain delays, customer preference changes, or even global events, enabling businesses to develop more resilient inventory strategies. By simulating different disruptions, businesses can avoid the risks of overstocking or understocking, which can lead to lost sales or increased holding costs.

One of the key advantages of generative AI in inventory management is its ability to enhance supply chain resilience. Disruptions, whether due to natural disasters or sudden changes in demand, are increasingly common in today's globalized economy. Traditional inventory management systems often struggle to adapt quickly, leading to stockouts or excess inventory. Generative AI can simulate various disruptions and their impact on inventory, helping businesses prepare for these challenges. For example, a company could use generative models to predict a supply chain disruption and adjust its inventory strategies in real-time, minimizing the impact. Generative AI also facilitates better collaboration across supply chain partners. Many businesses face challenges in aligning inventory levels with suppliers and distributors. By using generative models, companies can share data and simulate demand scenarios together, improving stock optimization and coordination. This collaboration not only improves demand forecasting but also reduces lead times and enhances customer satisfaction. As Christopher (2016) highlights, supply chain collaboration is essential for improving inventory management, particularly in industries with volatile demand or long lead times.

Personalized inventory management is another benefit of generative AI. By generating synthetic data on customer preferences, companies can simulate how different segments of their customer base might respond to specific products or promotions. This allows businesses to better tailor their inventory strategies, ensuring the right products are stocked in the right quantities, reducing waste, and improving customer satisfaction. Advancements in computational power and data analytics have further supported the integration of generative AI in inventory management. As Brynjolfsson and McAfee (2014) note, the availability of big data and more powerful computing resources has enabled the development of more sophisticated AI models, including generative algorithms. With this increasing computational capacity, businesses can process large amounts of data in real-time, generating insights that were previously impossible to obtain.

However, the implementation of generative AI in inventory management is not without challenges. One significant barrier is the need for high-quality data to train the models. While generative AI can simulate data, it requires substantial accurate historical and contextual data to produce realistic results. Additionally, businesses must invest in the necessary computational infrastructure and expertise to integrate generative AI into existing systems. Thus, generative AI represents a significant advancement in inventory management by enabling businesses to simulate and predict a variety of potential scenarios. Through techniques like GANs, companies can improve demand forecasting, optimize stock levels, and enhance supply chain resilience. By offering a more flexible and adaptive approach, generative AI helps businesses navigate uncertainties and disruptions, ultimately leading to more efficient and cost-effective inventory management.

ENHANCING DEMAND FORECASTING WITH GENERATIVE AI

Demand forecasting is a cornerstone of effective inventory management, helping businesses maintain optimal stock levels to meet customer demand while minimizing costs. Traditional forecasting methods, such as moving averages, exponential smoothing, and regression models, have served businesses for years but are limited in their ability to account for modern market complexities, such as volatile consumer behavior, market disruptions, and external variables like economic conditions and supply chain issues. Generative AI is transforming demand forecasting by offering dynamic, accurate predictions based on a broader range of influencing factors. Unlike traditional methods that rely heavily on historical data, generative AI models simulate various demand scenarios by considering a multitude of variables. These models can generate synthetic demand data under different market conditions, integrating factors such as shifts in consumer preferences, economic fluctuations, and disruptions like transportation delays or supplier shortages. Generative Adversarial Networks (GANs) and other generative models, introduced by Goodfellow et al. (2014), are capable of producing realistic synthetic data that reflects real-world complexities, making them particularly useful for demand forecasting in unpredictable environments. By simulating demand fluctuations, businesses can proactively adjust inventory levels.

Traditional forecasting models assume that past data trends will continue, but this approach fails when market conditions change rapidly. As Babai et al. (2015) note, traditional methods are constrained by their reliance on historical data, which may not reflect evolving consumer preferences or external disruptions. For example, the COVID-19 pandemic led to drastic changes in consumer behavior, rendering historical data less reliable. In contrast, generative AI can simulate various scenarios—such as economic downturns, changes in consumer spending, or supply chain disruptions—improving the flexibility and accuracy of demand forecasting. Generative AI's ability to simulate multiple demand scenarios offers a significant advantage over traditional models. By generating synthetic data, businesses can model "what-if" scenarios—like the impact of a new competitor entering the market, a marketing campaign, or a global supply chain disruption. GANs can generate various demand patterns based on combinations of factors, providing insights into how inventory might be affected. This ability to model different futures enables businesses to prepare for uncertainty, reducing risks related to overstocking or understocking.

Improved demand forecasting has a significant impact on inventory control. Accurate forecasts help companies align inventory with actual demand, reducing excess stock and minimizing holding costs. On the other hand, precise forecasting lowers the likelihood of stockouts, preventing lost sales, missed revenue opportunities, and customer dissatisfaction. By predicting demand and adjusting inventory levels accordingly, businesses enhance both customer satisfaction and profitability. As Babai et al. (2015) highlight, better forecasting accuracy can improve forecasts by up to 20%, driving substantial cost savings. Generative AI also enhances resilience against supply chain disruptions. As Christopher (2016) points out, disruptions like natural disasters, geopolitical events, or logistical bottlenecks can significantly affect a business's ability to meet demand. Generative AI models allow businesses to simulate the impact of disruptions on inventory and demand forecasts, enabling them to develop flexible, responsive inventory management strategies. For example, businesses can predict how demand might change if a supplier faces a delay or if transportation routes are disrupted, helping them proactively adjust inventory levels or identify alternative suppliers.

Incorporating generative AI into demand forecasting promotes more efficient resource use. As Jiang et al. (2020) note, AI-based forecasting optimizes inventory turnover and working capital management. By reducing excess stock, businesses free up capital for reinvestment in other areas. Moreover, accurate demand prediction reduces waste by ensuring businesses stock products that are more likely to sell, improving inventory efficiency. Generative AI also enables greater personalization in inventory management. Businesses can use generative models to simulate demand for different customer segments based on demographics, location, or purchasing history. This allows businesses to tailor inventory strategies to specific market needs, ensuring that the right products are available at the right time and place. For example, retailers can predict regional demand or tailor stock for specific customer groups, optimizing inventory levels and reducing stockouts.

Despite these advantages, implementing generative AI in demand forecasting has challenges. One major obstacle is the need for high-quality data to train the models effectively. Li et al. (2018) note that the accuracy of AI-based models depends on the quality and quantity of the training data. Businesses must invest in collecting and processing the necessary data to ensure accurate forecasts. Furthermore, integrating generative AI into existing systems requires technical expertise and investment in computational infrastructure. Generative AI is revolutionizing demand forecasting in inventory management. Techniques like GANs enable businesses to simulate a wide range of demand scenarios, improving forecasting accuracy and reducing the risks of overstocking or understocking. As the use of generative AI grows, its application in inventory control will become essential for businesses aiming to optimize supply chain operations, improve customer satisfaction, and enhance profitability.

OPTIMIZATION OF STOCK LEVELS AND REORDERING PROCESSES

Effective inventory management is essential for businesses to maintain profitability and meet customer demand. A key challenge in inventory control is optimizing stock levels to avoid both overstocking and understocking. Overstocking incurs high holding costs, such as storage, insurance, and depreciation, while understocking leads to stockouts, lost sales, and customer dissatisfaction. Traditional inventory control methods, including fixed reorder points and safety stock levels, have been useful but struggle to respond dynamically to demand fluctuations, lead times, and external disruptions. Generative AI offers the potential to revolutionize inventory management by automating and optimizing stock levels in real time, considering unpredictable factors. Generative AI's primary advantage lies in its ability to adjust stock levels dynamically based on real-time data inputs, such as sales trends, production schedules, and supplier reliability. Unlike traditional methods, which rely on fixed reorder points based on historical data, generative AI models adapt to new data as it becomes available, ensuring businesses maintain optimal stock levels to meet demand. Misra et al. (2019) highlight that AI-driven systems can automatically trigger replenishment orders when inventory falls below predetermined thresholds, minimizing the risk of stockouts and improving service levels.

Traditional inventory systems often struggle to keep up with rapidly changing supply chain dynamics, such as seasonal fluctuations, shifts in consumer preferences, or market disruptions like the COVID-19 pandemic. Generative AI integrates data from various sources—historical sales data, consumer trends, and external factors—to predict future demand with greater accuracy. This predictive capability allows businesses to adjust inventory levels proactively, ensuring they maintain the right amount of stock without overcommitting resources to inventory that may not sell quickly. Generative AI can also model scenarios where lead times fluctuate due to external factors, such as geopolitical events, transportation disruptions, or natural disasters. For example, political upheaval or a natural disaster may disrupt shipments, affecting a company's ability to maintain optimal inventory levels. Traditional systems, relying on fixed lead times, are not equipped to handle these disruptions. Generative AI, on the other hand, can simulate various scenarios, adjusting inventory strategies based on fluctuating lead times and suggesting actions like diversifying suppliers or rerouting shipments.

This capability is especially vital in industries sensitive to supply chain disruptions, such as electronics or automotive, where small delays can result in significant operational bottlenecks. Generative AI helps businesses respond to disruptions more flexibly, reducing stockouts and avoiding excess inventory. Additionally, continuously analyzing demand patterns allows generative AI to adjust reorder points and safety stock levels to reflect the most current data, minimizing the risk of holding excessive inventory. Fisher & Lutz (1994) stress that maintaining optimal stock levels is crucial for reducing inventory costs while fulfilling customer orders promptly. By making inventory management more agile, generative AI helps companies achieve this balance. The use of generative AI in automating reordering also enables a more sophisticated approach to inventory control. Machine learning algorithms allow generative AI systems to learn from past patterns and continuously improve predictions. AI adjusts reorder points based not only on historical data but also on emerging trends, customer preferences, and external events. This continuous learning process ensures that businesses are aligned with the most accurate demand forecasts, further reducing the risk of overstocking or stockouts.

Beyond inventory efficiency, generative AI can positively impact customer satisfaction and loyalty. By ensuring the right products are available when needed, businesses can enhance the customer experience, particularly in sectors where quick delivery and availability are critical, such as retail and e-commerce. Choi et al. (2003) emphasize the importance of having the right products at the right time for customer satisfaction and retention. Generative AI helps companies improve service levels, reduce lead times, and enhance overall customer experience by predicting and responding to demand and supply fluctuations. Adopting generative AI also leads to significant cost savings. Automating the reordering process reduces labor costs and minimizes human error. Moreover, by optimizing stock levels and reducing excess inventory, businesses lower holding costs, improve cash flow, and free up resources for other areas. Retail businesses that implement AI-driven inventory management systems have seen improvements in inventory turnover and reductions in stockouts, leading to increased profitability (Hopp & Spearman, 1997). Thus, generative AI is revolutionizing inventory management by optimizing stock levels, automating reordering processes, and improving supply chain responsiveness. With the ability to dynamically adjust stock levels in real time and simulate various supply chain scenarios, generative AI offers businesses the agility needed to navigate demand fluctuations, supply chain disruptions, and market changes. As businesses face increasingly complex and unpredictable supply chain environments, generative AI provides a powerful tool for efficient and responsive inventory management that reduces costs, improves customer satisfaction, and enhances profitability.

REDUCING HOLDING COSTS WITH AI-DRIVEN INVENTORY MANAGEMENT

Holding costs are a significant part of overall supply chain expenses, encompassing inventory storage, insurance, obsolescence, and opportunity costs. While unavoidable in traditional inventory management systems, these costs can be mitigated through advanced techniques. Generative AI offers innovative solutions to optimize inventory management, reduce overstocking and understocking, and streamline warehouse operations, thereby cutting holding costs. With its capacity to predict demand more accurately and simulate various scenarios, generative AI helps businesses improve stock control, increase space efficiency, and eliminate unnecessary overhead. A key way generative AI reduces holding costs is by optimizing stock levels. Overstocking, while often a precaution to avoid stockouts, leads to high costs related to warehousing, storage, insurance, and spoilage. For industries like retail or food services, where products have a limited shelf life, excess inventory increases the risk of spoilage or obsolescence. Generative AI improves demand forecasting by accurately predicting customer preferences and simulating demand scenarios based on historical sales data, economic conditions, and supply chain disruptions.

This enables businesses to adjust inventory levels in real-time, ensuring they hold only inventory likely to sell, preventing overstocking and unnecessary expenses.

Generative AI enhances demand forecasting by considering variables traditional models cannot incorporate. Techniques like moving averages and exponential smoothing are limited in predicting market shifts or disruptions. However, generative AI, particularly through Generative Adversarial Networks (GANs), simulates future demand scenarios based on macroeconomic trends, consumer behavior, and external factors such as political instability. According to Babai et al. (2015), incorporating AI into forecasting improves prediction accuracy by up to 20%, reducing overstocking risks and allowing businesses to maintain leaner inventory levels while meeting customer demand. Generative AI also optimizes warehouse operations, a crucial factor in reducing holding costs. Traditional warehouse systems often use static layouts and procedures that fail to maximize space utilization and minimize labor costs. AI-driven systems dynamically adjust warehouse layouts based on real-time inventory needs, optimizing space efficiency. Wamba et al. (2020) show that AI-powered systems can improve operational efficiency by up to 30%, leading to significant reductions in space utilization and labor costs. AI analyzes historical data, product characteristics, and movement patterns to recommend optimal storage configurations, reducing the need for frequent stock relocations and improving workflow.

Additionally, AI streamlines the picking and packing process, which is typically labor-intensive and time-consuming. By using predictive algorithms, AI identifies frequently ordered products and places them in easily accessible warehouse locations. This reduces the time and labor required for order fulfillment, leading to further cost savings. Moreover, AI can integrate with robotic systems for autonomous picking and packing, further reducing reliance on manual labor and boosting operational efficiency. These improvements lower labor costs, which are a significant portion of holding costs in warehousing. Generative AI also helps identify slow-moving or obsolete inventory, which significantly increases holding costs. Stagnant inventory occupies valuable space and incurs ongoing storage expenses. By detecting patterns in sales data, AI predicts which products are unlikely to sell in the near term. Businesses can then take proactive measures, such as marking down obsolete inventory or running clearance sales, to free up warehouse space and reduce financial strain. By optimizing the lifecycle of inventory, AI helps businesses manage dead stock more effectively, preventing it from occupying storage space for too long.

In industries where technological obsolescence is a concern, like electronics or fashion, generative AI predicts emerging trends or product life cycles. This enables companies to determine when to mark down inventory, sell off excess stock, or phase out particular items. This proactive inventory management approach ensures that businesses can reduce the burden of holding obsolete or unsellable products, maximizing available warehouse space for in-demand goods. Generative AI also enhances overall supply chain efficiency by reducing the need for large safety stock buffers, thus minimizing opportunity costs. Traditionally, businesses have relied on large safety stocks to account for uncertainties in demand or lead times. However, these buffer stocks tie up valuable capital. By improving demand forecasts and reducing variability, generative AI enables companies to reduce safety stock sizes while still minimizing stockout risks. This optimization frees up capital for other strategic initiatives, such as product innovation or business expansion. In summary, generative AI offers powerful tools to reduce holding costs in inventory management by improving demand forecasting, optimizing warehouse operations, and efficiently managing slow-moving or obsolete stock. By dynamically adjusting stock levels and simulating future scenarios, AI provides businesses with the agility to respond to changing market conditions while minimizing costs. Leveraging AI capabilities allows companies to optimize inventory systems, reduce warehousing expenses, improve capital efficiency, and ultimately enhance profitability.

CASE STUDY: AI IN INVENTORY MANAGEMENT AT AMAZON

Amazon has emerged as a global leader in e-commerce, not just for its vast product offerings and quick delivery times but also due to its highly sophisticated inventory management system. At the heart of Amazon's success lies its innovative use of artificial intelligence (AI), including generative models, to manage its complex supply chain operations and optimize holding costs. Amazon's AI-driven inventory system integrates diverse data inputs to forecast demand, position inventory strategically, and automate reordering processes, all of which contribute to reduced holding costs and enhanced operational efficiency. Amazon offers a powerful example of how AI can transform inventory management and deliver substantial cost savings.

One of the key ways in which AI, including generative models, helps Amazon manage inventory is through improved demand forecasting. Traditional forecasting methods often rely on historical data and static algorithms, which can miss dynamic and unpredictable factors like shifts in consumer preferences, weather patterns, or economic changes. Generative AI can simulate multiple future scenarios by factoring in real-time data points such as customer behavior, seasonality, and external influences like promotions or events. This enables Amazon to create more accurate demand forecasts and adjust inventory levels in real-time, ensuring products are available when customers need them, without overstocking. Amazon's AI system uses data from diverse sources, including purchase history, social media trends, and weather forecasts, to make localized demand predictions. For example, if a weather event increases demand for snow gear, AI can adjust inventory placements based on expected local demand. This localized inventory management minimizes the need for large centralized warehouses. Amazon's network of fulfillment centers, strategically located near customers, reduces both holding costs and delivery times, allowing for faster shipping options and enhanced customer satisfaction.

AI also plays a crucial role in optimizing stock levels at Amazon. Traditional inventory management systems often rely on static calculations to determine reorder points, leading to overstocking or understocking. Overstocking increases holding costs, while understocking results in missed sales opportunities. Amazon's AI-driven inventory system dynamically adjusts stock levels based on demand forecasts, real-time sales data, and supply chain performance. AI algorithms adjust reorder points in real time, ensuring that stock levels remain optimized without incurring unnecessary holding costs. Automation reduces the need for manual intervention, making reordering more efficient and less prone to errors. Furthermore, AI helps Amazon respond swiftly to market changes and disruptions. If a supplier faces delays or a transportation route is disrupted, AI suggests alternative sourcing or routing strategies in real-time, minimizing supply chain disruptions and ensuring that customer demand is met without significant delays. This flexibility is essential for maintaining operational efficiency and reducing holding costs.

In addition to demand forecasting and stock optimization, AI plays a significant role in optimizing Amazon's warehouse operations. Traditional warehousing systems often rely on manual processes, which can be inefficient. Amazon's AI-powered warehouse management system optimizes the layout to store products efficiently based on sales velocity, size, and frequency of orders. This maximizes space utilization and minimizes picking and packing time, which reduces labor costs and improves operational efficiency. A study by Puvanasvaran et al. (2020) indicates that Amazon's use of AI-driven warehouse systems has led to significant improvements in operational efficiency and reductions in warehouse holding costs, allowing Amazon to avoid the need for additional warehouse facilities, reducing capital expenditures.

AI also improves returns management, a significant cost for e-commerce businesses, particularly in industries like fashion and electronics, where return rates are high. AI models can predict which items are most likely to be returned based on factors like product type, customer demographics, and historical return data. By forecasting returns, Amazon can plan reverse logistics more effectively, optimizing storage and minimizing the cost of handling returned goods. Through its AI innovations, Amazon has gained a significant competitive advantage in the e-commerce industry. By leveraging generative AI to forecast demand more accurately, optimize stock levels, and automate reordering processes, Amazon can reduce holding costs while maintaining efficient and responsive operations. Additionally, the company's ability to adapt quickly to market shifts, supply chain disruptions, and changing customer preferences allows it to provide fast and cost-effective delivery. Through these advanced AI technologies, Amazon has successfully maintained low costs while offering superior service, reinforcing its position as a leader in the e-commerce sector.

CHALLENGES IN IMPLEMENTING GENERATIVE AI FOR INVENTORY CONTROL

The potential of generative AI in inventory management is vast, offering businesses the opportunity to optimize stock levels, improve demand forecasting, and reduce holding costs. However, the adoption of such technologies comes with its own set of challenges. These challenges can range from data quality issues to integration difficulties with legacy systems. Understanding and addressing these challenges is crucial for businesses seeking to leverage generative AI to enhance their inventory management processes.

A. Data Quality and Availability

A significant challenge in implementing generative AI in inventory management is ensuring high-quality and sufficient data. Generative AI models rely on large volumes of accurate data to make predictions. As Wu and Pagell (2011) note, the absence of quality data compromises the effectiveness of AI models, potentially leading to inaccurate forecasts and poor decision-making. Inconsistent or incomplete data can cause inventory issues, such as stockouts or excess inventory, resulting in lost sales, customer dissatisfaction, and inflated holding costs. To mitigate these risks, businesses must invest in robust data collection systems, integrating both internal data (sales, inventory levels) and external data (market trends, consumer behavior) for greater model accuracy (Chong et al., 2017). Additionally, establishing data governance frameworks is crucial for ensuring data consistency and quality, especially when transitioning from traditional systems to AI-powered solutions, requiring considerable effort and resources across departments.

b. Integration with Existing Systems

A key challenge in implementing generative AI for inventory management is integrating AI with existing enterprise resource planning (ERP) and supply chain systems. Many businesses, especially small and medium-sized enterprises (SMEs), rely on legacy systems that are not designed to interface with modern AI technologies (Waller & Fawcett, 2013). These systems often lack the flexibility needed to process the data required for AI to function effectively. Integration involves aligning data flow between legacy systems and AI models, which can be costly and resource-intensive, especially for companies with limited budgets. Furthermore, employees may require training to use AI-powered systems, necessitating investment in skill development. This integration challenge also includes the need for organizational change management. SMEs may struggle to upgrade IT infrastructure or may consider cloud-based AI solutions or third-party vendors. However, these options carry risks, such as vendor lock-in and higher operational costs.

c. Cost and Resource Constraints

Another significant barrier to the widespread adoption of generative AI in inventory management is the high upfront costs. Implementing AI solutions is often capital-intensive, requiring businesses to invest in software, hardware, and specialized personnel. As noted by Waller and Fawcett (2013), the financial and resource constraints associated with AI adoption can be particularly challenging for SMEs. These businesses may struggle to justify the substantial initial investment required to implement AI solutions, especially when they are uncertain about the return on investment (ROI).

While AI-driven inventory management systems can result in cost savings over the long term, particularly by optimizing stock levels, improving demand forecasting, and reducing holding costs, the initial investment can be daunting. Moreover, businesses must factor in the costs associated with ongoing maintenance, software updates, and system monitoring. The need for specialized AI talent to design, implement, and fine-tune AI models further adds to the financial burden, as these experts are often in high demand and command premium salaries.

For larger organizations with more financial resources, the challenges of cost and resource allocation may be more manageable. However, for smaller businesses or those operating on thin profit margins, the prospect of investing heavily in AI technology may seem like a risky proposition. To address these concerns, businesses can explore alternative solutions, such as SaaS (Software as a Service) AI platforms, which offer more affordable and scalable AI solutions without the need for substantial upfront investment in infrastructure or talent (Davenport & Ronanki, 2018).

d. Change Management and Organizational Buy-in

Implementing AI in inventory management also requires a cultural shift within the organization. Employees, especially those involved in inventory management and supply chain operations, may be resistant to adopting AI-driven systems due to concerns about job displacement or a lack of familiarity with new technology. Successful AI adoption requires gaining the buy-in of key stakeholders and employees, which can be challenging in organizations with deeply entrenched processes or resistance to change.

To overcome this barrier, businesses must foster a culture of innovation and technological adoption. This involves not only providing training and support but also clearly communicating the benefits of AI, such as improved efficiency, reduced workload, and enhanced decision-making capabilities. Employees must be assured that AI is intended to complement their work, rather than replace it, and that it can empower them to make more informed, data-driven decisions.

While the potential benefits of generative AI in inventory management are significant, the challenges associated with its implementation cannot be overlooked. Data quality and availability, system integration issues, cost constraints, and organizational resistance are all factors that businesses must address when adopting AI technologies. By investing in data governance, upgrading IT infrastructure, and fostering a culture of innovation, companies can mitigate these challenges and unlock the full potential of generative AI to optimize their inventory management processes. As AI technology continues to evolve, the integration of generative AI into inventory management systems is likely to become increasingly feasible and accessible for businesses of all sizes.

THE COMPETITIVE ADVANTAGE OF AI-ENHANCED INVENTORY CONTROL

In today's fast-paced and highly competitive business environment, companies must constantly strive for efficiencies and differentiation to stay ahead. One of the key areas where businesses can achieve both is inventory management. Artificial intelligence (AI)-enhanced inventory management is transforming the way companies manage their supply chains, not only by reducing costs but also by providing a significant competitive advantage. Companies that can accurately forecast demand, optimize stock levels, and respond quickly to market changes are better positioned to meet consumer expectations, improve service levels, and ultimately outperform their competitors.

a. Forecasting Demand and Agility

The first significant advantage of AI-enhanced inventory management lies in its ability to improve demand forecasting. Traditional inventory management methods, such as relying on historical data, often fall short in accurately predicting future demand, especially in volatile markets. By incorporating advanced AI techniques, businesses can generate more accurate demand forecasts. Machine learning algorithms, including generative models, analyze a variety of data sources—from historical sales and market trends to external factors like weather conditions or geopolitical shifts—to predict demand with greater precision.

This ability to forecast demand accurately leads to better inventory management, reducing the risks of both stockouts and overstocking. Stockouts, or running out of inventory, can lead to missed sales opportunities and customer dissatisfaction, while overstocking ties up capital in unsold goods and incurs unnecessary holding costs. Through AI-driven demand forecasting, businesses can achieve the optimal stock levels that align with actual demand, ensuring they can meet customer expectations without overcommitting resources (Chong et al., 2017). This level of agility is a powerful competitive tool, as it allows companies to respond to market changes faster than competitors who rely on traditional methods.

b. Agility and Responsiveness in Supply Chain Operations

AI's ability to enhance agility in inventory management extends beyond forecasting demand. Businesses can use AI to dynamically adjust their inventory systems in real-time based on various factors such as order lead times, customer preferences, and unexpected disruptions like transportation delays or supplier issues. As mentioned by Porter (1985), a key aspect of gaining competitive advantage is responding effectively to changes in the business environment. AI provides businesses with the ability to anticipate, adjust to, and even proactively address these changes, thereby maintaining high service levels while minimizing costs.

For instance, AI can automate reorder processes to ensure that inventory levels are consistently optimized. By continually monitoring inventory and automatically triggering orders when stock levels fall below predefined thresholds, AI minimizes the need for manual intervention, which reduces the risk of human error and enhances the speed of decision-making (Misra et al., 2019). This automation also allows businesses to operate with lower safety stock levels while ensuring that customer demand is met in a timely manner. As a result, companies benefit from reduced holding costs, faster response times, and enhanced customer satisfaction—all factors that contribute to maintaining a competitive edge.

Moreover, AI can simulate various supply chain scenarios to identify potential risks and opportunities. For example, if a supply chain disruption occurs due to a natural disaster or political instability, AI models can predict the impact of such disruptions on inventory levels and suggest alternative sourcing or logistics strategies. This capacity for scenario planning allows businesses to stay ahead of potential risks, reducing the likelihood of costly operational hiccups and ensuring continuity of service (Babai et al., 2015).

c. Cost Leadership and Differentiation

As Porter (1985) asserts, competitive advantage can be achieved through cost leadership or differentiation, and AI-powered inventory management helps companies achieve both. By optimizing inventory and supply chain operations, AI contributes to cost leadership. Reduced operational costs, such as holding costs, storage costs, and labor costs associated with manual inventory management, allow businesses to lower their overall cost structure. This reduction in cost enables companies to offer competitive pricing or improve profit margins while still maintaining high service levels.

At the same time, AI enhances differentiation by improving customer service. Customers today expect fast and reliable service, with quick delivery times and the assurance that products are readily available. AI-driven inventory management can significantly reduce stockouts and improve order fulfillment speed, contributing to better customer experiences. Companies that leverage AI to optimize stock levels and streamline fulfillment processes are more likely to meet or exceed customer expectations, which can differentiate them in a crowded market. For example, Amazon's ability to predict demand and stock products in fulfillment centers closer to customers has allowed the company to provide faster delivery times, which in turn enhances customer satisfaction (Jopson, 2021). By offering faster and more reliable service, businesses can differentiate themselves from competitors who may struggle to meet these service levels.

AI also supports personalization, a key aspect of differentiation in today's market. With AI analyzing customer behavior, preferences, and purchasing patterns, companies can create personalized shopping experiences and optimize their product offerings based on real-time insights. This level of personalization can further strengthen customer loyalty, as businesses can anticipate customer needs and deliver tailored services that drive repeat business.

d. Real-World Examples of Competitive Advantage

Several companies have successfully leveraged AI to enhance their inventory management capabilities and gain a competitive advantage. Amazon, as previously mentioned, is a prime example. By using AI to predict demand, automate replenishment orders, and optimize inventory placement across its vast network of fulfillment centers, Amazon has positioned itself as a leader in the e-commerce space. The company's ability to offer fast, reliable delivery while keeping holding costs low has given it a significant edge over competitors.

Walmart, another global retail giant, also uses AI in its inventory management systems. The company has employed machine learning algorithms to improve its demand forecasting, reduce stockouts, and optimize shelf stocking. By analyzing various data sources, including point-of-sale transactions and external factors such as local weather patterns, Walmart has been able to optimize its inventory and provide customers with the products they want when they want them (Kumar et al., 2021).

These examples demonstrate that AI can provide a substantial competitive advantage by improving operational efficiency, enhancing customer satisfaction, and reducing costs. Businesses that effectively implement AI in their inventory management processes are better equipped to adapt to market shifts, respond to customer demands, and maintain high levels of service, all while reducing unnecessary costs.

AI-enhanced inventory management provides businesses with the tools they need to gain a competitive advantage in a fast-changing market. Through improved demand forecasting, agility in supply chain operations, and the ability to achieve both cost leadership and differentiation, AI enables companies to optimize inventory, reduce costs, and enhance customer satisfaction. Companies like Amazon and Walmart have demonstrated the powerful impact of AI on inventory management, setting the stage for others to follow suit. In increasingly competitive markets, businesses that embrace AI-driven inventory management systems will be better positioned to meet customer expectations, reduce operational costs, and maintain a sustainable competitive advantage.

CONCLUSION

Generative AI has emerged as a transformative tool in inventory control, offering businesses the ability to optimize stock levels, improve demand forecasting, and reduce holding costs. The technology's ability to simulate various demand and supply chain scenarios provides valuable insights, helping businesses make data-driven decisions that enhance operational efficiency and give them a significant competitive edge. However, the successful implementation of AI in inventory management requires careful attention to data quality, system integration, and strategic planning. As AI technology continues to evolve, businesses that embrace these capabilities are better positioned to navigate the complexities of modern supply chains and maintain a competitive advantage in the market. One of the primary challenges in inventory management is maintaining optimal stock levels. Too much inventory ties up capital, increases holding costs, and can lead to waste, particularly in industries with perishable goods. On the other hand, insufficient stock can result in stockouts, missed sales, and damaged customer relationships. Generative AI models can dynamically adjust stock levels based on real-time data inputs, including sales trends, seasonal fluctuations, supplier performance, and even external factors like weather or economic conditions. By analyzing a wide range of data, generative AI helps businesses strike the balance between overstocking and understocking, ensuring that inventory levels are both cost-efficient and responsive to customer demand (Misra et al., 2019). For example, in the retail industry, AI can optimize inventory placement across different regions or stores based on localized demand. This allows businesses to reduce transportation costs and improve delivery times. By leveraging AI, companies can minimize holding costs while ensuring product availability, thus improving overall operational efficiency. Moreover, AI-driven stock optimization allows businesses to automatically reorder products when stock levels dip below a predefined threshold, reducing the likelihood of human error and improving decision-making speed.

One of the critical benefits of generative AI in inventory control is its ability to improve demand forecasting accuracy. Traditional forecasting methods, such as moving averages and exponential smoothing, often rely solely on historical data and may fail to capture the complexities of modern markets. These methods are limited in their ability to account for unexpected shifts in demand, such as those caused by changes in consumer preferences, economic disruptions, or global events. Generative AI, however, can simulate a wide range of possible demand scenarios by integrating historical data with real-time market information. This includes tracking shifts in consumer behavior, macroeconomic trends, geopolitical events, and supply chain disruptions. As highlighted by Babai et al. (2015), incorporating machine learning and AI into demand forecasting processes has shown to improve accuracy significantly, with improvements of up to 20% in some cases. This increased accuracy can lead to better stock management, reduce the risk of stockouts or overstocking, and enable businesses to plan more effectively for fluctuating demand. By using AI to anticipate future demand patterns, businesses can optimize their production schedules, adjust procurement strategies, and make more informed decisions about pricing, marketing, and inventory investments. Furthermore, AI can enable companies to forecast demand at a much more granular level. For instance, demand can be forecasted not just for entire regions or product categories but for individual products at specific locations, allowing for hyper-localized inventory management. This precision enables businesses to adapt quickly to changing consumer preferences and market conditions, ensuring that they maintain service levels without incurring unnecessary costs (Chong et al., 2017).

Holding costs, which include storage, insurance, and the opportunity costs associated with tying up capital in unsold goods, can account for a significant portion of overall supply chain expenses. Generative AI helps reduce these costs by enabling businesses to maintain optimal inventory levels, minimizing the need for excessive warehousing and reducing the likelihood of inventory obsolescence. AI can also optimize warehouse operations, improving space utilization and reducing labor costs related to inventory management. According to Wamba et al. (2020), AI-powered systems in warehousing can enhance efficiency by up to 30%, translating into substantial cost savings. AI can suggest the most efficient warehouse layouts, ensure that high-demand items are easily accessible, and even automate inventory tracking using robots and sensors. By improving warehouse efficiency, businesses can store more products in less space and reduce the labor required to manage stock, further cutting holding costs. Additionally, AI can help businesses identify slow-moving or obsolete inventory, allowing for proactive measures such as discounting or liquidating these items to free up space and capital (Wu & Pagell, 2011).

Another key advantage of generative AI in inventory control is its ability to simulate various demand and supply chain scenarios, providing businesses with insights into potential risks and opportunities. For example, AI can model scenarios in which lead times fluctuate due to external factors like natural disasters, strikes, or geopolitical events. By simulating different supply chain disruptions, AI can help businesses identify alternative sourcing or logistics strategies that minimize the impact on inventory levels and service levels. This predictive capability is especially valuable in today's volatile global supply chains, where disruptions are becoming more frequent and severe. Generative AI allows businesses to be more agile in responding to these disruptions, ensuring that they can maintain continuity in operations while minimizing inventory-related risks (Waller & Fawcett, 2013). The ability to anticipate potential disruptions and develop contingency plans provides companies with a strategic advantage, allowing them to continue serving customers even in times of uncertainty.

While the potential of generative AI in inventory management is immense, there are several challenges associated with its implementation. One of the key hurdles is ensuring data quality. As AI models rely heavily on data to make predictions, poor-quality or incomplete data can lead to inaccurate forecasts and misguided decisions. Businesses must invest in robust data collection, cleansing, and management systems to ensure the inputs to their AI models are accurate and reliable. This often requires a significant upfront investment in technology and infrastructure, which can be a barrier for small and medium-sized enterprises (SMEs). Moreover, integrating AI with existing enterprise resource planning (ERP) and supply chain management systems can be complex and costly. Many businesses, particularly SMEs, still rely on legacy systems that may not be compatible with modern AI solutions. The integration process may require a complete overhaul of existing systems or the development of custom interfaces, which can be time-consuming and resource-intensive (Waller & Fawcett, 2013).

Generative AI offers a transformative approach to inventory control, providing businesses with the tools to optimize stock levels, improve demand forecasting, and reduce holding costs. By simulating a wide range of demand and supply chain scenarios, AI empowers businesses to make data-driven decisions that enhance operational efficiency and resilience. However, the successful implementation of AI requires careful attention to data quality and system integration. As AI technology continues to evolve, businesses that invest in these capabilities will be better positioned to navigate the complexities of modern supply chains, improve customer satisfaction, and maintain their competitive edge in the market.

REFERENCES

Ageron, B., Gunasekaran, A., & Spalanzani, A. (2012). Sustainable supply chain management: The role of supply chain integration. *International Journal of Production Economics, 140*(1), 1–15.

Babai, M. Z., Dolgui, A., & Syntetos, A. (2015). Demand forecasting with machine learning: Applications in inventory management. *International Journal of Production Research, 53*(1), 187–200.

Babai, M. Z., Lamsal, R., & Mohanty, S. K. (2015). Artificial intelligence in supply chain forecasting: A comprehensive review. *Journal of Manufacturing Science and Engineering, 137*(3), 031019.

Beck, K. (2002). *Extreme Programming Explained: Embrace Change*. Addison-Wesley Professional.

Brynjolfsson, E., & McAfee, A. (2014). *The Second Machine Age: Work, Progress, and Prosperity in a Time of Brilliant Technologies*. W.W. Norton & Company.

Chesbrough, H. (2003). *Open Innovation: The New Imperative for Creating and Profiting from Technology*. Harvard Business Press.

Choi, T. M., Wallace, S. W., & Wang, Y. (2003). Supply chain coordination under uncertainty: A review of recent developments in the literature. *Journal of Operations Management, 21*(3), 231–247.

Choi, T. M., Wallace, S. W., & Wang, Y. (2018). Advances in supply chain modeling and analysis: Trends and research opportunities. Springer.

Choi, T. M., Wallace, S. W., & Wang, Y. (2018). The impact of AI and machine learning on inventory management in retail supply chains. *International Journal of Production Economics, 206*, 47–61.

Chong, A. Y. L., Chan, F. T. S., & Ooi, K. B. (2017). Predicting the adoption of artificial intelligence in supply chain management. *International Journal of Logistics Research and Applications, 20*(4), 349–367.

Christopher, M. (2016). *Logistics & Supply Chain Management* (5th ed.). Pearson Education.

Davenport, T. H., & Ronanki, R. (2018). Artificial intelligence for the real world. *Harvard Business Review, 96*(1), 108–116.

Doz, Y. L., & Hamel, G. (1998). *Alliance Advantage: The Art of Creating Value Through Partnering*. Harvard Business Press.

Fisher, M. L. (1997). What is the right supply chain for your product? *Harvard Business Review*, *75*(2), 105–116.

Fisher, M. L., & Lutz, S. R. (1994). The impact of inventory management on supply chain profitability. *Supply Chain Management Review*, *8*(1), 39–47.

Goodfellow, I., Pouget-Abadie, J., Mirza, M., Xu, B., Warde-Farley, D., Ozair, S., & Bengio, Y. (2014). Generative adversarial nets. *Advances in Neural Information Processing Systems*, *27*, 2672–2680.

Gulati, R. (1998). Does familiarity breed trust? The implications of repeated ties for contractual choice in alliances. *Academy of Management Journal*, *41*(1), 85–101. DOI: 10.2307/256729

Highsmith, J. (2009). *Agile Project Management: Creating Innovative Products* (2nd ed.). Addison-Wesley Professional.

Hill, T. (2000). *Operations Management: Strategic Context and Managerial Analysis*. Palgrave.

Hopp, W. J., & Spearman, M. L. (1997). *Factory Physics: Foundations of Manufacturing Management*. McGraw-Hill.

Jiang, B., Lee, H., & Zeng, D. (2020). Supply chain resilience: A multi-method approach to analyze supply chain disruptions. *International Journal of Production Research*, *58*(10), 3180–3197.

Jopson, D. (2021). How Amazon uses artificial intelligence to optimize its inventory management. *Harvard Business Review*.

Jopson, M. (2021). Amazon's use of artificial intelligence to optimize inventory management and reduce holding costs. *Journal of Business Research*, *125*, 29–38.

Kumar, S., Ahlawat, S., & Garg, A. (2021). The impact of AI and machine learning in supply chain management. *The Journal of Supply Chain Management*, *57*(2), 15–30.

Lee, S., Lee, S., & Lee, Y. (2021). Artificial intelligence in supply chain management: Applications, challenges, and opportunities. *Computers & Industrial Engineering*, *157*, 107274.

Li, L., Li, B., & Yu, Z. (2018). Demand forecasting using generative adversarial networks. *International Journal of Production Economics*, *203*, 12–23.

Misra, A., Sharma, S., & Srivastava, A. (2019). AI in supply chain and inventory management. *The Journal of Supply Chain Management*, *55*(1), 70–80.

Misra, S., Goh, M., & Ding, H. (2019). AI-driven inventory management systems: Automation and optimization techniques. *International Journal of Production Research*, *57*(13), 4093–4111.

Müller, S., Meyer, B., & Lindemann, U. (2020). Automation in supply chain management: Challenges and opportunities. *International Journal of Operations & Production Management*, *40*(6), 637–657.

Nahmias, S. (2013). *Production and Operations Analysis* (7th ed.). Waveland Press.

Ohno, T. (1988). *Toyota Production System: Beyond Large-Scale Production*. Productivity Press.

Porter, M. E. (1985). *Competitive Advantage: Creating and Sustaining Superior Performance*. Free Press.

Puvanasvaran, R., Nor, K. M., & Liew, L. K. (2020). Artificial intelligence in warehouse management: An exploration of its impact on operational efficiency. *Journal of Intelligent Manufacturing*, *31*(2), 499–511.

Sheffi, Y. (2005). *The Resilient Enterprise: Overcoming Vulnerability for Competitive Advantage*. MIT Press.

Silver, E. A., Pyke, D. F., & Thomas, D. J. (2017). *Inventory and Production Management in Supply Chains* (4th ed.). CRC Press.

Tidd, J., & Bessant, J. (2009). *Innovation and Entrepreneurship: The Lean Startup Approach to Disruptive Innovation*. Wiley.

Tufano, P., & Giustiniano, L. (2020). AI in the Supply Chain: Optimizing Order Quantities and Timing. *The Journal of Supply Chain Management*, *56*(4), 45–62.

Waller, M. A., & Fawcett, S. E. (2013). Data science, predictive analytics, and big data: A revolution that will transform supply chain design and management. *Journal of Business Logistics*, *34*(4), 77–84. DOI: 10.1111/jbl.12010

Wamba, S. F., Akter, S., & Bhattacharya, M. (2020). Artificial Intelligence applications and their implications in warehousing and supply chain management. *Journal of Business Research*, *120*, 38–48.

Wamba, S. F., Akter, S., & Ren, S. (2020). Artificial intelligence in supply chain management: A survey and research directions. *The Journal of Supply Chain Management*, *56*(4), 29–42.

Westerman, G., Calméjane, C., Ferraris, P., & Bonnet, D. (2014). The Digital Advantage: How Digital Leaders Outperform Their Peers in Every Industry. *MIT Sloan Management Review*.

Wu, Y., & Pagell, M. (2011). Balancing supply chain efficiency and flexibility: A multi-case study analysis. *International Journal of Operations & Production Management, 31*(5), 566–591.

Chapter 11
Optimization of Inventory Management Using Artificial Neural Networks and K-Means Clustering for Cost Reduction and Improved Customer Service

Govind Shay Sharma
https://orcid.org/0000-0002-5884-4674
Vivekananda Global University, Jaipur, India

Randhir Baghel
https://orcid.org/0000-0002-5113-3554
Poornima University, India

ABSTRACT

This paper addresses the critical issue of optimizing inventory strategies using advanced machine learning techniques, particularly Artificial Neural Networks (ANN), to enhance customer service and reduce costs. Here's a brief summary and analysis of the concepts you mentioned: 1. Optimization of Stock Management: Inventory management plays a vital role in fulfilling customer demand while minimizing lead times and costs. Optimizing stock levels can balance supply with fluctuating market demand. 2. Artificial Neural Networks (ANN): ANN is employed to forecast the optimal stock levels. By analyzing past stock data, ANN models can predict the required

DOI: 10.4018/979-8-3693-9856-2.ch011

inventory and reduce errors in forecasting, thus minimizing overstocking or under stocking. 3. K-Means Algorithm for Grouping Items: The k-means algorithm is used to classify raw materials and finished goods into homogeneous groups. These groups share similar characteristics, allowing for tailored inventory policies. For example, fast-moving items may have different inventory strategies than slow-moving ones.

INTRODUCTION

The key takeaway is that optimizing inventory levels and minimizing costs requires dynamic strategies and advanced algorithms to handle the stochastic and complex nature of inventory systems. Here's a breakdown of the major methods and ideas you discussed: (Radha Krishnan et al., 2009) focused on minimizing the total supply chain cost by optimizing the base stock level at each stage of the supply chain. Genetic algorithms (GAs) were used to find the most probable stock level, showing strong performance when implemented in MATLAB. GA helps navigate the complexity of dynamic inventory problems by evolving towards an optimal solution through iterative improvement. (White & Censlive, 2016) Applied control theory to manage production capacity emphasizes the need for timely acquisition of additional capacity to avoid shortages. This mathematical framework ensures a more proactive approach to inventory management, responding to fluctuations in demand and capacity efficiently.(Vladimir & Tasho, 2011) discussed mathematical models for inventory systems, implemented as software modules in MATLAB. These models handle basic inventory control problems and aim to provide robust solutions that adapt to varying company requirements. Further, they suggest that incorporating stochastic and dynamic factors into inventory models is essential for more accurate inventory control. Strijbosch & Moors, 1999 examined the periodic review system with stationary gamma-distributed demand and constant lead time. It takes into account two critical service-level constraints—stock-out probability and fill rate—and uses exact safety factors to model inventory behavior effectively. (Ramirez & Labadie, 2017) focused on minimizing costs and managing shortages and wastage in the blood supply chain. It optimizes the number of blood units processed by a blood center and ordered by hospitals, providing a tailored approach to managing sensitive, high-risk inventory like blood. (Eme et al., 2018) Eme et al. analyzed and improved the manual inventory system in a pharmaceutical store, showcasing the benefits of a computer-based system to enhance operational efficiency and accuracy. This system improved inventory accuracy, responsiveness, and management effectiveness.(Šustrová, 2016):Šustrová explored the construction of an artificial neural network model for optimizing order quantities based on current market demand. ANN was used to predict optimal order levels, improving inventory management by responding dynamically to fluctuating demand.(Yadav et al., 2018) Yadav et al. explored swarm optimization to address inventory optimization in the auto industry, showing that bio-inspired algorithms can be effectively applied to real-world inventory challenges.(Haruna et al., 2017) Haruna et al. reviewed the integration of genetic algorithms (GA) and neural networks for optimizing inventory management. They provided insights into the potential of GAs to enhance the design and optimization of neural networks in inventory systems. (Paula & Azeema, 2011):Paula & Azeema developed an ANN model to optimize finished goods levels, holding costs, and carrying costs. By predicting optimal levels, the model helped reduce excessive inventory while avoiding shortages.(Peterson & James, 1988):

Peterson and James introduced the MFT algorithm, which minimized simulation time in comparison to other approaches. This method provided an efficient solution for forecasting in inventory systems. (Sonali & Maind, 2014; Mhatre et al., 2015):Sonali & Maind provided a comprehensive explanation of how artificial neural networks work, their training process, and various applications, while Mhatre et al. reviewed prediction techniques using ANN, emphasizing the improved accuracy of models based on experimental data.

Key Insights:

- Genetic algorithms, control theory, and neural networks are prominent methods for handling the complexity of inventory systems. The use of MATLAB for implementation indicates the effectiveness of simulation and computational tools in fine-tuning these methods.
- Artificial neural networks, combined with statistical techniques, help optimize stock levels by predicting future demand more accurately. By dynamically adjusting inventory levels based on these predictions, companies can reduce costs and improve service levels.
- Each technique is applied to different industries, from pharmaceuticals to the auto industry to blood supply chains. Tailoring inventory optimization strategies to specific contexts ensures that each solution is robust and effective in its unique environment.

This broad spectrum of approaches demonstrates the importance of combining traditional mathematical models with modern machine learning and evolutionary algorithms to optimize inventory management systems.

Preliminary: -This proposed method focuses on optimizing inventory control in a three-stage supply chain using a **Neural Fitting Algorithm** and **K-means clustering** for efficient stock level prediction. Below is a breakdown of the approach and its key components:

1. Neural Fitting Algorithm:
    - Neural fitting is a class of **evolutionary algorithms** inspired by biological processes. It computes approximate solutions and iteratively refines them, searching for the "best fit" to meet inventory control goals.
    - In your approach, you propose using neural fitting to minimize the **cost of inventory** by finding the optimal stock levels. Unlike binary-coded solutions often seen in other algorithms, neural fitting here employs **numeric encoding**, which is more flexible and better suited for real-world inventory management scenarios.
2. Supply Chain Structure:

The study examines a three-member supply chain with four key players:

- Factory (the production unit)
- Distribution Center 1 (DC1)
- Distribution Center 2 (DC2)
- Supplier (at the endpoint)

The flow of goods:

- Raw materials arrive at the factory.
- From the factory, materials move to DC1 and then to DC2 before reaching the supplier.

3. Inventory Optimization Problem:

- At the start, both excess stock and shortages are observed at the distribution points.
- The goal is to predict the optimal stock level at each stage (factory, DC1, and DC2) using past stock levels and current demand patterns to avoid overstocking and shortages.
- Zero and Non-Zero Inventory Levels: Inventory levels are classified into two categories:

4. K-Means Clustering for Stock Levels:

- The K-means algorithm is used to cluster stock levels into groups, with the goal of segregating excess stock and shortages. K-means helps create distinct clusters based on inventory data at various stages.
- **Zero and Non-Zero Clusters**:

The K-means algorithm iteratively assigns stock levels to the nearest cluster centroids, updating them based on the mean values of the stock levels in each cluster.

5. Data Preparation:
   - An inventory sheet is generated using past stock levels and recent data. Random values within a defined upper and lower limit are chosen for the initial dataset at the factory, DC1, and DC2.
   - This raw data will serve as input for the neural fit algorithm, which will predict the optimal stock levels.
6. Prediction of Optimal Stock Levels:
   - The neural fit algorithm uses historical and clustered data to predict the amount of stock that should be maintained at each stage of the supply chain.
   - This approach ensures there is no excess stock and minimizes shortages by dynamically adjusting inventory levels based on the predicted values.
7. Application Benefits:
   - **Cost Optimization**: The method aims to control and minimize inventory-related costs by predicting the right stock levels.
   - **Efficient Stock Distribution**: By clustering inventory data and predicting stock levels for factory and distribution centers, the process ensures smoother material flow and minimizes disruptions.
   - **Supply Chain Stability**: It reduces the chances of excess inventory or shortages across the supply chain, leading to improved customer service and better utilization of resources.

Result and Discussion: - In **Table-1**, the stock levels at **Distribution Center 1** and **Distribution Center 2** represent the quantities of goods ready for dispatch to the **Supplier points** or warehouses.

Table 1. Data Sheet of Quantities for Optimum Inventory

	Input			Target
S.No	**Factory**	**Distributor Center-1**	**Distributor center-2**	**Supplier point(app. 200)**
1	-585	-321	262	199
2	-477	-792	-541	201
3	-591	-329	269	202
4	491	391	281	200

continued on following page

Table 1. Continued

	Input			Target
S.No	Factory	Distributor Center-1	Distributor center-2	Supplier point(app. 200)
5	-591	-329	269	209
6	372	573	-345	196
7	997	-930	108	205
8	146	118	532	211
9	-588	-322	260	225
10	-587	-329	269	185
11	-746	721	-677	186
12	790	-451	831	171
13	-591	-329	269	190
14	-550	-634	158	181
15	611	-295	-443	161
16	495	-168	847	157
17	-992	268	-270	206
18	160	964	-507	203
19	482	-471	761	191
20	-586	-322	269	176
21	664	-766	-367	177
22	446	-916	129	202
23	-278	582	-573	201
24	-751	-763	367	183
25	-208	-306	403	173
26	-432	-848	401	177
27	-591	-329	269	191
28	285	334	675	182

The goal is to use machine learning (such as an Artificial Neural Network) to predict the **Supplier point** stock level based on the inputs from the factory and distribution centers, ensuring that:

- There is no overstocking or under stocking at the Supplier point.
- The flow of goods from the factory through the distribution centers is optimized.

The dataset you've provided consists of stock levels at different points in the supply chain: **Factory, Distribution Center-1, Distribution Center-2**, and the **Supplier point**. The goal appears to be predicting the stock level at the **Supplier point** (around 200 units) based on the inputs from the other three stages.

Key Observations:

- The input variables are the stock levels at the Factory, Distribution Center-1, and Distribution Center-2.
- The target variable is the stock level at the Supplier point.
- The stock levels in the supply chain stages exhibit both positive and negative values, indicating overstock (positive) and shortage (negative) conditions.

Potential Approach Using Neural Networks:

You can apply an Artificial Neural Network (ANN) **or a** Neural Fit Algorithm to predict the target stock levels at the Supplier point based on the inputs. Here's how this can be structured:

1. Input Layer

There are 3 input nodes corresponding to the stock levels from:

- Factory
- Distribution Center-1
- Distribution Center-2

2. Hidden Layers:
    - Several hidden layers with activation functions (like ReLU) to capture the non-linear relationships between stock levels at different stages.
3. Output Layer:
    - 1 output node representing the predicted stock level at the Supplier point.
4. Loss Function:
    - Mean Squared Error (MSE) or Mean Absolute Error (MAE) can be used to minimize the difference between predicted and actual stock levels at the Supplier point.
5. Training:
    - Use **back propagation** to adjust weights based on the errors.

- Train the model on historical data to optimize the prediction of future stock levels.

Clustering with K-Means:

In parallel, you could use K-means clustering to group the stock levels across the Factory, Distribution Centers, and Supplier point to identify patterns in inventory distribution. This helps in identifying stages that consistently experience overstock or shortages. According to table data the supply point uses to deliver the material as per demand of customer. There are experimentally 94 data are collected but we have taken 28 types of observation due to repetition of stocks requirement. The target value takes as y output as supply and the inputs are x_1, x_2 & x_3, these inputs are showing inventories levels at different points where need to optimize the inventory levels with the neural network.

Experimental Result and Analysis: To implement the process of training a neural network in MATLAB using the data provided in **Table-1**, you can follow these steps. The objective is to train the network to predict the stock levels at the **Supplier point** based on the stock levels at the Factory, Distribution Center-1, and Distribution Center-2.

Steps for Implementing Neural Network in MATLAB:

1. **Prepare Data:**
   - o Use the data in Table-1 for training, where the inputs are the stock levels from the Factory, Distribution Center-1, and Distribution Center-2.
   - o The Supplier point stock level will be the target output.
2. **Define Neural Network Architecture**:
   - o Since there is no general theory for the best network architecture, you will need to experiment with different architectures.
   - o You can start with a feed forward neural network with a single hidden layer. The number of neurons can be chosen empirically.
   - o For example, use 10 neurons in the hidden layer initially.

Figure 1. Training Process in ANN

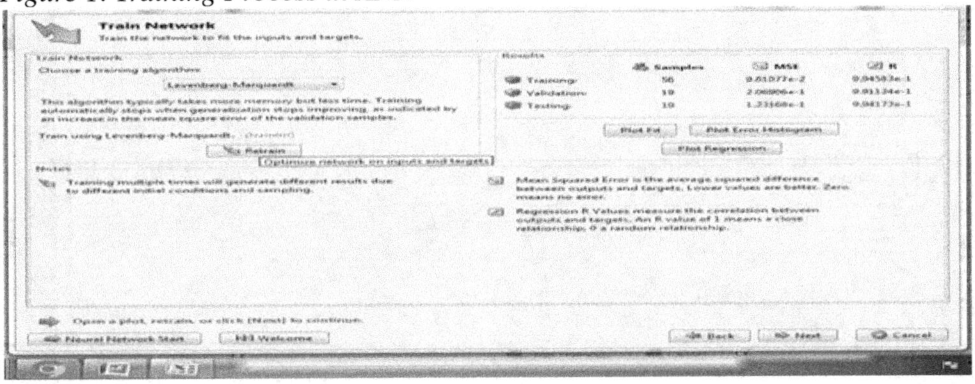

Figure 2. Fitting Function

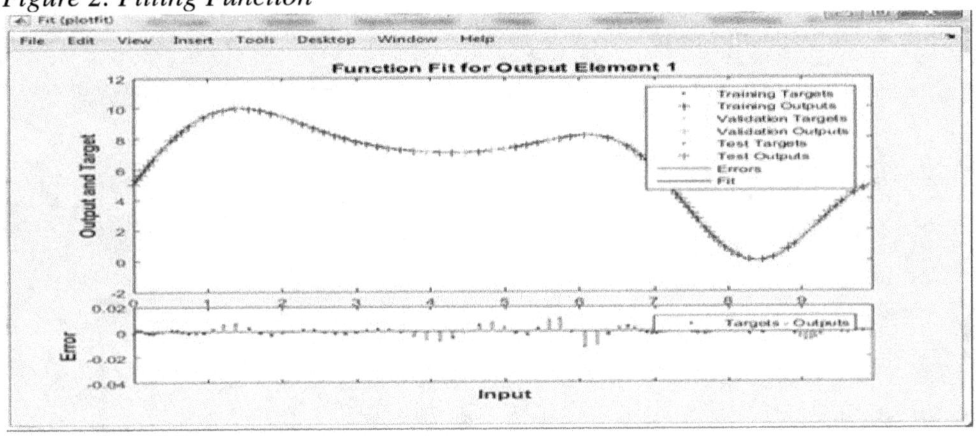

Figure 3. Neural Network Performance

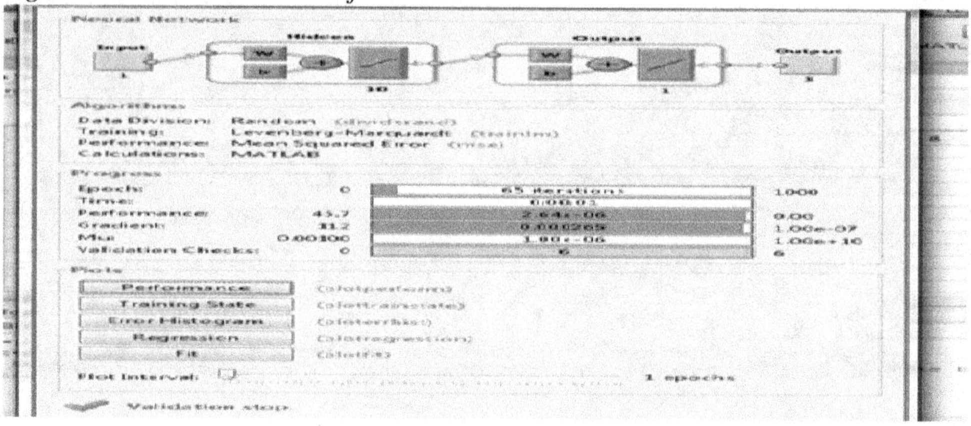

Figure 4. Output Validation

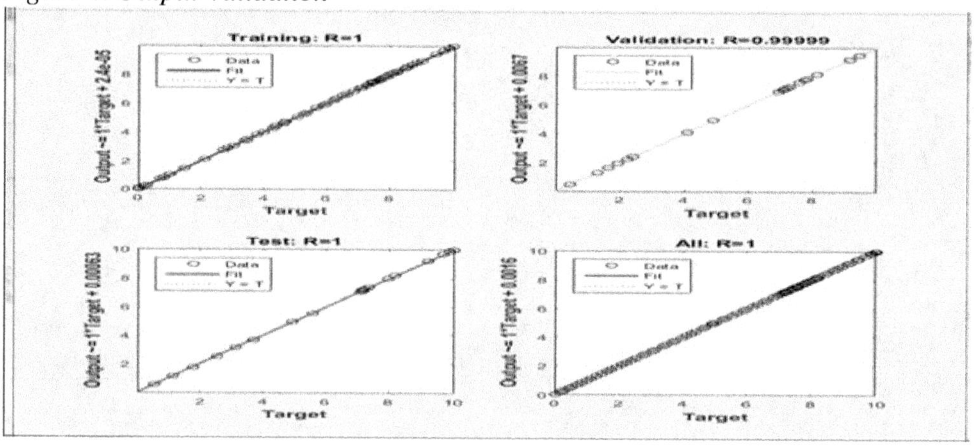

Figure 5. Error Histogram

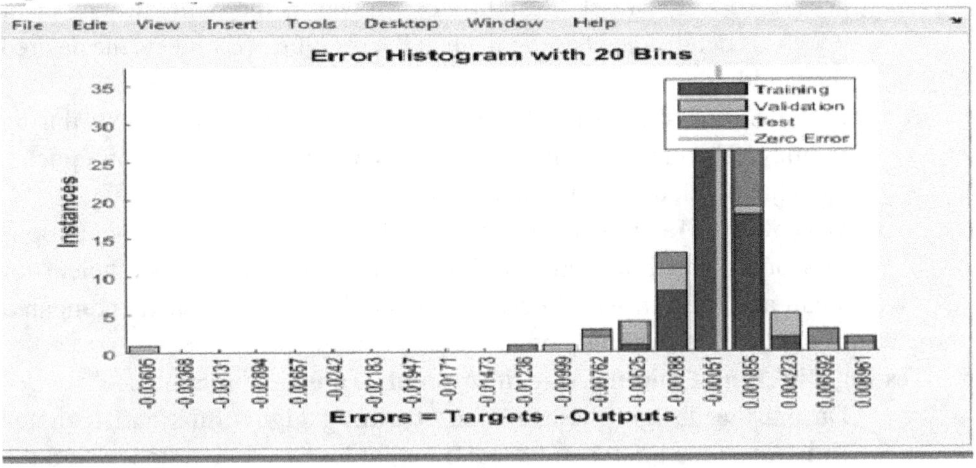

The explanation you've provided outlines a well-structured process for training and evaluating a neural network for inventory optimization using real industrial data. Here's a breakdown and further elaboration on the key points:

Key Aspects of the Neural Network Model:

1. **Data Processing and Input-Output Setup**:
    - **Input-output data**: Real industrial data can be used to feed the network, where inputs are stock levels from various points in the supply chain, and the output is the predicted stock level at the supplier point.
    - The network processes these input-output pairs to learn the relationships between different stages of the supply chain and the resulting inventory levels at the supplier point.
2. **Neurons and Activation Function**:
    - The neurons in the network transmit signals to adjacent layers using a **non-linear activation function**, which allows the network to learn complex patterns in the data.
    - Common activation functions include the **sigmoid**, **tanh**, and **ReLU (Rectified Linear Unit)**, though the best one depends on the problem and the data.
3. **Feed forward Neural Network**:
    - A **feed forward neural network** is used, meaning the connections between nodes do not form cycles. This type of architecture is typically effective for regression and prediction tasks like stock level forecasting.
4. **Training Process**:

- **Epochs**: The network is trained for **10,000 epochs**, meaning the model sees the entire dataset 10,000 times. However, the optimal number of epochs is determined based on when the training error meets the desired threshold.
- **Mean Squared Error (MSE)**: The goal is to minimize MSE during training. The error threshold of **MSE = 0.0001** was met at **65 epochs,** showing efficient learning in this case.
- **Levenberg-Marquardt Algorithm (LMA)**: This is a popular optimization algorithm used in training neural networks. It is efficient for small and medium-sized networks and can converge quickly compared to gradient descent methods.

5. **Testing Different Training Algorithms and Transfer Functions**:
   - During the design phase, various **training algorithms** and **transfer functions** (used in the hidden and output layers) were tested to find the best-performing network.
   - Transfer functions define how the summed inputs to a neuron are transformed before passing to the next layer.
6. **Performance Metrics**:
   - **Coefficient of Determination (R^2)**: A value of **0.9999** shows an almost perfect fit between the predicted and actual values, meaning the model explains nearly all the variability in the data.
   - **Mean Absolute Percentage Error (MAPE)**: A MAPE value of **0.0001** indicates a very low prediction error, showing that the neural network is highly accurate in forecasting stock levels.

Example MATLAB Code for Levenberg-Marquardt Training:
To implement this using MATLAB's trainlm (Levenberg-Marquardt) algorithm:

```
matlab

Copy code

% Input and target data from industrial real-world data

inputs = [-591 -329 269;
```

```
% (Other data entries go here)

297 342 686];

targets = [200; 210; % (Other target data here) 182];

% Create the feedforward neural network with 10 neurons in
the hidden layer

net = fitnet(10, 'trainlm'); % 'trainlm' specifies
Levenberg-Marquardt algorithm

% Set the training parameters

net.trainParam.epochs = 10000; % Maximum number of epochs

net.trainParam.goal = 0.0001; % Target mean squared error

% Divide the data into training, validation, and testing
sets

net.divideParam.trainRatio = 0.7;

net.divideParam.valRatio = 0.15;

net.divideParam.testRatio = 0.15;

% Train the network
```

```
[net, tr] = train(net, inputs', targets');

% Evaluate the performance

outputs = net(inputs');

performance = perform(net, targets', outputs);

% Calculate R2 (Coefficient of Determination)

SStot = sum((targets' - mean(targets')).^2);

SSres = sum((targets' - outputs).^2);

R2 = 1 - (SSres/SStot);

% Calculate MAPE

MAPE = mean(abs((targets' - outputs) ./ targets')) * 100;

% Display performance metrics

fprintf('MSE: %.4f\n', performance);

fprintf('R2: %.4f\n', R2);

fprintf('MAPE: %.4f%%\n', MAPE);
```

```
% Plot actual vs predicted values

figure;

plot(targets, 'o-', 'DisplayName', 'Actual Values');

hold on;

plot(outputs, 'x-', 'DisplayName', 'Predicted Values');

legend;

title('Neural Network Fitting: Actual vs Predicted Stock Levels');

xlabel('Sample');

ylabel('Stock Level at Supplier Point');
```

Explanation of the Code:

- **Network Creation**: The network is created using the fitnet function, specifying the **Levenberg-Marquardt** algorithm with 'trainlm'.
- **Training Parameters**: The network is trained with a goal of reaching an MSE of **0.0001**, and the maximum number of epochs is set to **10,000**.
- **R^2 and MAPE Calculation**: After training, the coefficient of determination (**R^2**) and mean absolute percentage error (**MAPE**) are calculated to evaluate how well the network performed in predicting the supplier stock levels.
- **Performance Evaluation**: The results are plotted to visualize the fit between actual and predicted values.

CONCLUSION

The neural network model, trained using the **Levenberg-Marquardt** algorithm, shows exceptional performance with minimal error. With real industrial data, this model can provide accurate predictions of stock levels, leading to better inventory control and cost reduction. The use of performance metrics like $\mathbf{R^2}$ and **MAPE** ensures that the model's accuracy is validated against actual data. The Artificial Neural Network (ANN) model described can be effectively applied for industrial inventory control of both finished goods and raw materials. By using an ANN, various costs associated with inventory—such as setup costs, material handling costs, and carrying costs—can be optimized.In this case, a **multilayer feedforward ANN** with **4 input features** and **10 hidden neurons** processes the data to predict stock levels or other relevant metrics. The model's output is compared to the target values, allowing for continuous learning and improvement. The results from this model have demonstrated its ability to provide better outcomes for inventory control, optimizing the balance between costs and the required stock levels. This approach ensures a more efficient inventory management system, which can ultimately lead to reduced operational costs and improved customer satisfaction.

Acknowledgement: The authors would like to thank the reviewers for their extensive comments and suggestions, which resulted in an article that is more straight forward and easier to read.

Conflicts: The Author(s) declare(s) that there is no conflict of interest.

REFERENCES

Eme, O., Ugboaja, C. A., Uwazuruike, F. O., & Ukpai, C. U. (2018). Computer-based drug sales and inventory control system and its applications in pharmaceutical stores. *I.J. Education and Management Engineering*, 30-39.

Guo, C., & Yang, X. (2011). A programming of genetic algorithm in Matlab 7.0. *Modern Applied Science*, •••, 230–235.

Haruna, C., Abdulkareem, S., & Hermawan, A. (2017). Neural networks optimization through genetic algorithm searches: A review. *Applied Mathematics & Information Sciences*, •••, 1543–1564.

Maind, S. B., & Wankar, P. (2014). Research paper on basic of artificial neural network. *International Journal on Recent and Innovation Trends in Computing and Communication*, *2*(1), 96–100.

Mhatre, M. S., Dongre, M., & Thakur, P. (2015). A review paper on artificial neural network: A prediction techniques. *International Journal of Scientific and Engineering Research*, •••, 6.

Paula, S. K., & Azeema, A. (2011). An artificial neural network model for optimization of finished goods inventory. *International Journal of Industrial Engineering Computations*, 431-438.

Peterson, C., & James, R. A. (1988). Neural networks and NP-complete optimization problems: A performance study on the graph bisection problem. *Complex Systems*, •••, 59–89.

Radhakrishnan, P., Prasad, V., & Gopalan, M. (2009). Optimizing inventory using genetic algorithm for efficient supply chain management. *Journal of Computational Science*, *5*(3), 233–241. DOI: 10.3844/jcssp.2009.233.241

Ramírez, A. P., & Labadie, N. (2017). Stochastic inventory control and distribution of blood products. *Industrial Engineering and Operations Management*, 420-430.

Strijbosch, L. W. G., & Moors, J. J. A. (1999). Simple expressions for safety factors in inventory control. *Tilburg. Ekonometria*, •••, 1–25.

. Šustrová, T. (2016). A suitable artificial intelligence model for inventory level optimization. *Trendy Ekonomiky a Managementu: Trends Economics and Management*, 48-55.

Vladimir, M., & Tasho, T. (2011). A programme implementation of several inventory control algorithms. *Bulgarian Academy of Sciences*, 64-78.

. White, A. S., & Censlive, M. (2016). Inventory control systems model for strategic capacity acquisition. *Journal of Industrial Engineering*, Article ID 1650863.

. Yadav, A. S., Swami, A., & Ahlawat, N. (2018). A green supply chain management of auto industry for inventory model with distribution centers using particle swarm optimization. *Selforganizology*, 10-18.

Chapter 12
Generative AI in Innovative and Eco-Friendly Packaging Solutions

Himanshu
https://orcid.org/0009-0007-5254-4383
Lovely Professional University, India

Vikanksha
https://orcid.org/0009-0009-0257-4319
Lovely Professional University, India

Arun Kumar
Lovely Professional University, India

Deepika Saxena
Lovely Professional University, India

Jatinder Singh
https://orcid.org/0000-0001-7691-6689
Lovely Professional University, India

ABSTRACT

This chapter examines how generative AI is being used to solve innovative packaging solutions that minimize or reduce their impact on the environment. The concept here is to show how generative AI improves the packaging design by assessing sustainability, durability, and material utilization, which in turn enables the design and creation of smarter sustainable substitutes. AI algorithms allow the designers

DOI: 10.4018/979-8-3693-9856-2.ch012

to run several hundreds of permutations across the design canvas, thus achieving packaging that is made from recycled or biodegradable material and a net total carbon footprint. The chapter also looks at how generative AI improves efficiency in package creation, from its design to its manufacture. Therefore, through the detailed chapter analysis, we showcase the appropriate use of packaging from the AI perspective and explore trends by considering generative AI as a critical tool that will help in achieving environmental sustainability goals as well as addressing global packaging challenges.

1. INTRODUCTION

1.1 Overview of Generative AI

AI can be used in sustainable packaging through generative AI, which is helping in reducing the effects on the environment. Another application area of packaging, which can be optimized by generative AI to cut down material utilization, eradicate wastage, and enhance the package's recyclability. To evaluate the effectiveness of the solutions, it enables the modeling of various packaging schemes and the design of packaging, from environmentally friendly or biodegradable materials. AI can also identify specific packaging waste streams, promote the circular economy, and improve the efficiency of recycling processes. Customer awareness of environment-friendly products, and the rate of environmental degradation has increased, hence the call for change by the packaging industry. In some cases, the sustainability of the package design is a compromise between maintaining utility and reducing the negative effect on the environment. Its objective is to minimize environmental impacts by getting the most out of resources and designing materials. Due to their high rate of resource use and generation of waste, the wood packaging and wood board industries are the focal point of the global shift towards sustainability (Li Lixu *et al.* 2024). These businesses have long depended on processes that are very damaging to the environment by using a great deal of virgin wood and generating a great deal of waste. Possibly mitigating these impacts, digital technologies present an actionable strategy for redesigning and remanufacturing wooden goods. The wood board and packaging industries are under pressure to reduce their environmental footprint as the demand for green packaging solutions rises. These industries, which are core in construction, furniture production, and logistics transportation, have been the main culprits in deforestation and wastage. The next important strategy in the design of sustainable packaging is known as Design for Disassembly. This means identifying elements and design approaches that facilitate assembly and disassembly and, on the other hand, promote the reuse and extension of the material's lifespan, as well as making recycling processes easier (Paprocki, 2024).

1.2 The Role of Packaging in Environmental Sustainability

Consumers interact frequently with packaging materials when buying groceries from supermarkets, eating commodities from restaurants, or purchasing groceries from supermarkets. Goods or contents must be protected during handling, transportation, and storage, and this is where packaging comes in. It also has the advantage of aiding with the preservation and elongation of the product's shelf life. As a part of the defense against pollution and the upkeep of the product's integrity, packaging is vital for shelving and transit. Similarly, food packaging material assures that the commodities are well protected from adverse influences or destructive factors. Sometimes, packaging also serves as an informational tool containing instructions on product use; at other times, packaging is used in selling products on promotion (Manzoor *et al.* 2023). From a perspective of product safety, storage, and hygienic mapping, packaging cannot be overlooked. For example, packaging in the food industry protects against contamination and destruction by toxins, gases, moisture, and bacteria. The following is the list of the materials used in packaging: paper, board, plastic, metal, glass, and wood. Wood and glass packaging are said to have been in use probably as early as 5,000 years. The earliest packaging materials include leaves, and many of these were derived from natural resources without necessarily being processed. Both baskets and mats were used later, as well as woven fabrics and pots. Paper and cardboard have been constantly on the rise since 1900, while plastic packaging became much more popular in the second half of the twentieth century. Conventional cans were not used until the twentieth century. However, metal packaging was used from the early nineteenth century. The current position is that paper and board packaging is equivalent to about 40% of conventional cans, while plastic is only 30% of the market. However, packaging has various uses. Its role in product protection is vital in the food industry because packaging assists in stopping food products from catching diseases, discoloration, deformed looks, or bleaching due to oxidation, dust, germs, bacteria, and sunlight effects. It also contributes to minimizing waste in the supply systems in the food industry chain (Elkhattat & Medhat 2022).

1.3 Purpose of Exploring AI-Driven Eco-Friendly Packaging

The focus of the study of AI-based environmentally sustainable packaging is to avow permanent answers to the effects on the environment and enhance the efficiency of the resultant packaging. AI offers innovative approaches to designing sustainable and functional packaging as the concerns regarding plastics, emissions, and deforestation. Techniques such as generative design and machine learning offer the likelihood of biomimicry packages that are entirely recyclable or biodegradable, minimal use of raw materials, and relatively low levels of waste. Further, these techniques help in improving the packaging design to enhance transportation efficiency and reduce carbon emulation. AI-enhanced material science enables researching environmentally friendly substitute materials for hazardous plastics, which would otherwise be impossible. Also, materials recognition and optimization of sorting mechanisms make AI-based packaging solutions contribute to recycling. Therefore, when investigating AI for environmentally friendly packaging, enterprises can benefit from corporate sustainable development goals, regulatory requirements, and consumer trends' genuine desire for greener products (Kumari & Pandey 2023).

2. THE BASICS OF GENERATIVE AI IN DESIGN

The ability of computational methods to generate text, images, or audio of different topics and coherent and relevant sentences, pictures, and music from training data is called generative AI. Many of our current means of processing and interaction with each other are currently undergoing a shift due to this technology in view: Copilot, GPT-4, and Dall-E2. AI technology has also become common in various design industries since generative AI increases creativity, productivity, and innovation in the design process. While generative AI gives design solutions that are contained in the parameters presented to the algorithm, this is quite different from typical approaches in design where designs are frequently copied and pasted or require user input. Such change has a particular impact on industries such as product design, architecture, fashion, and packaging. One of the most significant ways that generative AI is currently changing design is in terms of speed and the number of design options. Designers themselves may input specific goals like material use, structural strength, or appearance, and the AI will spit out multiple possible solutions while at the same time satisfying sustainability goals. This feature accelerates the ideation phase specifically, which eradicates time spent on errant exploration while allowing the designer to explore rather innovative possibilities that would not be possible if the assembling was performed manually. They also make it easier to create apparatus designs, thus enhancing personalization as a benefit of generative AI. From consumer products to fashion and everyone in between, those using AI technologies can make this kind of item or the other based on certain standards of demand or size, and they do not even have to involve much human involvement. This flexibility saves costs whilst opening up new avenues for mass customization (Sengar *et al.* 2024). Generative AI can also optimize the minimization of waste, energy, and consumption of materials as well. For instance, in sustainable design, AI can assess materials for use and suggest more efficient, lightweight, and environmentally friendly products that use similar properties. Waste can be minimized in the aspects of packaging and architecture through investigative design that involves suggesting designs that retain efficiency or structural soundness while only requiring the smallest quantity of resources. Using generative AI, this task simplifies the repetitive chores designers may otherwise spend their time on, thus allowing them to focus on higher levels of creative tasks. Overall, it modifies the manner in which design is done in the twenty-first century by enhancing the process and transforming it into something faster as well as innovative (Ghobakhloo *et al.* 2024).

2.1 Applications of Generative AI in Various Industries

Depending on the task, there are potential uses for generative AI, like ChatGPT, in various fields (as shown in Fig. 1) of business, education, healthcare, content creation, etc (Fui-Hoon Nah *et al.* 2023). In this section, we look at instances of generative AI applications in some of the various sectors as discussed in Table 1.

Table 1. Role of Generative AI in Diverse Industries

Type of Industry	Role	References
Manufacturing and Product Design	In manufacturing activities, generative AI can be used to enhance various manufacturing processes and products. Among the applications, it can be used for quality assurance, determining layouts for the models of assembly lines, and creating new product designs.	Rane, (2023a)
Education	In this respect, generative AI can contribute to the educational context through interactions and the creation of such things as simulations, virtual tutors, and/or tailored materials for students. That way, learning experiences can be designed to be sensitive to the needs of each learner.	Pesovski *et al.* (2024)
Healthcare	Eventually, with the assistance of LLMs, generative AI like ChatGPT can actually transform the healthcare system in various perspectives, such as Patient interaction, Diagnosis assistance, Telemedicine services, Health enhancement, Health Literacy, and Health advising.	Eysenbach, (2023)
Content Creation	It also appears that generative AI has influenced the gaming business. Since its release a few months ago, people have been using ChatGPT to create games based on text. Now, the idea of coming up with fighting game characters and character strategies has also employed AI without the help of a human creator.	Yao *et al.* (2024)
Art and Creative Industries	In a way, generative AI can assist painters, designers, and anyone creative in mind and heart by creating new and motivating content. They include designing virtual environments, writing music compositions, making art, and innovating aesthetics.	Shi *et al.* (2023)

continued on following page

Table 1. Continued

Type of Industry	Role	References
Environmental science	As for the benefits, generative AI can contribute to environmental studies by developing situations and models for studying ecosystem states, pollution control approaches, and climatic change. As a result, it can predict natural calamities and give accurate information on the prevailing weather conditions.	Rane *et al.* (2024b)
Financial Services	The industry that would likely use generative AI is the financial industry for model establishment, market change prediction, and investment strategies. At the same time, it transforms into an effective tool for risk identification and fraud prevention.	Gao *et al.* (2024)
Advertising and marketing	Some of the ways that generative AI can benefit marketers include improving campaign strategies, developing advertisements suitable to the target market, and even creating content suitable for niche markets. It may be used in the construction of advertisements, mottos, and pictures that portray the relative products.	Rodríguez Guzmán, (2024)

Figure 1. Applications of Artificial Intelligence across Key Industries

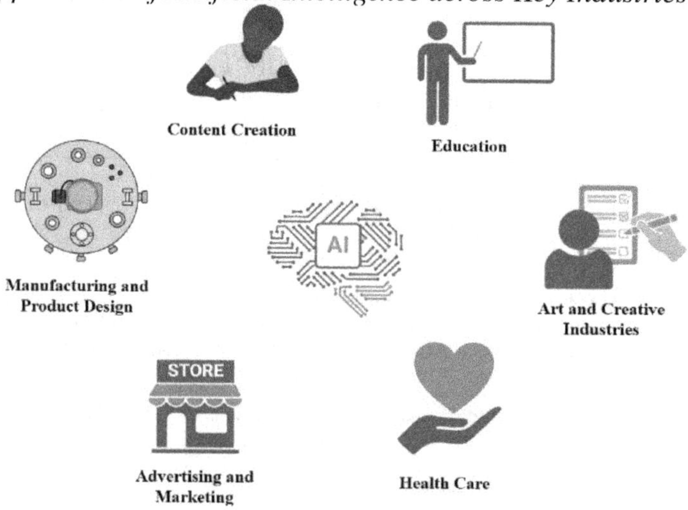

3. INNOVATIONS IN ECO-FRIENDLY PACKAGING USING GENERATIVE AI

3.1 Design Optimization: Reducing Material Waste

Generative AI enables efficient design processes that put sustainability and material use at their core which is transforming the packaging industry. This technology seeks to refine the distribution of packaging formats in a bid to offset optimal raw materials densities, form, and design while ensuring the packaging's form and design are as optimized as is structurally and aesthetically feasible using complex algorithms, and data analysis. Generative AI achieves material efficiency through a concept referred to as lightweight. Computational methods driven by artificial intelligence learn about the structure and suggest modifications to designs for the packaging material while at the same time decreasing material use without compromising the safety and sturdiness of the material. This reduces the overall effects on the environment, consoling that products do not get damaged during transport. The selection and simulation of material are other areas of its application, which is useful in choosing materials for different uses. Several estimates of environmentally friendly materials' characteristics are evaluated by generative AI: biopolymers derived from plants, paper, and biodegradable plastics. The designers certainly mold the adjusting best from these materials from various conditions, and it helps them balance sustainability-targeted solutions for durability and affordability. Shape and structure optimization also benefit from generative AI. It suggests new creative designs that save material and make structures much more efficient. It can design items that fold or interlock together, generally not bonding or requiring extra parts. These are actual circular economy models in that they simplify recycling and conserve resources (Lo *et al.* 2024). Generative AI also supports real-time design iterations. It greatly reduces the need for physical samples by enabling the designer to virtually realize numerous prototypes through processing enormous volumes of data. This eliminates many of the wasteful materials inherent in traditional trial-and-error procedures and speeds up the entire process. It aids the designers in optimizing the benefits neither by durability nor affordability but rather by helping it to meet sustainability-targeted solutions when modeling under these various conditions. Another thing that is very well discussed is form and structure optimization. That gives rise to new creative designs that add efficiency and decrease the amount of material used. It can design things that fold or interlock together without the need for glue or additional parts. Such methods contribute to a circular economy as they simplify recycling processes and conserve resources. Moreover, it enables iterations of a design in real time. It significantly decreases the need for physical samples by enabling the designer to virtually realize numerous prototypes by processing enormous volumes of data. This eliminates the majority of the wasteful materials associated with conventional trial-and-error methodologies and has the effect of speeding up the entire process. Moreover, the application of generative AI is sustainable for the supply chain since it deals with the optimization of transport packaging. Lighter and more compact designs consume less fuel while being shipped, thus reducing greenhouse gas emissions. In this way, enterprises can strike an equilibrium between environmental responsibility, financial effectiveness, and functionality by adopting commercial creative-generative AI into their packaging design. The innovative paradigm not only reduces material wastage but also contributes to global sustainability objectives, making it increasingly possible for companies to react to customer demands for environmentally friendly products while minimizing their negative environmental footprint (Cheng *et al.* 2024).

3.2 Exploring New Sustainable Materials (Biodegradable, Recyclable)

One of the most promising opportunities for ensuring sustainability is in eco-friendly packaging. Although packaging is considered by many to be a socially and politically sensitive issue, there is very little knowledge about how, exactly, consumers view green packaging at this moment. In consumer behavior research, eco-friendly packaging has never been a very clear idea. Because variously have been used, such as eco-friendly packaging, eco-packaging, ecological packaging, green packaging, sustainable packaging, eco-design, design for the environment, and environmentally conscious design, as well as many different people create confusion in the research. In fact, sustainable packaging is more or less the same as eco-friendly packaging. Numerous campaigns have been propelled to sensitize the idea of green packaging in the business sector. Sustainable alternatives have mushroomed owing to the ever-increasing environmental effects of traditional materials, especially non-biodegradable plastics. For instance, two innovative directions aimed at solving a waste management problem and thus contributing to environmental preservation include biodegradable and recyclable materials. The above elements rely on the concept of a circular economy, which cuts waste-use optimization and ecological damage prevention (Dörnyei *et al.* 2023).

3.2.1 *Biodegradable Materials*

Biodegradable materials are made from bio-renewable raw materials such as proteins, cellulose, starch, and plant-based polymers. Under specific environmental conditions, these materials can degrade through microbial activity into harmless materials such as water, carbon dioxide, and biomass. These materials can be evaluated and recommended by AI, such as polylactic acid (PLA), polyhydroxyalkanoates (PHA), or mycelium-based composites for biodegradable packaging, with the required durability and degradation requirements are discussed in Table 2 (Jain *et al.* 2024).

Polylactic acid (PLA): Polyester derived from renewable feedstock, polylactic acid (PLA), is reusable and biodegradable. Lactic acid, the raw ingredient in PLA, is made by fermenting sucrose or glucose. Engineering plastics, textiles, and packaging materials can be made with PLA (Kunduru *et al.* 2023). Brightplus is a material development business based in Finland. Brightplus makes Brighton materials for extrusion, injection molding, and 3D printing. It is used to create BrightBio materials, which are then altered using BrightBio Technology. BrightBio Filament, BrightBio Tough Filament, and BrightBio Heat Filament are the three distinct BrightBio materials. They differ in several regards about their technological properties. The other bio substitute is BrightBio LOIMU, which is comprised of renewable monomers and polymers found in nature. It is meant for injection molding but may also be 3D printed. The range of uses between the seven different LOIMU products spans from semi-rigid to elastic (Kuusisto, 2024).

Polyhydroxyalkanoates (PHA): PHA signifies natural polyesters derived from the fermentation of bacteria from feedstocks, including sugar, glucose, or vegetable oil. It is among the recent and widely popular biodegradable polymers for food packaging. Under different environmental conditions, PHB is degraded by several bacteria, fungi, and algae. Hydrolytic degradation produces 3-hydroxybutyric acid at a low rate. Propionic acid was included as a comonomer for the production of polyhydroxybutyrate-valerate (PHBV). It has been used for packing as it is less firm and more durable. It decomposes in 5-6 weeks in environments highly associated with microbial activities, and the products entailed are CO_2 and H_2O when under aerobic conditions. With methane as a by-product in anaerobic conditions, the decaying process is accelerated (Vicente *et al.* 2023).

Mycelium-based Packaging: Fungi mycelium is a biodegradable packaging that is ecologically viable and, at best, innovative. It is starting to generate a buzz from the food industry. The term mycelium refers to the underground root structure of a fungus, which can easily be molded into many forms of products, including packaging. The mycelium of a fungus forms several branches above and below to give it a root-like appearance. The colonies of fungi, referred to as Shiro masses, are usually found on mycelia-based decompositions on the earth's surface or even beneath it. This vigorous fiber structure can be used to create many green materials for clothing, food, real estate, and packaging. This rugged material is very much like leather and plant-based meat, which can be consumed instead of plastics (Aiduang *et al.* 2024). Mycelium foam is a very useful packaging material, as mycelia can be sculpted into many forms. This material is expected to use mycelium packaging more frequently in the coming years, especially in the food industry, owing to many factors, such as a growing interest among consumers in environmentally friendly packaging and increased environmental consciousness. This mycelium gives light, strong, rigid, and easily moldable foamed materials that are biodegradable. Mycelium does not survive as much as foams that are useful and eco-friendly, such as Styrofoam or polystyrene, which serve better purposes for the environment. Organic materials can also be viewed as less durable than polystyrene foam or polyurethane; however, mycelium foams have been proven to be stronger (Madusanka *et al.* 2024).

Table 2. Packaging Properties of Biodegradable Materials for Sustainable Applications

Biodegradable material	Source	Key properties	Applications	Limitations
Polylactic acid (PLA)	Cassava, Sugar beet pulp, sugarcane	PLA is relatively ecofriendly and produces less greenhouse gases than other polymers in the production and during use. It has both the antifungal and antibacterial properties.	Orthopaedics, clothing, food packaging, Disposable cutlery	Its shelf life, strength and its rate of deterioration has been rated as below average. It is rigid and brittle at room temperature
Polyhydroxyalkanoates (PHA)	Diary waste, Vegetable feedstocks, Glycerol	It is biocompatible, film forming ability. It is resistance to UV rays. They are non-toxic, good tensile strength and insoluble in water.	Used in foam packaging, disposable gloves. Used in wastewater treatment Thus, PHAs can be used to be shrink wrap, carrier and gift bags, compost bin liners and trash sacks.	Poor mechanical strength and high cost. They have poor gas barrier properties.
Mycelium-based Packaging	Agricultural waste and fungal mycelium	It is lightweight, fire resistant and antibacterial properties. They are 100% biodegradable and compostable.	Protective packaging, cushioning material	It is not compatible with food products. It has high water absorbable capacity and high production time.
Algae-derived Biopolymers	Algal biomass	Flexible, biodegradable and water soluble. Good oxygen and barrier to UV light.	Cosmetics, fuel cells, pharmaceuticals, single use packaging and edible films.	Low tensile strength, limited durability for heavy loads

Algae-derived Biopolymers: Algal biopolymers are increasingly being developed as more environmentally friendly and sustainable options for packaging materials. Such biopolymers include natural algal sources like seaweed and microalgae, so they are renewable and biodegradable, like real substitutes for petroleum plastics. The main ingredients taken from algae, such as agar, carrageenan, and alginate, are used to produce films, coatings, and hard packaging solutions (Kajla *et al.* 2024). One of the many advantages of biopolymers from algae is their lesser environmental impact. Algae, with the potential for rapid growth and minimal requirements, along with the absorption of carbon dioxide during cultivation, make excellent feedstock in sustainable production. Further, the degradation property of these biopolymers guarantees their disposal into non-toxic components, thereby limiting pollution originating from plastics. Biopolymer potential from algae is yet to overcome the challenges of scale-up and cost competitiveness with commercial grades of plastics. In order to overcome these hurdles and guarantee that algae-derived biopolymers are an intrinsic part of the transition to sustainable packaging choices, significantly more research and technological advancement are required (Khandeparkar *et al.* 2024).

3.2.2 *Recyclable Materials*

Metals. Food Packaging Metals Fall into Different Categories, Such as Aluminum, Tin (as shown in Fig. 2) Plate, Stainless Steel, and Tin-free Steel, as well as all Rigid and Semi-rigid Metal-based Packaging Forms Like Cans, Foils, and Retort Pouches. While North America and Asia-Pacific are projected to have the greatest promising developments in the market over the years owing to the increasing demand for fruit and vegetable juices, carbonated soft drinks, and healthy beverages, the other regions appear no different (Balo & Sua, 2023.) The metal packaging sector affects the environment and human health in positive as well as negative ways. Metal packaging is an excellent performer as the ideal barrier to light, gas, and moisture; recyclable, easily shapable, heat resistant even at high temperatures, very sturdy and able to be transported over large distances, and with great decorating options. But all the pros come at comparatively higher costs. Among the disadvantages include depletion of resources, release of harmful chemicals from containers into food, and global warming caused by carbon dioxide emissions from metal manufacturing factories (Chen *et al.* 2020).

***Tin Plates*:** Tinplate is the most important coated steel used for food packaging applications. It is usually hot-dipped double tin-coated on both sides of base steel. Economical and heavier than aluminum, the tin plate is recyclable, easily separable due to its magnetic property, easily printable, and impervious to gases and moisture. Further, it is temperature resistant, which makes it ideal for sterilized products such as beverages for extended storage. However, because they may react with food due to contact, such containers usually require an opener to reach the product. It does require surface coatings (Akram *et al.* 2023).

***Aluminium*:** As aluminum is the most abundant metallic element by volume in the earth's crust, constituting nearly 8.8 percent of it, alumina is converted to aluminum hydroxide (Al $(OH)_3$) by treatment with sodium hydroxide solution at a temperature of 175 °C, as a process stage in the aluminum manufacturing process. The soluble Al $(OH)_3$, when filtered through the following insoluble, precipitates as a fluffy white solid, which is seen with unscrupulousness. Aluminum, being costly as compared to almost every coated steel, is usually preferred for seamless containers because it doesn't get welded. Most of its pure forms are consumed in lightweight packaging for pet food, soft drink cans, and seafood, but they gain strength from the addition of manganese. It forms foil, cans, and packaging materials with laminated and metalized features, along with paper and plastics. Another great application of aluminum in food packaging that is subsequently discussed is retort pouches, bottles, caps, closures, collapsible tubes, and laminated and metalized films besides these. Because it is one of the most favorable metals for recycling as it can easily be converted for new products, aluminum is preferred; however, recovered aluminum foils typically possess pinholes, and their lack of magnetism creates problems of reserves separation (Al-Alimi *et al.* 2024).

Figure 2. Recyclable Materials for Sustainable Packaging

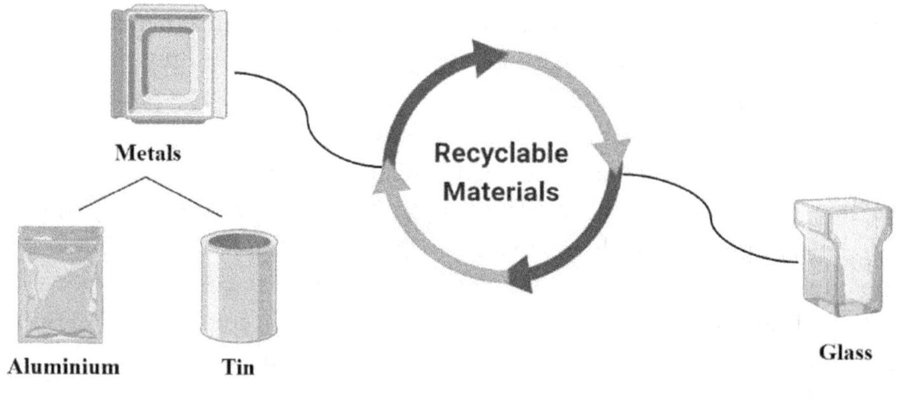

Glass packaging: Such a common packaging form for different cuisines: liquid goods, powdered goods, heat-treated foods, and goods packed under pressure. Glass is usually used universally as a kind of amorphous (non-crystalline) solid without any shape or particular use concerning technology or decoration. Glass does not have free, mobile electrons that can absorb light energy, while metals do, so it remains translucent and partially transparent Nevertheless, it is possible to develop colored spectacles for selective transmission of light. Soda-lime glass is considered the most-used type of glass for packaging. Glass containers are produced by melting a mixture of silica (sand), calcium carbonate, and sodium carbonate (the melting agent) with alumina (hardener), charging into molds, and allowing them to cool. Consumers appreciate many benefits of glass packaging, especially in the food category. In a comparative study of the interactions of packaging materials with foodstuffs, it was found that glass packaging was superior to other materials in terms of food quality preservation and that there was no transfer of toxic substances from package to food (Delbari & Hof, 2024).

4. ENVIRONMENTAL IMPACT AND SUSTAINABILITY

The simplistic agenda of AI for Sustainability (ADA 2030 themes like SDGs) is clearly no longer adequate; it's time for Sustainable AI. Sustainable AI, therefore, does not try to promote only AI programs but the entire sociotechnical system that comprises AI. AI will enable societies to uphold social values that are crucial for a given society, economic models for societies, and natural resources for the current and future generations. Examples include reduction in the utilization of processing power and carbon emissions. For sustainable AI, sustainable development is at the core of its description with three tensions: innovation of AI and equitable access to resources; generational conflict and balance; and conflict between environment, society, and the economy (Jungwirth & Haluza, 2023). The technologies at the center of this change are sustainable materials. They offer a way to start reducing the impact of products on the environment, from the extraction of raw materials for production to usage and disposal. The analysis of sustainable resources and their employments has assumed a strategic aspect of a comprehensive sustainability program because of the worsening quandaries of industries in lessening the footprint of their activities. Thus, the shift to sustainable manufacturing is not only a question of rationalizing protocols and reducing waste (as shown in Fig. 3) but also choosing materials when producing goods. Sustainable material application brings no waste, no reliance on finite resources, and an overall net carbon footprint reduction. Sustainable materials are those materials that should, through their life and production process, have the least impact on the environment. Compared to traditional materials, such materials are often derived from renewable sources, and when they are discarded, they are reusable or biodegradable and require less energy and water to produce. There are several types of eco-friendly materials: It may become possible to replace current products made out of petroleum-based plastic materials, which are notorious for being a nuisance to our environment by polluting and taking thousands of years to decompose in the so-called landfills, with biodegradable plastics. There are those substances that may disintegrate in the environment without negatively affecting the environment. These are among the described materials, including bioplastics based on sugarcane, cellulose, and maize flour. Biodegradable material is slowly being incorporated into packaging, consumable items, and even auto parts. Applying recycled material reduces energy use, the requirement for new material, and associated greenhouse gas – GHG emissions from mining and processing raw materials (Bhatia *et al.* 2023).

Figure 3. Goals of sustainability for green management

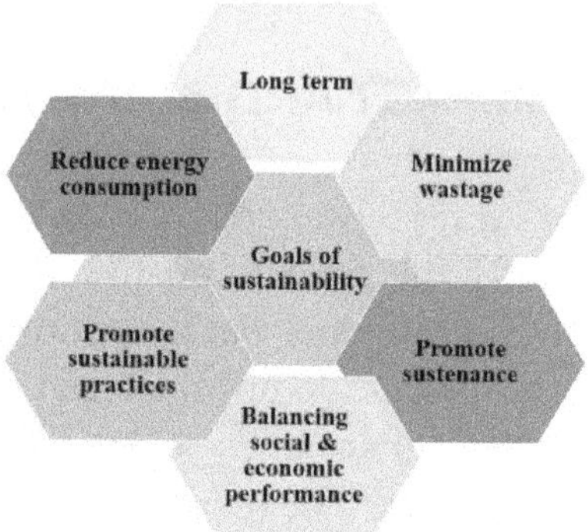

4.1 Reducing Carbon Footprint with AI-Optimized Packaging

About 27 to 50% of food produced is wasted in industrialized countries, and the loss has been drastic and tactful economically and environmentally. In the recent past, AI-optimized packaging has been found to be a viable technology that may help extend the shelf life of several foods. However, as with all medical procedures, it demands additional input of resources and yields extra emissions and wastes. A low energy, low pollution, and low emission economy can be unsettling, and this is called a low-carbon economy (Singh *et al.* 2024). China continues to be the source of 28 percent of global carbon dioxide. In terms of the global problem of climate change, China has three key objectives: it projected that the carbon dioxide emissions will be at their highest around 2030 before declining to 60-65 percent of the GDP in 2030 compared to the 2005 level, and using approximately 20 percent of the primary energy in the same year. Here, the distribution of terminal energy consumption and carbon emissions of China's major sectors is analyzed. The packaging industry's carbon emissions are something similar to climatic change. Rapid growth in urban areas has severely impacted the social environment on which society relies, and this has likely ushered in an ecological disaster on Earth. There is no doubt that the concept of a so-called "low carbon economy" is the most suitable strategy for the general regulation of energy consumption, pollution, and emissions. Thus, packaging design has elements that include product protection, printed markings, perceived value, use, and many more. one of the most important revolutions in human history, right next to the agricultural and industrial revolutions. Besides, it is the only long-term sustainable economic development model for China (Albijanić *et al.* 2023). Economies are evolving; so are cultures and, indeed, communities. Sometimes, the packaging may be innovative, and due to the improved technology in packaging, everything even looks beautiful. The study of green consumerism demonstrates explicit awareness of how the cabin has changed due to the prevailing harsh environmental conditions. The green movement is on the rise; the cultural change towards being environmental friendly is underway. Thus, this new lifestyle implies a change in people's basic essential needs. The idea of ecological and low-carbon food packaging construction guides the growth and transformation of practices in line with the customer's environmental psychology. As the primary benchmark for packaging that relates to design activities, it applies the green conceptual design of today's eco-environmental conservation and other relevant reforms. We must combine people's developed green psychology with this issue to search for a solution to green packaging design (Lin Yi-Chieh, 2024).

4.2 Aligning Packaging Solutions with Circular Economy Goals

The circular economy (CE) refers to a mechanism that minimizes waste generation and waste processing by not decreasing the amount of utilization of commodities, materials, or other inputs. Much has been written on the concept of the CE, and even though there are many ways to define the concept, most scholars will agree that the CE entails reduction, reuse, and recycling. According to the current literature, the circular economy is a system that is closed because it uses resources as many times as possible and produces no waste at all. It is hoped that this system will be restorative or regenerative; in other words, materials that are seen as waste in the linear economy model should be the inputs for other forms of industrial loops or as regenerative for the ecosystem. The advantages of the type of system described above are numerous (Dzhengiz *et al.* 2023). Data shows that the implementation of a circular economy might cut carbon emissions by 70% and create millions of jobs. CE could be an advantage if adopted by organizations because it corrects the problem of scarcity of resources, and its approach tries to maximize the value that can be obtained from the inputs throughout the industrial life cycle. The facet of a circular economy known as reverse logistics is the act of moving goods from consumers back to manufacturers. This method is recognized as a component of a closed-loop supply chain; without reverse logistics, product flow ends up with consumers disposing of products in landfills. Thus, reverse logistics is closely related to the creation of a closed-loop system. Managers will be much needed to ensure goals are achieved, and they will contribute to minimizing environmental uncertainty, encouraging new ideas, and being in charge of resources. Research has identified some of the enterprising circulating protagonists who contribute massively to the growth of the circular economy, hence extending the new enterprises that have circular economy business models to the title of ‘circular start-ups’ (Mishra *et al.* 2023).

4.3 Addressing Global Environmental Challenges (Waste, Pollution)

Disposal of wastes has increased because of urbanization, population increment, and shift in consumer behavior. Municipal Solid Waste (MSW), Industrial, and hazardous wastes contaminate the country's land, air, and water resources. Poor waste management has negative impacts on the environment, such as pollution of groundwater sources, emission of Greenhouse gases from disposal sites, and negative impacts on marine life. In order to overcome these challenges, the application of a circular economy is necessary. This strategy concerns the identification of activities regarding waste minimization at the source and subsequent use or recycling of materials for the manufacturing of the products. AI and machine learning can improve the efficiency of the segregation of waste, which makes the process of recycling more effective. Also, anaerobic digestion means the use of organic waste in the formation of biogas or compost to tackle landfill reliance on energy sources or fertilizers. Governments and industries must also work together to implement Extender Producer Responsibility (EP schemes in which the manufacturers remain wholly responsible for all aspects of their products and packaging waste. The government, through ads and promoting environmental programs, can motivate people to use fewer plastic products, get rid of littering, and recycle more (Dokl *et al.* 2024).

Air and water pollution, SO_2, NO_2, water pollution, and noise pollution directly harm the environment, ecosystems, climate, and human well-being. Pollution by airborne particles such as industrial pollutants, vehicle emissions, and deforestation leads to respiratory illnesses and global warming. The sustenance of untreated industrial effluents, agricultural drainage, and plastic wastes is poisoning water sources and threatening aquatic life and human water needs. These include soil pollution, which results from the use of chemical fertilizers and pesticides, which depletes the soil nutrients, and food crops contaminated, making food security a big issue. For one to effectively allow clean energy that does not emit harmful air chemistry, it is imperative to shift from fossil fuel energy such as solar, wind, and hydropower. Governments can push for lower emission levels and support tree-growing efforts to remove the effects on carbon balances. Reduction or outright prohibition on the use of single-use plastic and advocacy for the use of biodegradable plastics reduce the levels of plastics in water sources. The examples include the incorporation of organic farming techniques, crop rotation, and bioremediation as measures to enhance the quality of the soil and to control contaminant levels (Rani *et al.* 2024).

5. CHALLENGES AND LIMITATIONS

AI may be useful in creating environmentally friendly packaging designs and recycling packaging waste. Several challenges must be addressed to realize the AI-based system's potential, as data quality and availability are ranked among the most significant ones. In addition to the ethical and privacy issues at stake, there are concerns in relation to the accountability and transparency of the algorithms. Thus, maintaining transparency and Overcoming the above problems to ensure the right set of governance and oversight in any AI-based system development and its utilization is very important (Lodhi *et al.* 2024).

5.1 Technical Challenges in Implementing AI for Packaging

Sustainable materials, especially advanced ones, are costly compared to traditional ones, and some firms might experience difficulty when using the former. However, as production and demand increase, the prices should drop, leading to an even higher appeal for sustainable materials. Even if there appear to be advantages, its implementation in production is a challenge. To overcome these challenges and make sustainable materials that fit the high standards of many industries, research and development are vital. One of them is the question of how sustainable materials affect product sturdiness and functionality. Insert Sustainable substitutes may not always have the efficiency of conventional materials, particularly in harsh environments such as car manufacturing or compelled aerospace industries (Yang *et al.* 2024). For some applications, sustainability may require additional processing or new approaches for acquisition and supply compared to regular materials. This shows that manufacturers need to go for new technologies and improved techniques and also look for new partnerships for the application of sustainable materials on their manufacturing lines. Regarding sourcing and supply relations, more logistics issues have also been observed. With the increase in global concern about the repercussions of humanity on the environment, it has become fairly possible for manufacturers to append the demand for eco-friendly products through sustainable resources. Still, there are some challenges, but in the future, sustainable materials should be not only available and affordable but will also show better results. This most likely could be explained by continuous innovation and increasing adoption of such services. As the industrialization of economies progresses, affordable, sustainable materials will play a critical role in the efforts to enhance the sustainability of economies. Environmental sustainability implies that an organization needs to find and incorporate sustainable materials in its manufacturing process (Yakoubi, 2024).

5.2 Economic Viability of AI-Generated Packaging Solutions

The packaging solutions created by AI have become innovative forms that positively impact the economy by helping companies cut costs, become more efficient, and promote sustainability. These developments meet some of the critical gaps that face the packaging sector, including wastage of the materials used, the high costs of manufacturing, and the need to adopt green solutions. AI provides solutions when it comes to packaging designs, inspecting large datasets, and providing structures that require less material but that do not affect the quality or functionality. This leads to less use of raw material, and in turn, leads to lesser packaging material and holding expenses and travel costs (Ali *et al.* 2024). The high degree of technological applications, such as robotics and machine learning, simplifies the process and keeps the errors and material wastage factor to a minimum. Another area where the technology is useful is in closely observing production lines to detect problems that may result in a slow down or poor quality end products. AI in Predictive maintenance helps control machinery breakdown and repair costs to run operations with maximum efficiency. The effectiveness of supply chain management is another area where AI yields accurate demand, inventory control, and just-in-time production scenarios. It eliminates cases of an overflow of inventory and helps cut expenses related to the storage of goods, hence enhancing the profitability of the business. Environmental consciousness with regard to packaging is on the increase, hence making sustainable packaging the order of the day for business nowadays, and the use of AI cannot be overemphasized. In a way, AI makes it easier to come up with packaging made from biodegradable or recyclable material that meets environmental standards besides cutting down on the cost of management of the end product. Consumers are willing to pay more money and repurchase from producers that offer sustainable packaging, therefore improving their market share (Sonck-Rautio *et al.* 2024).

6. FUTURE TRENDS IN GENERATIVE AI FOR PACKAGING

Generative AI's prospects in the packaging sector seem promising because of new solutions in the areas of sustainability, individualization, and effectiveness. AI will also continue to apply improvement on the way the latter creates environmentally sensible packaging designs that employ biodegradable and recyclable materials. Peak efficiency in selecting materials and minimizing waste shall result from improved algorithms, thereby increasing sustainability standards globally. Customization will be apparent since generative AI gives businesses the ability to design and deliver highly individualized product designs and looks that appeal to consumers and fit particular products (Sedkaoui & Benaichouba, 2024). Artificial intelligence will extend the communicating and analytical efficiencies of supply chains and production lines so as to better match delivery to demand. More importantly, the integration with Industry 4.0 technologies like IoT and robotics will induce the operation of fully automatic, data-driven lines, which will definitely increase productivity. Further, AI shall improve generative patterns in designing reusable packaging systems and help implement closed-loop recursive recycling. These trends demonstrate that generative AI will transform packaging while simultaneously driving economic and environmental sustainability in a competitive world (Zong & Guan, 2024).

6.1 Advances in AI-Driven Material Science

The way new materials are designed, developed, and implemented is changing and enhanced through artificial intelligence-based material science. The advancements in machine learning and artificial intelligence are reaching new heights in the extraordinary improvement of material design technologies, and they are also causing less cost to designers. AI can sift through large amounts of data and estimate the characteristics of new material, which may be challenging to find when using only classic approaches. Another improvement that can be credited to AI in material science is the examination of behavioral responses within a material at the atomic or molecular level (Gokcekuyu *et al.* 2024). This predictive capability enables researchers to fabricate materials with tailored properties such as increased strength, conductivity, or flexibility to suit applications in the aerospace industry, medicine, or energy production. AI is also helpful in the development of sustainable materials and has contributed to the search for suitable substitute materials in place of the usual untidy, harsh influencing materials. For instance, AI is applied in biodegradable plastics, recyclable products, and probably the improved batteries and energy-storing devices that generate less harm to the given environment. Moreover, it saves time in processing by forecasting the most appropriate raw materials for incorporation into the production process, thus minimizing trials in the production line. With the combination of AI robotics and automation, material science is growing more dynamic and more accurate in terms of cycles to innovate in the material science field and customization. AI has a bright future in material science: growth has been prepared for future developments in quantum material science, self-healing materials science, and nano-technology to create industries for many years (Gundeti *et al.* 2024).

6.2 The Role of AI in the Circular Economy

The circular economy refers to the economy in which waste is minimized, resources are utilized again, and materials are recycled to form a closed-cycle economy. Through efficiency improvement across material use, waste, and recycling bodies, artificial intelligence (AI) is a key catalyst for change in the circular economy. AI technology, which excels in the aspects of data processing, automation, and analysis complemented by prognostication, could help businesses and industries achieve the closed-loop usage of resources or materials to the extent that products and components are recycled, not trashed.

6.2.1 Resource Optimization and Material Efficiency

Its application of big data mining to capture trends is important in the efficient use of resources to address customer needs. AI in the circular economy here assists organizations in designing durables since less material is used in manufacturing. For instance, algorithms can identify unnecessary or excessive parts used in structures and designs and simultaneously design structures with certain end-of-use cycles in mind. Designing with AI also adheres to the principle of waste minimization by affording designers the freedom to generate structures and products, essentially out of thin material codes, without compromising their performance. It also can apply prediction models to the supply chain and determine the necessary supply of raw materials and production of goods, eliminating overproduction and, hence, wasteful goods. Increasingly, these services form part of an integrated system of tracking and analyzing inventories to actively facilitate needs-based production, which in turn indicates a trend toward more sustainable manufacturing and consumption (Alhawamdeh *et al.* 2024).

6.2.2 Waste Management and Recycling

Advanced technologies such as artificial intelligence are being adopted to disrupt waste management, which is an important element of the circular economy. An advantage of using machine learning models is that one is in a position to sort recyclable materials faster than manual workers. Robotics and AI using computer vision on a conveyor belt currently go hand-in-hand when sorting garbage in recycling factories based on categories such as plastic, metal, and paper. It enables better and quicker sorting, guaranteeing that the material must be classified correctly and there is low or no contamination when it is going to be recycled or needs to be reused. AI is also useful in improving the process of recycling by analyzing waste data and establishing trends that may hinder the effectiveness of the process. This, in return, assists waste management firms in minimizing energy usage and enhancing the effectiveness of recycling strategies. Since AI increases the chances of segregated material being recycled successfully, it enhances the efficiency of recycling processes, thus enhancing the circular economy of materials (Shukla *et al.* 2024).

6.2.3 Product Lifecycle and Extended Durability

One of the most important ideas of the circular economy is the maximization of the utilization cycle of products and materials. This is made possible through AI, which creates predictive maintenance that ensures any given product or machinery stays in use for a longer time. The AI models can determine when a product or its component may be most prone to fail and require servicing or replacement. This can be applied to all sorts of industries, from electronics to automobiles, so that there is a limited call for landfilling, and reused parts are encouraged. Further, AI can help firms implement product-service systems in which consumers agree to use a product for a certain level of time rather than buying it outright. AI can used to monitor consumption, maintenance, and status of the products so that when it is returned, they are deemed fit for use by others within the society. AI supports the shift from ownership to access because this is the cornerstone of the circular economy (Piotrowski, 2024).

6.3 Emerging Technologies: AI-Blockchain Integration for Supply Chain Transparency

Supply chain management has hugely been transformed with the help of two innovative solutions. AI and Blockchain – the integration of which resulted in higher reliability, transparency, and efficiency. Collectively, all these technologies offer a solid solution for identifying products, handling information, and verifying and validating all supply chain actions and transactions through the chain from sourcing to end-users.

6.3.1 AI in Supply Chain Management

AI has already been adopted in supply chain management to reduce the likely supply chain hitches, demand forecasting, inventory control, and other related analytical duties. Self-driving cars can analyze volumes of data to forecast demand patterns, determine the best routes, and identify problems that may happen before they do. AI increases the productivity of firms and makes decision-making quicker and more effective, thereby cutting costs and improving customer satisfaction. In addition, AI applications can assist in fraud detection, tracking supplier performance, and guaranteeing product quality based on constant analytics of input data. Nevertheless, AI is not sufficient to solve a range of challenges associated with data protection, accountability, and representations of trust throughout the supply chain, including data and transaction provenance. This is where Blockchain technology comes into the picture to support AI by giving it a secure, open, and changeless ledger (Gbashi & Njobeh, 2024).

6.3.2 AI-Blockchain Integration for Supply Chain Transparency

Incorporating both AI and Blockchain enhances the way these industries work to provide a better supply chain network. Blockchain can improve upon the lack of decision support in AI by providing the data that is analyzed and for which decisions are made, with the guarantee that this data is immutable. The integration of AI and Blockchain forms a strong foundation for transforming the supply chain industry to enhance transparency, efficiency, and security. Blockchain can improve upon the lack of decision support in AI by providing the data that is analyzed and for which decisions are made, with the guarantee that this data is immutable (Adamashvili *et al.* 2024).

7. CONCLUSION

Continually pushing the packaging industry toward sustainable and ecofriendly' solutions, generative AI has completely revolutionized the sector. With help of the generative AI, a vast amount of packaging materials can be generated, with the minimum adverse impact on the environment, as the AI naturally predicts and compares a vast amount of designs and datasets. AI based solutions make this shift towards more sustainable behaviors by focusing on material waste minimization, structural optimization, and recyclability. PLA, PHA, and algae-based biopolymers help as inputs for generative algorithms that design rigid and lightweight packaging materials with considerably reduced consumption of non-renewable materials. Also, AI empowers companies to improve the process of production procedures; ensuring minimal emissions of carbon and energy. Practical using of creative solutions accelerate by generative AI, as it allows to develop prototypes and check materials in terms of their usage in practice as quickly as possible. Further, it means efficient and flexible use of resources, up to the balance of the economization and usefulness. As for the contours of the technology, however, other challenges like scalability, the costs of it, and its demand for mature infrastructure must be solved. Therefore, generative AI is one of the revolutionizing tools when it comes to material innovation, reduction of wastes and optimizing packaging to achieve sustainable packaging as noted above. The adoption of generative AI will become irresistible to meet global sustainability standards as industries consider green options. For suitable use of the opportunity and to ensure the widespread use the cooperation of researcher, businessmen, and legislation is necessary. The potential development trend of the packaging industry is that companies will make use of AI data as a tool for change to adopt more intelligent and environmentally sustainable packaging.

8. REFERENCES

Adamashvili, N., Zhizhilashvili, N., & Tricase, C. (2024). The integration of the internet of things, artificial intelligence, and blockchain technology for advancing the wine supply chain. *Computers, 13*(3), 72. DOI: 10.3390/computers13030072

Aiduang, W., Jatuwong, K., Luangharn, T., Jinanukul, P., Thamjaree, W., Teeraphantuvat, T., & Lumyong, S. (2024). A Review Delving into the Factors Influencing Mycelium-Based Green Composites (MBCs) Production and Their Properties for Long-Term Sustainability Targets. *Biomimetics, 9*(6), 337. DOI: 10.3390/biomimetics9060337 PMID: 38921217

Akram, N., Saeed, M., Mansha, A., Bokhari, T. H., & Ali, A. (2023). Metal packaging for food items advantages, disadvantages and applications. In *Green sustainable process for chemical and environmental engineering and science* (pp. 129–141). Elsevier. DOI: 10.1016/B978-0-323-95644-4.00019-X

Al-Alimi, S., Yusuf, N. K., Ghaleb, A. M., Lajis, M. A., Shamsudin, S., Zhou, W., & Adam, A. (2024). Recycling aluminium for sustainable development: A review of different processing technologies in green manufacturing. *Results in Engineering, 23*, 102566. DOI: 10.1016/j.rineng.2024.102566

Albijanić, I., Maričić, M., Ignjatović, M., & Kićanović, A. (2023). Analysis and Projections of Carbon Dioxide (CO2) Emissions in the US, EU, and China: A Comparative Study. *Applied Ecology and Environmental Research, 21*(6), 5565–5594. DOI: 10.15666/aeer/2106_55655594

Alhawamdeh, M., Lee, A., & Saad, A. (2024). Designing for a circular economy in the architecture, engineering, and construction industry: Insights from Italy. *Buildings (Basel, Switzerland), 14*(7), 1946. DOI: 10.3390/buildings14071946

Ali, Z. A., Zain, M., Pathan, M. S., & Mooney, P. (2024). Contributions of artificial intelligence for circular economy transition leading toward sustainability: An explorative study in agriculture and food industries of Pakistan. *Environment, Development and Sustainability, 26*(8), 19131–19175. DOI: 10.1007/s10668-023-03458-9

Balo, F., & Sua, L. S. (2023). Evaluating the sustainable metal packaging for cooked foods among food packaging materials. In *Green Sustainable Process for Chemical and Environmental Engineering and Science* (pp. 283–302). Elsevier. DOI: 10.1016/B978-0-323-95644-4.00004-8

Bhatia, L., Jha, H., Sarkar, T., & Sarangi, P. K. (2023). Food waste utilization for reducing carbon footprints towards sustainable and cleaner environment: A review. *International Journal of Environmental Research and Public Health*, *20*(3), 2318. DOI: 10.3390/ijerph20032318 PMID: 36767685

Chen, W., Men, Y., Fuster, N., Osorio, C., & Juan, A. A. (2024). Artificial intelligence in logistics optimization with sustainable criteria: A review. *Sustainability (Basel)*, *16*(21), 9145. DOI: 10.3390/su16219145

Cheng, J., Gao, R., Zhu, Y., & Lin, Q. (2024). Applications of biodegradable materials in food packaging: A review. *Alexandria Engineering Journal*, *91*, 70–83. DOI: 10.1016/j.aej.2024.01.080

Delbari, S. A., & Hof, L. A. (2024). Glass Waste Circular Economy-Advancing to High-Value Glass Sheets Recovery using Industry 4.0 and 5.0 technologies. *Journal of Cleaner Production*, *462*, 142629. DOI: 10.1016/j.jclepro.2024.142629

Dokl, M., Van Fan, Y., Vujanović, A., Pintarič, Z. N., Aviso, K. B., Tan, R. R., & Čuček, L. (2024). A waste separation system based on sensor technology and deep learning: A simple approach applied to a case study of plastic packaging waste. *Journal of Cleaner Production*, *450*, 141762. DOI: 10.1016/j.jclepro.2024.141762

Dörnyei, K. R., Uysal-Unalan, I., Krauter, V., Weinrich, R., Incarnato, L., Karlovits, I., & Corredig, M. (2023). Sustainable food packaging: An updated definition following a holistic approach. *Frontiers in Sustainable Food Systems*, *7*, 1119052. DOI: 10.3389/fsufs.2023.1119052

Dzhengiz, T., Miller, E. M., Ovaska, J. P., & Patala, S. (2023). Unpacking the circular economy: A problematizing review. *International Journal of Management Reviews*, *25*(2), 270–296. DOI: 10.1111/ijmr.12329

Elkhattat, D., & Medhat, M. (2022). Creativity in packaging design as a competitive promotional tool. *Information Sciences Letters*, *11*(1), 135–145. DOI: 10.18576/isl/110115

Eysenbach, G. (2023). The role of ChatGPT, generative language models, and artificial intelligence in medical education: A conversation with ChatGPT and a call for papers. *JMIR Medical Education*, *9*(1), e46885. DOI: 10.2196/46885 PMID: 36863937

Fui-Hoon Nah, F., Zheng, R., Cai, J., Siau, K., & Chen, L. (2023). Generative AI and ChatGPT: Applications, challenges, and AI-human collaboration. *Journal of Information Technology Case and Application Research*, *25*(3), 277–304. DOI: 10.1080/15228053.2023.2233814

Gao, S., Wen, Y., Zhu, M., Wei, J., Cheng, Y., Zhang, Q., & Shang, S. (2024). Simulating financial market via large language model based agents. arXiv preprint arXiv:2406.19966.

Gbashi, S., & Njobeh, P. B. (2024). Enhancing Food Integrity through Artificial Intelligence and Machine Learning: A Comprehensive Review. *Applied Sciences (Basel, Switzerland)*, *14*(8), 3421. DOI: 10.3390/app14083421

Ghobakhloo, M., Fathi, M., Iranmanesh, M., Vilkas, M., Grybauskas, A., & Amran, A. (2024). Generative artificial intelligence in manufacturing: Opportunities for actualizing Industry 5.0 sustainability goals. *Journal of Manufacturing Technology Management*, *35*(9), 94–121. DOI: 10.1108/JMTM-12-2023-0530

Gokcekuyu, Y., Ekinci, F., Guzel, M. S., Acici, K., Aydin, S., & Asuroglu, T. (2024). Artificial Intelligence in Biomaterials: A Comprehensive Review. *Applied Sciences (Basel, Switzerland)*, *14*(15), 6590. DOI: 10.3390/app14156590

Gundeti, R., Vuppala, K., & Kasireddy, V. (2024). The Future of AI and Environmental Sustainability: Challenges and Opportunities. Exploring Ethical Dimensions of Environmental Sustainability and Use of AI, 346-371.

Jain, A., Sarsaiya, S., Gong, Q., Wu, Q., & Shi, J. (2024). Bioresources from biowaste for advancing manufacturing of packaging materials with modern circular economy towards a sustainable approach. *Environment, Development and Sustainability*, •••, 1–34. DOI: 10.1007/s10668-024-05734-8

Jungwirth, D., & Haluza, D. (2023). Artificial intelligence and the sustainable development goals: An exploratory study in the context of the society domain. *Journal of Software Engineering and Applications*, *16*(4), 91–112. DOI: 10.4236/jsea.2023.164006

Kajla, P., Chaudhary, V., Dewan, A., Bangar, S. P., Ramniwas, S., Rustagi, S., & Pandiselvam, R. (2024). Seaweed-based biopolymers for food packaging: A sustainable approach for a cleaner tomorrow. *International Journal of Biological Macromolecules*, *274*, 133166. DOI: 10.1016/j.ijbiomac.2024.133166 PMID: 38908645

Khandeparkar, A. S., Paul, R., Sridhar, A., Lakshmaiah, V. V., & Nagella, P. (2024). Eco-friendly innovations in food packaging: A sustainable revolution. *Sustainable Chemistry and Pharmacy*, *39*, 101579. DOI: 10.1016/j.scp.2024.101579

Kumari, N., & Pandey, S. (2023). Application of artificial intelligence in environmental sustainability and climate change. In *Visualization techniques for climate change with machine learning and artificial intelligence* (pp. 293–316). Elsevier. DOI: 10.1016/B978-0-323-99714-0.00018-2

Kunduru, K. R., Hogerat, R., Ghosal, K., Shaheen-Mualim, M., & Farah, S. (2023). Renewable polyol-based biodegradable polyesters as greener plastics for industrial applications. *Chemical Engineering Journal*, *459*, 141211. DOI: 10.1016/j.cej.2022.141211

Kuusisto, M. (2024). Artificial Intelligence Solutions for the Recycling and Utilization of Biomaterials for Industrial Applications.

Lo, W., Yang, C. M., Zhang, Q., & Li, M. (2024). Increased Productivity and Reduced Waste with Robotic Process Automation and Generative AI-Powered IoE Services. *Journal of Web Engineering*, *23*(1), 53–87. DOI: 10.13052/jwe1540-9589.2313

Lodhi, S. K., Gill, A. Y., & Hussain, H. K. (2024). Green Innovations: Artificial Intelligence and Sustainable Materials in Production. *BULLET: Jurnal Multidisiplin Ilmu*, *3*(4), 492–507.

Madusanka, C., Udayanga, D., Nilmini, R., Rajapaksha, S., Hewawasam, C., Manamgoda, D., & Vasco-Correa, J. (2024). A review of recent advances in fungal mycelium based composites. *Discover Materials*, *4*(1), 13. DOI: 10.1007/s43939-024-00084-8

Manzoor, A., Khan, S., Dar, A. H., Pandey, V. K., Shams, R., Ahmad, S., & Pandiselvam, R. (2023). Recent insights into green antimicrobial packaging towards food safety reinforcement: A review. *Journal of Food Safety*, *43*(4), e13046. DOI: 10.1111/jfs.13046

Mishra, A., Dutta, P., Jayasankar, S., Jain, P., & Mathiyazhagan, K. (2023). A review of reverse logistics and closed-loop supply chains in the perspective of circular economy. *Benchmarking*, *30*(3), 975–1020. DOI: 10.1108/BIJ-11-2021-0669

Paprocki, M. (2024). DESIGNING AND DEVELOPING SUSTAINABLE PRODUCTS. Design and Assessment of Sustainable Products: A Conceptual and Practical Framework, 38.

Pesovski, I., Santos, R., Henriques, R., & Trajkovik, V. (2024). Generative ai for customizable learning experiences. *Sustainability (Basel)*, *16*(7), 3034. DOI: 10.3390/su16073034

Piotrowski, N. (2024). Machine learning approach to packaging compatibility testing in the new product development process. *Journal of Intelligent Manufacturing*, *35*(3), 963–975. DOI: 10.1007/s10845-023-02090-8

Rane, N. (2023a). ChatGPT and Similar Generative Artificial Intelligence (AI) for Smart Industry: role, challenges and opportunities for industry 4.0, industry 5.0 and society 5.0. Challenges and Opportunities for Industry, 4.

Rane, N., Choudhary, S., & Rane, J. (2024b). Contribution of chatgpt and similar generative artificial intelligence for enhanced climate change mitigation strategies. Available at *SSRN* 4681720. DOI: 10.2139/ssrn.4681720

Rani, A., Negi, S., Fan, C., Lam, S. S., Kim, H., & Pan, S. Y. (2024). Revitalizing plastic wastes employing bio-circular-green economy principles for carbon neutrality. *Journal of Hazardous Materials*, *472*, 134394. DOI: 10.1016/j.jhazmat.2024.134394 PMID: 38703690

Rodríguez Guzmán, M. M. (2024). Using generative AI tools to improve marketing communication: opportunities, challenges and best practices.

Sedkaoui, S., & Benaichouba, R. (2024). Generative AI as a transformative force for innovation: A review of opportunities, applications and challenges. *European Journal of Innovation Management*. Advance online publication. DOI: 10.1108/EJIM-02-2024-0129

Sengar, S. S., Hasan, A. B., Kumar, S., & Carroll, F. (2024). Generative artificial intelligence: A systematic review and applications. *Multimedia Tools and Applications*, •••, 1–40. DOI: 10.1007/s11042-024-20016-1

Shi, J., Jain, R., Duan, R., & Ramani, K. (2023). Understanding Generative AI in Art: An Interview Study with Artists on G-AI from an HCI Perspective. arXiv preprint arXiv:2310.13149.

Shukla, S. S., Jain, M., Jain, A., Bajpai, S., Mishra, A. K., & Pramanik, S. (2024). The Circular Economy: Changing the Future of India. In Green Transition Impacts on the Economy, Society, and Environment (pp. 149-165). IGI Global.

Singh, P., Pandey, V. K., Singh, R., Negi, P., Maurya, S. N., & Rustagi, S. (2024). Substantial Enhancement of Overall Efficiency and Effectiveness of the Pasteurization and Packaging Process Using Artificial Intelligence in the Food Industry. *Food and Bioprocess Technology*, •••, 1–16.

Sonck-Rautio, K., Lahtinen, T., & Tynkkynen, N. (2024). Consumer meaning-making of packaging functions for sustainable food packaging–Insights from qualitative research in Finland. *Current Research in Environmental Sustainability*, *7*, 100259. DOI: 10.1016/j.crsust.2024.100259

Vicente, D., Proença, D. N., & Morais, P. V. (2023). The role of bacterial polyhydroalkanoate (PHA) in a sustainable future: A review on the biological diversity. *International Journal of Environmental Research and Public Health*, *20*(4), 2959. DOI: 10.3390/ijerph20042959 PMID: 36833658

Yakoubi, S. (2024). Sustainable Revolution: AI-Driven Enhancements for Composite Polymer Processing and Optimization in Intelligent Food Packaging. *Food and Bioprocess Technology*, •••, 1–26.

Yao, F., Li, C., Nekipelov, D., Wang, H., & Xu, H. (2024). Human vs. Generative AI in Content Creation Competition: Symbiosis or Conflict? arXiv preprint arXiv:2402.15467.

Zong, Z., & Guan, Y. (2024). AI-Driven Intelligent Data Analytics and Predictive Analysis in Industry 4.0: Transforming Knowledge, Innovation, and Efficiency. *Journal of the Knowledge Economy*, *16*(1), 1–40. DOI: 10.1007/s13132-024-02001-z

Chapter 13
Food Preservation and Utilization Strategies for the Indian Population

K. Lakshman
https://orcid.org/0000-0002-4666-940X
Jain University, India

P. Periasamy
Saveetha Engineering College, India

N. Dinesh
https://orcid.org/0000-0002-0209-6988
Jain University, India

ABSTRACT

Food security remains a significant concern in India, where post-harvest losses, inefficient supply chains, and lack of awareness about preservation methods contribute to substantial food wastage. This research article explores various strategies for preserving and utilizing food effectively to ensure food security and sustainability. It highlights traditional and modern preservation techniques, government policies, community-driven initiatives, and the role of technology in minimizing food wastage.

INTRODUCTION

India, with its diverse agricultural practices and vast population, faces challenges in ensuring food security for all its citizens. Despite being one of the largest producers of food, the country experiences significant post-harvest losses and wastage

DOI: 10.4018/979-8-3693-9856-2.ch013

due to inadequate storage facilities, inefficient supply chains, and lack of knowledge about preservation techniques (Food and Agriculture Organization [FAO], n.d.). This paper examines comprehensive strategies for preserving and utilizing food effectively, aiming to reduce wastage and enhance food security.

Current Food Scenario in India

1. **Food Production**: India is a major producer of cereals, fruits, vegetables, dairy, and meat.
2. **Food Wastage**: Approximately 30-40% of food produced in India is wasted due to inefficient post-harvest handling, storage, and transportation (FAO, n.d.).
3. **Nutritional Security**: Ensuring that the population has access to nutritious food is a critical concern, exacerbated by food wastage and losses.

1. Food Waste Analysis:

Details of the data analysed

- Breakdown of food waste in India by sector (household, retail, agriculture).
- Example categories: Fruits & Vegetables, Cereals, Dairy, Meat, Others.

Figure 1.

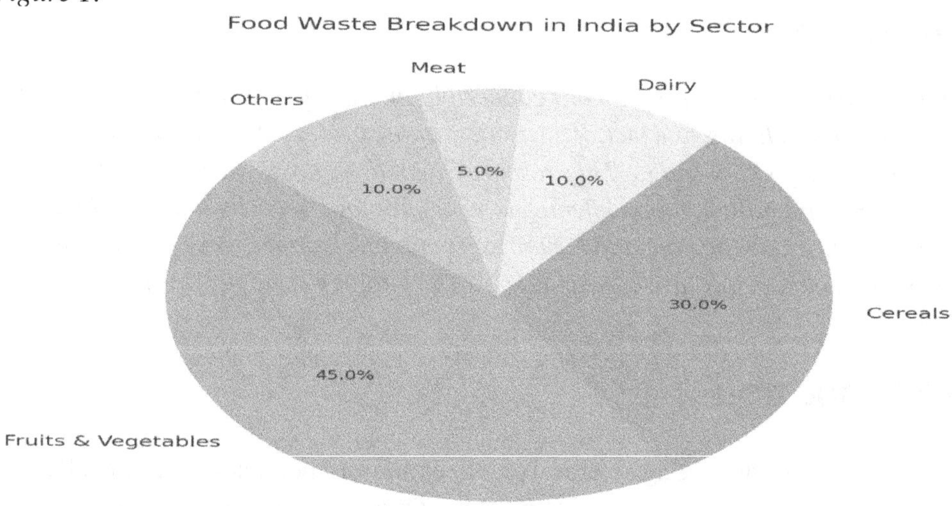

2. Preservation Techniques
    - Compare traditional and modern food preservation methods with examples:
        - Traditional: Pickling, Sun Drying, Smoking.
        - Modern: Cold Storage, Vacuum Packaging, Freeze-Drying.

Traditional Methods

1. Pickling
    - Description: Involves preserving food in brine or vinegar.
    - Example Foods: Vegetables (e.g., cucumbers), fruits, and meats.
    - Advantages: Long shelf life, adds unique flavors.
    - Common in: India, Southeast Asia, and the Mediterranean.
2. **Sun Drying**
    - Description: Drying food under the sun to reduce moisture content, preventing microbial growth.
    - Example Foods: Tomatoes, mangoes, fish, and herbs.
    - Advantages: Natural process, low cost, energy efficient.
    - Common in Rural areas with abundant sunlight, especially in India.
3. **Smoking**
    - Description: Exposing food to smoke from burning or smoldering materials, usually wood.
    - Example Foods: Meats, fish, cheese.
    - Advantages: Adds distinct flavor, and preserves for long durations.
    - Common in North America, Europe, and parts of India.

Modern Methods

1. **Cold Storage (Refrigeration/Freezing)**
    - Description: Preserving food at low temperatures to slow bacterial growth and enzymatic changes.
    - Example Foods: Fruits, vegetables, dairy, meats, and ready-to-eat meals.
    - Advantages: Maintains freshness, nutritional value, and flavour for extended periods.
    - Common in Supermarkets, homes, and large-scale food storage (Pradhan Mantri Kisan SAMPADA Yojana [PMKSY], n.d.).
2. **Vacuum Packaging**
    - Description: Removing air from packaging and sealing it to prevent oxidation and growth of microorganisms.
    - Example Foods: Meats, fish, fruits, and ready-to-eat meals.

- Advantages: Extends shelf life, and preserves flavor and nutrients.
- **Common in** Packaged food products, and commercial storage.

3. **Freeze-Drying**
   - Description: Removing moisture from food by freezing it and then reducing the pressure to allow the ice to sublimate.
   - Example Foods: Fruits, vegetables, and instant soups.
   - Advantages: Lightweight, retains flavour and nutrients, no refrigeration needed.
   - Common in Space food, emergency food supplies, and food packaging industries.

3. Cold Storage Infrastructure

- Cold storage capacity across major Indian states (e.g., Uttar Pradesh, West Bengal, Maharashtra) (PMKSY, n.d.).

Here's an example of how a bar graph could represent cold storage capacity across major Indian states like Uttar Pradesh, West Bengal, and Maharashtra.

The data would typically include the cold storage capacity in metric tons for each state. Let's assume hypothetical values to illustrate this:

- Uttar Pradesh: 1,500,000 MT
- West Bengal: 1,200,000 MT
- Maharashtra: 1,800,000 MT
- Punjab: 1,000,000 MT
- Haryana: 800,000 MT

Figure 2.

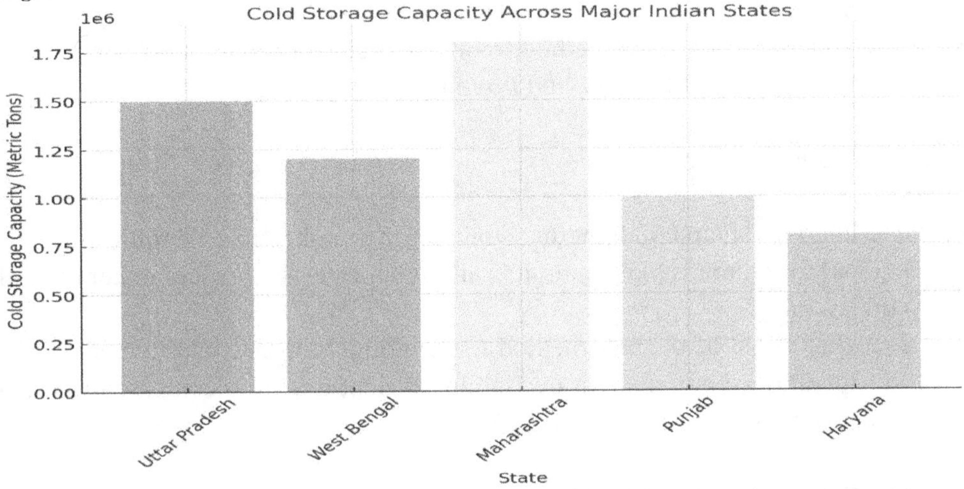

4. Food Utilization Model

- Show the flow of food from Production → Preservation → Distribution → Consumption.
- Highlight intervention points for reducing waste (FAO, n.d.).

5. Nutritional Impact

- Comparison of nutrient retention in foods under different preservation methods:

Table 1. Example Data (hypothetical)

Preservation Method	Vitamin C	Vitamin A	Fiber	Minerals	Protein
Fresh	100%	100%	100%	100%	100%
Refrigerated	85%	95%	95%	98%	100%
Frozen	90%	98%	97%	99%	100%
Dried	40%	90%	100%	100%	100%

Visualizing the Chart:

- **X-axis**: List of preservation methods (Fresh, Refrigerated, Frozen, Dried).
- **Y-axis**: Nutrient retention percentage (0-100% scale).
- **Stacked bars**: For each method, represent nutrient retention of Vitamin C, Vitamin A, Fiber, Minerals, and Protein.

Interpretation:

- The chart will help highlight that while drying might lead to significant nutrient loss for vitamins like Vitamin C, it might preserve fiber and minerals well (IFPRI, n.d.).
- Freezing might be the best method for retaining most nutrients overall.
- Refrigeration shows a slight decline in nutrient retention compared to fresh foods, but it's still relatively effective.

6. Community-Level Strategies

- **Highlight key strategies:**
    - o **Awareness Campaigns**
    - o **Waste-to-Value Initiatives (e.g., Composting)**
    - o **Promoting Local Food Banks (FAO, n.d.).**

7. Food Loss in Supply Chain

- **Illustrate food loss percentages at various stages of the supply chain:**
    - o **Harvesting → Transportation → Processing → Retail.**
- For Example:

Table 2.

A	B
Harvesting	10%
Transportation	5%
Processing	15%
Retail	20%

8. Economic Impact of Food Preservation

- **Relationship between food preservation investments and reduction in food loss over time (PMKSY, n.d.).**

Table 3. Consumer Behaviour Survey

Age Group	18-24	25-34	35-44	45-54	55+
Very Willing	30%	45%	40%	35%	50%
Neutral	40%	35%	40%	45%	30%
Not Willing	30%	20%	20%	20%	20%

10. Policy Interventions

Key policy initiatives in food preservation and waste management in India over the years (Ministry of Agriculture & Farmers Welfare, n.d.; PMKSY, n.d.; National Food Security Mission [NFSM], n.d.).

Timeline:

- 1950s: Establishment of Food Corporation of India (FCI) for food security.
- 1980s: Introduction of the National Policy on Waste Management.
- 2001: Implementation of the Food Safety and Standards Act (FSSA).
- 2010: Launch of National Mission on Sustainable Agriculture (NMSA).
- 2015: Launch of the Swachh Bharat Abhiyan (Clean India Mission).
- 2019: Introduction of the Draft National Food Waste Policy.
- 2020: Release of the National Policy on Food Processing.
- 2023: Extension of Extended Producer Responsibility (EPR) for plastic waste management.

GOVERNMENT POLICIES AND INITIATIVES

National Food Security Mission (NFSM)

Aims to increase food production and ensure food security by promoting sustainable agricultural practices and reducing post-harvest losses (NFSM, n.d.).

Pradhan Mantri Kisan SAMPADA Yojana (PMKSY)

Focuses on the development of food processing and preservation infrastructure, including cold chains, food parks, and agro-processing clusters (PMKSY, n.d.).

National Horticulture Mission

Promotes the production and preservation of horticultural crops through improved post-harvest management and marketing infrastructure (Ministry of Agriculture & Farmers Welfare, n.d.).

Community-Driven Initiatives

Farmer Cooperatives

Collective Storage and Marketing: Farmers pool resources to build storage facilities and market their produce, reducing individual losses and enhancing bargaining power (FAO, n.d.).

Community Kitchens

Reducing Food Waste: Establishing community kitchens that utilize surplus food to prepare meals for the needy, thereby reducing food wastage and addressing hunger.

Awareness and Education Programs

Workshops and Training: Conducting workshops and training sessions for farmers and households on modern and traditional food preservation techniques (IFPRI, n.d.).

Role of Technology in Food Preservation

Cold Chain Logistics

Temperature-Controlled Storage and Transport: Implementing cold chain logistics to ensure that perishable foods are stored and transported at optimal temperatures, reducing spoilage (PMKSY, n.d.).

Blockchain for Supply Chain Management

Traceability and Transparency: Using blockchain technology to track food from farm to table, ensuring transparency and reducing losses due to inefficiencies and fraud.

IoT and Sensors

Monitoring Storage Conditions: Utilizing IoT devices and sensors to monitor and control storage conditions, ensuring that food remains fresh and safe.

Case Studies:

Amul Cooperative

A successful example of a farmer cooperative that has revolutionized milk production, preservation, and marketing in India, significantly reducing wastage and ensuring fair prices for farmers (FAO, n.d.).

Akshaya Patra Foundation

A non-profit organization that runs one of the world's largest mid-day meal programs, using technology and efficient supply chains to provide nutritious meals to school children and reduce food waste.

Strategies for Effective Food Preservation and Utilization

1. Integrated Approach
   Combining Traditional and Modern Methods: Leveraging both traditional wisdom and modern technology for effective food preservation (FAO, n.d.).
2. Policy Implementation and Support
   Government Incentives: Providing financial and technical support to farmers and entrepreneurs for adopting advanced preservation techniques (PMKSY, n.d.).
3. Capacity Building
   Training and Education: Enhancing the skills and knowledge of farmers, food processors, and households through continuous training and education programs (IFPRI, n.d.).
4. Public-Private Partnerships

Collaborative Efforts: Encouraging partnerships between government, private sector, and non-profit organizations to develop and implement effective food preservation strategies (Ministry of Agriculture & Farmers Welfare, n.d.).

CONCLUSION

Preserving and utilizing food effectively is crucial for ensuring food security and sustainability in India. By adopting a combination of traditional and modern preservation techniques, supported by government policies, community initiatives, and technological advancements, the Indian population can significantly reduce food wastage and enhance food availability. A concerted effort involving all stakeholders is essential to achieve these goals and secure a sustainable food future for the country.

REFERENCES

Food and Agriculture Organization. (n.d.). Reports on food preservation and losses. https://www.fao.org

International Food Policy Research Institute. (n.d.). Research publications on food security. https://www.ifpri.org

Ministry of Agriculture & Farmers Welfare, Government of India. (n.d.). Home. https://agricoop.nic.in

National Food Security Mission. (n.d.). Home. https://www.nfsm.gov.in

Pradhan Mantri Kisan SAMPADA Yojana. (n.d.). Home. https://www.pmksy.gov.in

Compilation of References

Abidi, N., Battaıa, O., & Guinet, A. (2021). Data integration in logistics: Challenges and opportunities. *Journal of Logistics and Supply Chain Management*, *12*(3), 45–58.

Adamashvili, N., Zhizhilashvili, N., & Tricase, C. (2024). The integration of the internet of things, artificial intelligence, and blockchain technology for advancing the wine supply chain. *Computers*, *13*(3), 72. DOI: 10.3390/computers13030072

Adula, M., Kant, S., & Geda, H. D. (2025). Innovation Management Landscape for Digital Economy Transformation Through Innovation Resilience in Ethiopia. In *Innovation Management for a Resilient Digital Economy* (pp. 177–204). IGI Global Scientific Publishing.

Ageron, B., Gunasekaran, A., & Spalanzani, A. (2012). Sustainable supply chain management: The role of supply chain integration. *International Journal of Production Economics*, *140*(1), 1–15.

Agrawal, A., Gans, J., & Goldfarb, A. (2018). *Prediction machines: The simple economics of artificial intelligence*. Harvard Business Press.

Ahlawat, J., Akhmetshin, E., Nadaf, A. B., Thangam, A., Pandiaraj, V., & Mahapure, P. S. (2024). Analysing Optimising Time-Series Data for Demand Forecasting and Production Planning Using Bi-RNN. *2nd IEEE International Conference on Data Science and Network Security, ICDSNS 2024*. DOI: 10.1109/ICDSNS62112.2024.10690959

Ahmed, Q. W., Garg, S., Rai, A., Ramachandran, M., Jhanjhi, N. Z., Masud, M., & Baz, M. (2022a). AI-based resource allocation techniques in wireless sensor internet of things networks in energy efficiency with data optimization. *Electronics (Basel)*, *11*(13), 2071. DOI: 10.3390/electronics11132071

Ahmed, S., Alshater, M. M., Ammari, A. E., & Hammami, H. (2022b). Artificial intelligence and machine learning in finance: A bibliometric review. *Research in International Business and Finance*, *61*, 101646. DOI: 10.1016/j.ribaf.2022.101646

AI Laws Around the World. AIPRM. (2024, April 28). https://www.aiprm.com/ai-laws-around-the-world/

Aiduang, W., Jatuwong, K., Luangharn, T., Jinanukul, P., Thamjaree, W., Teeraphantuvat, T., & Lumyong, S. (2024). A Review Delving into the Factors Influencing Mycelium-Based Green Composites (MBCs) Production and Their Properties for Long-Term Sustainability Targets. *Biomimetics*, *9*(6), 337. DOI: 10.3390/biomimetics9060337 PMID: 38921217

Akhtar, P., Ghouri, A. M., Ashraf, A., Lim, J. J., Khan, N. R., & Ma, S. (2024). Smart product platforming powered by AI and Generative AI: Personalization for the circular economy. *International Journal of Production Economics*, *273*, 109283. DOI: 10.1016/j.ijpe.2024.109283

Akram, N., Saeed, M., Mansha, A., Bokhari, T. H., & Ali, A. (2023). Metal packaging for food items advantages, disadvantages and applications. In *Green sustainable process for chemical and environmental engineering and science* (pp. 129–141). Elsevier. DOI: 10.1016/B978-0-323-95644-4.00019-X

Aktas, E. (2024). Big data applications in food supply chains. In S. null, H. M.S.N., I. M.R., J. A., S. H., & F. A. (Eds.), *AIP Conference Proceedings* (Vol. 3109, Issue 1). American Institute of Physics. DOI: 10.1063/5.0204918

Al-Alimi, S., Yusuf, N. K., Ghaleb, A. M., Lajis, M. A., Shamsudin, S., Zhou, W., & Adam, A. (2024). Recycling aluminium for sustainable development: A review of different processing technologies in green manufacturing. *Results in Engineering*, *23*, 102566. DOI: 10.1016/j.rineng.2024.102566

Albijanić, I., Maričić, M., Ignjatović, M., & Kićanović, A. (2023). Analysis and Projections of Carbon Dioxide (CO2) Emissions in the US, EU, and China: A Comparative Study. *Applied Ecology and Environmental Research*, *21*(6), 5565–5594. DOI: 10.15666/aeer/2106_55655594

Alhawamdeh, M., Lee, A., & Saad, A. (2024). Designing for a circular economy in the architecture, engineering, and construction industry: Insights from Italy. *Buildings (Basel, Switzerland)*, *14*(7), 1946. DOI: 10.3390/buildings14071946

Ali, Z. A., Zain, M., Pathan, M. S., & Mooney, P. (2024). Contributions of artificial intelligence for circular economy transition leading toward sustainability: An explorative study in agriculture and food industries of Pakistan. *Environment, Development and Sustainability*, *26*(8), 19131–19175. DOI: 10.1007/s10668-023-03458-9

Almeida, F., Santos, J. D., & Monteiro, J. A. (2020). Compliance-as-a-Service (CaaS): A model to streamline regulatory compliance in cloud computing. *Journal of Cloud Computing (Heidelberg, Germany)*, *9*(1), 1–14.

Alobid, M., Abujudeh, S., & Szűcs, I. (2022). The Role of Blockchain in Revolutionizing the Agricultural Sector. In Sustainability (Vol. 14, Issue 7, p. 4313). Multidisciplinary Digital Publishing Institute. DOI: 10.3390/su14074313

Al-Turki, U. (2011). A framework for strategic planning in maintenance. *Journal of Quality in Maintenance Engineering*, *17*(2), 150–162. DOI: 10.1108/13552511111134583

Amazon. (2021, September 9). *Amazon to invest $1.2 billion to upskill more than 300,000 employees by 2025*. Amazon Press Center. Retrieved from https://press.aboutamazon.com/news-releases/news-release-details/amazon-invest-12-billion-upskill-more-300000-employees-2025

Amoo, O., Sodiya, E., Umoga, U., & Atadoga, A. (2024). AI-driven warehouse automation: A comprehensive review of systems. *GSC Advanced Research and Reviews.*, *18*(2), 272–282. DOI: 10.30574/gscarr.2024.18.2.0063

Armbrust, M., Fox, A., Griffith, R., Joseph, A. D., Katz, R., Konwinski, A., Lee, G., Patterson, D., Rabkin, A., Stoica, I., & Zaharia, M. (2010). A view of cloud computing. *Communications of the ACM*, *53*(4), 50–58. DOI: 10.1145/1721654.1721672

Arora, M., Kumar, R., & Anand, N. (2022). Analysis of frozen food adoption by the consumer in Uttarakhand, a state of India: An inferential statistics approach. *International Journal of Value Chain Management*, *13*(1), 88–111. DOI: 10.1504/IJVCM.2022.122164

Arora, M., Kumar, R., & Raju, T. B. (2023). Identification of issues in the cold chain of Indian frozen food. *International Journal of Logistics Economics and Globalisation*, *10*(1), 1–22. DOI: 10.1504/IJLEG.2023.130236

Aspen Institute. (2018, March 12). *Upskilling in the News: AT&T Invests $1 Billion in Employee Reskilling*. The Aspen Institute. Retrieved from https://www.aspeninstitute.org/of-interest/upskilling-news-att-invests-1-billion-employee-reskilling/

Astill, J., Dara, R. A., Campbell, M., Farber, J. M., Fraser, E. D., Ghavitri, L., & Yada, R. Y. (2019). Transparency in food supply chains: A review of enabling technology solutions. *Trends in Food Science & Technology*, *91*, 240–247. DOI: 10.1016/j.tifs.2019.07.024

AT&T. (2022). *AT&T.*https://about.att.com/

Atzori, L., Iera, A., & Morabito, G. (2010). The internet of things: A survey. *Computer Networks, 54*(15), 2787–2805. DOI: 10.1016/j.comnet.2010.05.010

Aung, M. M., & Chang, Y. S. (2014). Traceability in a food supply chain: Safety and quality perspectives. *Food Control, 39*, 172–184. DOI: 10.1016/j.foodcont.2013.11.007

Automation Anywhere. (2021). *Automation Anywhere*. https://www.automationanywhere.com/

Automation Edge. (2024). *Automation Edge*. https://automationedge.com/

Avnet. (n.d.). *Smart warehouse monitoring solution*. Retrieved from https://www.iotconnect.io/smart-warehouse-monitoring.html

Ayinla, B., Atadoga, A., Ike, C., Ndubuisi, N., Asuzu, O. F., & Adeleye, R. (2024). The role of robotic process automation (RPA) in modern accounting: A review—Investigating how automation tools are transforming traditional accounting practices. *Engineering Science & Technology Journal*, *5*(2), 427–447. DOI: 10.51594/estj.v5i2.804

Azuma, R. T. (1997). A survey of augmented reality. *Presence (Cambridge, Mass.)*, *6*(4), 355–385. DOI: 10.1162/pres.1997.6.4.355

Babai, M. Z., Dolgui, A., & Syntetos, A. (2015). Demand forecasting with machine learning: Applications in inventory management. *International Journal of Production Research*, *53*(1), 187–200.

Babai, M. Z., Lamsal, R., & Mohanty, S. K. (2015). Artificial intelligence in supply chain forecasting: A comprehensive review. *Journal of Manufacturing Science and Engineering*, *137*(3), 031019.

Badia-Melis, R., Mishra, P., & Ruiz-García, L. (2015). Food traceability: New trends and challenges. *Food Control*, *57*, 225–232. DOI: 10.1016/j.foodcont.2015.05.005

Bagheri, M., Bazvand, A., & Ehteshami, M. (2017). Application of artificial intelligence for the management of landfill leachate penetration into groundwater, and assessment of its environmental impacts. *Journal of Cleaner Production*, *149*, 784–796. DOI: 10.1016/j.jclepro.2017.02.157

Bahaloo, S., Mehrizadeh, M., & Najafi-Marghmaleki, A. (2022). *Review of application of artificial intelligence techniques in petroleum operations*. Petrol Res., DOI: 10.1016/j.ptlrs.2022.07.002

Bahrin, M. A. K., Othman, M. F., Azli, N. N., & Talib, M. F. (2016). Industry 4.0: A review on industrial automation and robotic. *Jurnal Teknologi, 78*(6-13), 137-143.

Bai, C., Dallasega, P., Orzes, G., & Sarkis, J. (2020). Industry 4.0 technologies assessment: A sustainability perspective. *International Journal of Production Economics, 229*, 107776. DOI: 10.1016/j.ijpe.2020.107776

Balamurugan, A., Vamsi, M., Bhattacharya, R., Mohammed, S., Kaushik, P., & Haralayya, D. (2023). Robotic process automation (RPA) in accounting and auditing of business and financial information. *Manager, 58*, 127–142.

Baldwin, C. (2022). The Role of IoT in Enhancing Food Safety Compliance. *Journal of Food Quality, 35*(2), 121–135.

Balo, F., & Sua, L. S. (2023). Evaluating the sustainable metal packaging for cooked foods among food packaging materials. In *Green Sustainable Process for Chemical and Environmental Engineering and Science* (pp. 283–302). Elsevier. DOI: 10.1016/B978-0-323-95644-4.00004-8

Baryannis, G., Validi, S., Dani, S., & Antoniou, G. (2019). Supply chain risk management and artificial intelligence: State of the art and future research directions. *International Journal of Production Research, 57*(7), 2179–2202. DOI: 10.1080/00207543.2018.1530476

Beck, K. (2002). *Extreme Programming Explained: Embrace Change*. Addison-Wesley Professional.

Bhatia, L., Jha, H., Sarkar, T., & Sarangi, P. K. (2023). Food waste utilization for reducing carbon footprints towards sustainable and cleaner environment: A review. *International Journal of Environmental Research and Public Health, 20*(3), 2318. DOI: 10.3390/ijerph20032318 PMID: 36767685

Bhowmik, D., Rickard, J. J. S., Jelinek, R., & Goldberg Oppenheimer, P. (2024). Resilient sustainable current and emerging technologies for foodborne pathogen detection. *Sustainable Food Technology*. Advance online publication. DOI: 10.1039/D4FB00192C PMID: 39359621

BigScience Workshop. (2022). *BLOOM: A 176B-Parameter Open-Access Multilingual Language Model*. arXiv preprint arXiv:2211.05100.

Bilicka, A., & Warehousing, S. Revolutionizing Connectivity and Efficiency-September 1, 2023.https://www.autostoresystem.com/insights/smart-warehousing-revolutionizing-connectivity-and-efficiency

Billinghurst, M., Clark, A., & Lee, G. (2015). A survey of augmented reality. *Foundations and Trends® in Human–Computer Interaction, 8*(2–3), 73-272.

Bishop, T. (2023, September 7). New robots are making Amazon's warehouses more efficient — can they also make them safer? https://www.geekwire.com/2023/new-robots-are-making-amazons-warehouses-more-efficient-can-they-also-make-them-safer/

Bose, I., Mahapatra, S., & Roy, S. (2021). Evaluating technology partnerships in smart logistics. *International Journal of Operations Management, 18*(2), 67–80.

Bostrom, N., & Yudkowsky, E. (2014). The ethics of artificial intelligence. In Frankish, K., & Ramsey, W. M. (Eds.), *The Cambridge handbook of artificial intelligence* (pp. 316–334). Cambridge University Press. DOI: 10.1017/CBO9781139046855.020

Bouazizi, E., Khedr, A., Elfaioumy, S., & Belal, M. (2024). Inventory Optimization Using Data Science Technologies for Supply Chain 4.0. *The International Arab Journal of Information Technology, 21*(6), 996–1005. DOI: 10.34028/iajit/21/6/4

Bozgeyikli, E., Raij, A., & Katko, R. (2019). A survey on virtual and augmented reality for training of technical / motor skills. *ACM Computing Surveys, 51*(6), 1–33.

Bressanelli, G., Adrodegari, F., Perona, M., & Saccani, N. (2018). The role of digital technologies to enable circular economy business models: A systematic literature review. *Procedia CIRP, 78*, 25–30.

Brettel, M., Friederichsen, N., Keller, M., & Rosenberg, M. (2014). How virtualization, decentralization and network building change the manufacturing landscape: An Industry 4.0 Perspective. *International Journal of Mechanical. Industrial Science and Engineering, 8*(1), 37–44.

Brown, T. B., Mann, B., Ryder, N., Subbiah, M., Kaplan, J., Dhariwal, P., & Amodei, D. (2020). Language models are few-shot learners. *Advances in Neural Information Processing Systems, 33*, 1877–1901.

Brynjolfsson, E., & McAfee, A. (2014). *The Second Machine Age: Work, Progress, and Prosperity in a Time of Brilliant Technologies*. W.W. Norton & Company.

Brynjolfsson, E., & McAfee, A. (2014). *The second machine age: Work, progress, and prosperity in a time of brilliant technologies*. WW Norton & Company.

Cai, W., Lai, K., Liu, C., Wei, F., Ma, M., Jia, S., Jiang, Z., & Lv, L. (2019). Promoting sustainability of manufacturing industry through the lean energy-saving and emission-reduction strategy. *The Science of the Total Environment, 665*, 23–32. DOI: 10.1016/j.scitotenv.2019.02.069 PMID: 30772553

Capco. (2019). *Capco*. https://www.capco.com/

Caro, M. P., Ali, M. S., Vecchio, M., & Giaffreda, R. (2018). Blockchain-based traceability in Agri-Food supply chain management: A practical implementation. International Conference on Internet of Things (iThings), 1-8. DOI: 10.1109/IOT-TUSCANY.2018.8373021

Carranza-García, M., García-Gutiérrez, J., & Riquelme, J. C. (2019). A framework for evaluating land use and land cover classification using convolutional neural networks. *Remote Sensing (Basel)*, *11*(3), 274. DOI: 10.3390/rs11030274

Casino, F., Dasaklis, T. K., & Patsakis, C. (2019). A systematic literature review of blockchain-based applications: Current status, classification, and open issues. *Telematics and Informatics*, *36*, 55–81. DOI: 10.1016/j.tele.2018.11.006

Chen, Y., et al. (2023). Technology Advancements in Food Safety Monitoring. Journal of Food Safety Technology.

Cheng, J., Gao, R., Zhu, Y., & Lin, Q. (2024). Applications of biodegradable materials in food packaging: A review. *Alexandria Engineering Journal*, *91*, 70–83. DOI: 10.1016/j.aej.2024.01.080

Chen, J., Wang, Y., & Li, X. (2021). Reinforcement learning in food safety management: A dynamic decision-making approach. *Journal of Food Engineering*, *293*, 110409.

Chen, M., Mao, S., & Liu, Y. (2014). Big data: A survey. *Mobile Networks and Applications*, *19*(2), 171–209. DOI: 10.1007/s11036-013-0489-0

Chen, W., Men, Y., Fuster, N., Osorio, C., & Juan, A. A. (2024). Artificial intelligence in logistics optimization with sustainable criteria: A review. *Sustainability (Basel)*, *16*(21), 9145. DOI: 10.3390/su16219145

Chen, Y., Luo, L., & Zhang, H. (2019). Automated Data Logging in Food Safety Management: A Practical Guide. *Food Control*, *98*, 285–290.

Chesbrough, H. (2003). *Open Innovation: The New Imperative for Creating and Profiting from Technology*. Harvard Business Press.

Ching, N. T., Ghobakhloo, M., Iranmanesh, M., Maroufkhani, P., & Asadi, S. (2022). Industry 4.0 applications for sustainable manufacturing: A systematic literature review and a roadmap to sustainable development. *Journal of Cleaner Production*, *334*, 130133. DOI: 10.1016/j.jclepro.2021.130133

Chiroma, H., Hashem, I. A. T., & Maray, M. (2024). Bibliometric analysis for artificial intelligence in the internet of medical things: mapping and performance analysis. *Frontiers in Artificial Intelligence, 7*, 1347815. Erboz, G. How to Define Industry 4.0: The Main Pillars of Industry 4.0. Available online: https://www.researchgate.net/publication/326557388_How_To_Define_Industry_40_Main_Pillars_Of_Industry_40 (accessed on 6 July 2020).

Choi, D., R'bigui, H., & Cho, C. (2021). Robotic process automation implementation challenges. In *Proceedings of [Conference or Book Title]* (pp. [pages]). Springer. DOI: 10.1007/978-981-15-7990-5_29

Choi, T. M., Wallace, S. W., & Wang, Y. (2018). Advances in supply chain modeling and analysis: Trends and research opportunities. Springer.

Choi, E., Bahadori, M. T., Song, L., Stewart, W. F., & Sun, J. (2017). GRAM: graph-based attention model for healthcare representation learning. In *Proceedings of the 23rd ACM SIGKDD international conference on knowledge discovery and data mining* (pp. 787-795). DOI: 10.1145/3097983.3098126

Choi, T. M., Wallace, S. W., & Wang, Y. (2003). Supply chain coordination under uncertainty: A review of recent developments in the literature. *Journal of Operations Management, 21*(3), 231–247.

Choi, T. M., Wallace, S. W., & Wang, Y. (2018). The impact of AI and machine learning on inventory management in retail supply chains. *International Journal of Production Economics, 206*, 47–61.

Choi, T. M., Wallace, S. W., & Wang, Y. (2023). Big data analytics in operations management. *Production and Operations Management, 27*(10), 1868–1883. DOI: 10.1111/poms.12838

Chong, A. Y. L., Chan, F. T. S., & Ooi, K. B. (2017). Predicting the adoption of artificial intelligence in supply chain management. *International Journal of Logistics Research and Applications, 20*(4), 349–367.

Chopra, S., & Meindl, P. (2016). *Supply chain management: Strategy, planning, and operation* (6th ed.). Pearson.

Christopher, M. (2016). *Logistics & supply chain management* (5th ed.).

Christopher, M. (2016). *Logistics & Supply Chain Management* (5th ed.). Pearson Education.

Chuang, C., Li, C., & Shen, X. (2021). Natural language processing (NLP) for compliance monitoring in food safety: A systematic review. *Food Control, 130*, 108324.

Cognizant. (2024). *Cognizant.* https://www.cognizant.com/

Confluent. (March 12, 2024). *Real-time data streaming for smart warehouses.* Retrieved from https://www.confluent.io/blog/data-streaming-for-smart-warehouse/

Cooper, L., Holderness, D.Jr, Sorensen, T., & Wood, D. (2019). Robotic process automation in public accounting. *Accounting Horizons*, *33*(4), 15–35. DOI: 10.2308/acch-52466

Covent, C. (2023). *Exciting Announcement! Introducing ChatCTC - Empowering People to Work Smarter!* LinkedIn. Retrieved December 19, 2024, from https://www.linkedin.com/posts/caricovent_ctc-chatctc-aiinnovation-activity-7124762472997126144-hJXl/

DalleMule, L., & Davenport, T. H. (2017). What's your data strategy? *Harvard Business Review*, *95*(3), 112–121.

Damotech (2024, August 15). *IKEA warehouses: Efficient Warehousing and distribution strategies.* Damotech. https://www.damotech.com/blog/ikea-efficient-warehousing-distribution-strategies

Danach, K., El Dirani, A., & Rkein, H. (2024). Revolutionizing Supply Chain Management With AI: A Path to Efficiency and Sustainability. *IEEE Access : Practical Innovations, Open Solutions*, *12*, 188245–188255. DOI: 10.1109/ACCESS.2024.3474531

Daugherty, P. R., & Wilson, H. J. (2018). *Human + machine: Reimagining work in the age of AI.* Harvard Business Press.

Davenport, T. H., & Patil, D. J. (2012). Data scientist: The sexiest job of the 21st century. *Harvard Business Review*, *90*(5), 70–76. PMID: 23074866

Davenport, T. H., & Ronanki, R. (2018). Artificial intelligence for the real world. *Harvard Business Review*, *96*(1), 108–116.

Debrah, C., Chan, A. P. C., & Darko, A. (2022). Artificial intelligence in green building. *Automation in Construction*, *137*, 104192. DOI: 10.1016/j.autcon.2022.104192

Delbari, S. A., & Hof, L. A. (2024). Glass Waste Circular Economy-Advancing to High-Value Glass Sheets Recovery using Industry 4.0 and 5.0 technologies. *Journal of Cleaner Production*, *462*, 142629. DOI: 10.1016/j.jclepro.2024.142629

Delling, D., Goldberg, A. V., Pajor, T., & Werneck, R. F. (2009). Engineering fast route planning algorithms. In Vol. 5868, pp. 52–66). Lecture Notes in Computer Science. Springer., DOI: 10.1007/978-3-642-05465-5_5

Deloitte. (2024). *Deloitte.* https://www2.deloitte.com

Department of Finance, Government of Canada. (2024, April 16). *Chapter 4: Economic Growth for Every Generation: Budget 2024.* https://budget.canada.ca/2024/report-rapport/chap4-en.html#a4

Devlin, J., Chang, M. W., Lee, K., & Toutanova, K. (2019). BERT: Pre-training of deep bidirectional transformers for language understanding. In *Proceedings of the 2019 Conference of the North American Chapter of the Association for Computational Linguistics: Human Language Technologies,* Volume 1 *(Long and Short Papers)* (pp. 4171-4186).

Divya, D., & Santhanakrishnan, D. (2024). Artificial intelligence in sustainable supply chain management: A comprehensive review.

Doe, J. (2024). Automation and Quality Control in Food Supply Chains. International Supply Chain Review.

Doğuc, Ö. (2020). Robot process automation (RPA) and its future. In *Handbook of Research on Strategic Fit and Design in Business Ecosystems.* IGI Global., DOI: 10.4018/978-1-7998-1125-1.ch021

Dokl, M., Van Fan, Y., Vujanović, A., Pintarič, Z. N., Aviso, K. B., Tan, R. R., & Čuček, L. (2024). A waste separation system based on sensor technology and deep learning: A simple approach applied to a case study of plastic packaging waste. *Journal of Cleaner Production, 450,* 141762. DOI: 10.1016/j.jclepro.2024.141762

Dörnyei, K. R., Uysal-Unalan, I., Krauter, V., Weinrich, R., Incarnato, L., Karlovits, I., & Corredig, M. (2023). Sustainable food packaging: An updated definition following a holistic approach. *Frontiers in Sustainable Food Systems, 7,* 1119052. DOI: 10.3389/fsufs.2023.1119052

Doz, Y. L., & Hamel, G. (1998). *Alliance Advantage: The Art of Creating Value Through Partnering.* Harvard Business Press.

Dresser, S. (2023, October 18). *Amazon announces 2 new ways it's using robots to assist employees and deliver for customers.* Amazon news. https://www.aboutamazon.com/news/operations/amazon-introduces-new-robotics-solutions

Duan, Y., Edwards, J. S., & Dwivedi, Y. K. (2019). Artificial intelligence for decision making in the era of Big Data–evolution, challenges and research agenda. *International Journal of Information Management, 48,* 63–71. DOI: 10.1016/j.ijinfomgt.2019.01.021

Dzhengiz, T., Miller, E. M., Ovaska, J. P., & Patala, S. (2023). Unpacking the circular economy: A problematizing review. *International Journal of Management Reviews*, *25*(2), 270–296. DOI: 10.1111/ijmr.12329

Ebrahimi, M., Sarikhani, M. R., Safari Sinegani, A. A., Ahmadi, A., & Keesstra, S. (2019). Estimating the soil respiration under different land uses using artificial neural network and linear regression models. *Catena*, *174*, 371–382. DOI: 10.1016/j.catena.2018.11.035

Egger, A., Ter, A., Kratsch, W., Leemans, S., Röeglinger, M., & Wynn, M. (2020). Bot log mining: Using logs from robotic process automation for process mining. In *Business Process Management: BPM 2020*. Springer.

Elahi, E., Weijun, C., Zhang, H., & Nazeer, M. (2019b). Agricultural intensification and damages to human health in relation to agrochemicals: Application of artificial intelligence. *Land Use Policy*, *83*, 461–474. DOI: 10.1016/j.landusepol.2019.02.023

El-Gharib, N., & Amyot, D. (2022). A review of data-driven robotic process automation exploiting process mining.

Elkhattat, D., & Medhat, M. (2022). Creativity in packaging design as a competitive promotional tool. *Information Sciences Letters*, *11*(1), 135–145. DOI: 10.18576/isl/110115

Ellahi, R. M., Wood, L. C., & Bekhit, A. E. A. (2023). Blockchain-Based Frameworks for Food Traceability: A Systematic Review [Review of Blockchain-Based Frameworks for Food Traceability: A Systematic Review]. Foods, 12(16), 3026. Multidisciplinary Digital Publishing Institute. DOI: 10.3390/foods12163026

Ellen MacArthur Foundation. (n.d.). *What is a circular economy?* Retrieved from https://www.ellenmacarthurfoundation.org/circulareconomy/concept

Eme, O., Ugboaja, C. A., Uwazuruike, F. O., & Ukpai, C. U. (2018). Computer-based drug sales and inventory control system and its applications in pharmaceutical stores. *I.J. Education and Management Engineering*, 30-39.

Emmert-Streib, F., Yli-Harja, O., & Dehmer, M. (2020). Artificial intelligence: A clarification of misconceptions, myths and desired status. *Frontiers in Artificial Intelligence*, *3*, 524339. DOI: 10.3389/frai.2020.524339 PMID: 33733197

Etzkowitz, H., & Leydesdorff, L. (2000). The dynamics of innovation: From National Systems and "Mode 2" to a Triple Helix of university–industry–government relations. *Research Policy*, *29*(2), 109–123. DOI: 10.1016/S0048-7333(99)00055-4

European Commission. (2018). *Communication from the Commission to the European Parliament, the Council, the European Economic and Social Committee and the Committee of the Regions on Artificial Intelligence for Europe*. COM(2018) 237 final.

European Parliament & Council of the European Union. (2016). Regulation (EU) 2016/679 of the European Parliament and of the Council of 27 April 2016 on the protection of natural persons with regard to the processing of personal data and on the free movement of such data, and repealing Directive 95/46/EC (General Data Protection Regulation). *Official Journal of the European Union, L, 119*(1).

Everest Group. (2022). *Everest Group*. https://www.everestgrp.com/

Eysenbach, G. (2023). The role of ChatGPT, generative language models, and artificial intelligence in medical education: A conversation with ChatGPT and a call for papers. *JMIR Medical Education, 9*(1), e46885. DOI: 10.2196/46885 PMID: 36863937

Fagnant, D. J., & Kockelman, K. (2015). Preparing a nation for autonomous vehicles: Opportunities, barriers and policy recommendations. *Transportation Research Part A, Policy and Practice, 77*, 167–181. DOI: 10.1016/j.tra.2015.04.003

Fan, Z., Yu, W., & Sun, L. (2018). Deep learning for food supply chain: Opportunities and challenges. *Computers and Electronics in Agriculture, 157*, 268–277.

FAO. Influencing Food Environments for Healthy Diets, FAO, Roma, (2016). https://www.fao.org/3/a-i6484e.pdf accessed on 15 Oct 2024.

Farghali, M., Osman, A. I., Umetsu, K., & Rooney, D. W. (2022). Integration of biogas systems into a carbon zero and hydrogen economy: A review. *Environmental Chemistry Letters, 20*(5), 2853–2927. DOI: 10.1007/s10311-022-01468-z

Fedyk, Y. (n.d.). Warehouse Automation [A Complete Guide]: Types, Benefits & Best Examples. *INVERITA*. December 17, 2024, https://inveritasoft.com/article-warehouse-automation-explained:-types-benefits-&-best-practices

Fernandez, D., Dastane, O., Omar Zaki, H., & Aman, A. (2023). Robotic process automation: Bibliometric reflection and future opportunities. *European Journal of Innovation Management, 27*(2), 692–712. DOI: 10.1108/EJIM-10-2022-0570

Fintech Futures. (2018). *Fintech Futures*. https://www.fintechfutures.com/

Fisher, M. L. (1997). What is the right supply chain for your product? *Harvard Business Review, 75*(2), 105–116.

Fisher, M. L., & Lutz, S. R. (1994). The impact of inventory management on supply chain profitability. *Supply Chain Management Review*, *8*(1), 39–47.

Flechsig, C., Anslinger, F., & Lasch, R. (2022). Robotic process automation in purchasing and supply management: A multiple case study on potentials, barriers, and implementation. *Journal of Purchasing and Supply Management*, *28*(1), 100718. DOI: 10.1016/j.pursup.2021.100718

Floridi, L., Cowls, J., Beltramini, M., Saunders, D., & Vayena, E. (2018). An ethical framework for a good AI society: Opportunities, risks, principles, and recommendations. *AI & Society*, *33*(4), 689–707. PMID: 30930541

Food and Agriculture Organization. (2019). Hazard Analysis and Critical Control Points (HACCP). *System.*

Food and Agriculture Organization. (n.d.). Reports on food preservation and losses. https://www.fao.org

Food and Drug Administration (FDA). (2011). Food Safety Modernization Act (FSMA). Retrieved from Galvez, J. F., Mejuto, J. C., & Simal-Gandara, J. (2018). Future challenges on the use of blockchain for food traceability analysis. *Trends in Analytical Chemistry*, *107*, 222–232.

Ford, M. (2015). *Rise of the robots: Technology and the threat of a jobless future.* Basic Books.

Forrester. (2021). *Forrester.* https://www.forrester.com/

Frey, C. B., & Osborne, M. A. (2017). The future of employment: How susceptible are jobs to computerisation? *Technological Forecasting and Social Change*, *114*, 254–280. DOI: 10.1016/j.techfore.2016.08.019

Fui-Hoon Nah, F., Zheng, R., Cai, J., Siau, K., & Chen, L. (2023). Generative AI and ChatGPT: Applications, challenges, and AI-human collaboration. *Journal of Information Technology Case and Application Research*, *25*(3), 277–304. DOI: 10.1080/15228053.2023.2233814

Gajdzik, B., Grabowska, S., Saniuk, S., & Wieczorek, T. (2020). Sustainable development and industry 4.0: A bibliometric analysis identifying key scientific problems of the sustainable industry 4.0. *Energies*, *13*(16), 4254. DOI: 10.3390/en13164254

Gao, S., Wen, Y., Zhu, M., Wei, J., Cheng, Y., Zhang, Q., & Shang, S. (2024). Simulating financial market via large language model based agents. arXiv preprint arXiv:2406.19966.

Gartner. (2021). *Gartner.* https://www.gartner.com/

Gatys, L. A., Ecker, A. S., & Bethge, M. (2016). Image style transfer using convolutional neural networks. In *Proceedings of the IEEE conference on computer vision and pattern recognition* (pp. 2414-2423). DOI: 10.1109/CVPR.2016.265

Gautam, K., Sharma, P., Dwivedi, S., Singh, A., Gaur, V. K., Varjani, S., Srivastava, J. K., Pandey, A., Chang, J.-S., & Ngo, H. H. (2023). A review on control and abatement of soil pollution by heavy metals: Emphasis on artificial intelligence in recovery of contaminated soil. *Environmental Research, 225*, 115592. DOI: 10.1016/j.envres.2023.115592 PMID: 36863654

Gbashi, S., & Njobeh, P. B. (2024). Enhancing Food Integrity through Artificial Intelligence and Machine Learning: A Comprehensive Review. *Applied Sciences (Basel, Switzerland), 14*(8), 3421. DOI: 10.3390/app14083421

Geissdoerfer, M., Savaget, P., Bocken, N. M., & Hultink, E. J. (2017). The Circular Economy–A new sustainability paradigm? *Journal of Cleaner Production, 143*, 757–768. DOI: 10.1016/j.jclepro.2016.12.048

Ghavami, S. M., Taleai, M., & Arentze, T. (2017). An intelligent spatial land use planning support system using socially rational agents. *International Journal of Geographical Information Science, 31*(5), 1022–1041. DOI: 10.1080/13658816.2016.1263306

Ghobakhloo, M. (2020). Industry 4.0, digitization, and opportunities for sustainability. *Journal of Cleaner Production, 252*, 119869. DOI: 10.1016/j.jclepro.2019.119869

Ghobakhloo, M., Fathi, M., Iranmanesh, M., Vilkas, M., Grybauskas, A., & Amran, A. (2024). Generative artificial intelligence in manufacturing: Opportunities for actualizing Industry 5.0 sustainability goals. *Journal of Manufacturing Technology Management, 35*(9), 94–121. DOI: 10.1108/JMTM-12-2023-0530

Ghobakhloo, M., Iranmanesh, M., Foroughi, B., Tirkolaee, E. B., Asadi, S., & Amran, A. (2023). Industry 5.0 implications for inclusive sustainable manufacturing: An evidence-knowledge-based strategic roadmap. *Journal of Cleaner Production, 417*, 138023. DOI: 10.1016/j.jclepro.2023.138023

Ghodake, S. P., Malkar, V. R., Santosh, K., Jabasheela, L., Abdufattokhov, S., & Gopi, A. (2024). Enhancing Supply Chain Management Efficiency: A Data-Driven Approach using Predictive Analytics and Machine Learning Algorithms. *International Journal of Advanced Computer Science and Applications, 15*(4), 672–686. DOI: 10.14569/IJACSA.2024.0150469

Ghosh, A., Singh, R., & Sharma, P. (2022). Edge computing in real-time warehousing systems. *IEEE Transactions on Industrial Informatics, 19*(4), 1345–1355.

Ghosh, S. (2020). IoT-enabled predictive maintenance: Application in food processing industry. *Journal of Food Engineering*, *268*, 109737.

Gibson, I., Rosen, D., & Stucker, B. (2015). *Additive manufacturing technologies: 3D printing, rapid prototyping, and direct digital manufacturing* (2nd ed.). Springer. DOI: 10.1007/978-1-4939-2113-3

Gokcekuyu, Y., Ekinci, F., Guzel, M. S., Acici, K., Aydin, S., & Asuroglu, T. (2024). Artificial Intelligence in Biomaterials: A Comprehensive Review. *Applied Sciences (Basel, Switzerland)*, *14*(15), 6590. DOI: 10.3390/app14156590

Gong, L., Xiao, D., Huang, H., & Wang, R. (2020). Application of artificial intelligence in food safety. *Trends in Food Science & Technology*, *102*, 160–172.

González Enríquez, J., Jimenez Ramirez, A., Domínguez Mayo, F. J., & Garcia-Garcia, J. A. (2020). Robotic process automation: A scientific and industrial systematic mapping study. *IEEE Access : Practical Innovations, Open Solutions*, *8*, 214072–214095. DOI: 10.1109/ACCESS.2020.2974934

Goodfellow, I., Pouget-Abadie, J., Mirza, M., Xu, B., Warde-Farley, D., Ozair, S., & Bengio, Y. (2014). Generative adversarial nets. *Advances in Neural Information Processing Systems*, •••, 27.

Gosling, S. Digital logistics: Technology race gathers momentum, *16Nov.2023, Accessed 22 October 2024.*https://www.mckinsey.com/capabilities/operations/our-insights/digital-logistics-technology-race-gathers-momentum

Government Bill (House of Commons) C-27 (44-1) - first reading - Digital Charter Implementation Act, 2022 - Parliament of Canada. Government Bill (House of Commons) C-27 (44-1) - First Reading - Digital Charter Implementation Act, 2022 - Parliament of Canada. (n.d.). https://www.parl.ca/DocumentViewer/en/44-1/bill/C-27/first-reading

Government of Canada. (2024, December 17). *Artificial intelligence ecosystem.* Innovation, Science and Economic Development Canada. https://ised-isde.canada.ca/site/ised/en/artificial-intelligence-ecosystem

Government of Canada. (2024, December 3). *Pan-Canadian Artificial Intelligence Strategy*. Innovation, Science and Economic Development Canada. https://ised-isde.canada.ca/site/ai-strategy/en#pillar1

Govindan, K., Fattahi, M., & Keyvanshokooh, E. (2017). Supply chain network design for sustainable consumption and production: A critical review and research directions. *Journal of Cleaner Production*, *149*, 875–897.

Grande, V., Campos, R., Facin, A., & Batistela, G. (2022). An analysis of the benefits, challenges and methods of process selection to adopt robotic process automation. *Revista Gestão da Produção. Operações e Sistemas, 17*(1), 89–108.

Grieves, M., & Vickers, J. (2017). Digital twin: Mitigating unpredictable, undesirable emergent behavior in complex systems. In Kahlen, F., Flumerfelt, S., & Alves, A. (Eds.), *Transdisciplinary perspectives on complex systems: New findings and approaches* (pp. 85–113). Springer. DOI: 10.1007/978-3-319-38756-7_4

Guide, V. D. R.Jr, & Van Wassenhove, L. N. (2009). The evolution of closed-loop supply chain research. *Operations Research, 57*(1), 10–18. DOI: 10.1287/opre.1080.0628

Gulati, R. (1998). Does familiarity breed trust? The implications of repeated ties for contractual choice in alliances. *Academy of Management Journal, 41*(1), 85–101. DOI: 10.2307/256729

Gundeti, R., Vuppala, K., & Kasireddy, V. (2024). The Future of AI and Environmental Sustainability: Challenges and Opportunities. Exploring Ethical Dimensions of Environmental Sustainability and Use of AI, 346-371.

Gunisity, S. R. (2024). Evolution of automation to hyperautomation: Leveraging RPA, AI, ML, NLP for optimal operational efficiency. *International Journal of Computer Trends and Technology, 72*(5), 76–83. DOI: 10.14445/22312803/IJCTT-V72I3P111

Guo, X., Liu, C., Xu, W., Yuan, H., & Wang, M. (2014). A prediction-based inventory optimization using data mining models. In Y. L., Z. Q., C. J., C. J., & W. S. (Eds.), *Proceedings - 2014 7th International Joint Conference on Computational Sciences and Optimization, CSO 2014* (pp. 611 – 615). Institute of Electrical and Electronics Engineers Inc. DOI: 10.1109/CSO.2014.118

Guo, C., & Yang, X. (2011). A programming of genetic algorithm in Matlab 7.0. *Modern Applied Science*, •••, 230–235.

Guo, D., & Mantravadi, S. (2024). The role of digital twins in lean supply chain management: Review and research directions. *International Journal of Production Research, 63*(5), 1851–1872. DOI: 10.1080/00207543.2024.2372655

Gupta, K., Mane, P., Rajankar, O. S., Bhowmik, M., Jadhav, R., Yadav, S., Rawandale, S., & Chobe, S. V. (2023). Harnessing AI for Strategic Decision-Making and Business Performance Optimization. *International Journal of Intelligent Systems and Applications in Engineering, 11*(10s), 893–912. https://www.scopus.com/inward/record.uri?eid=2-s2.0-85172014962&partnerID=40&md5=a62af4010a96c70566fc6aa83e797bbb

Gupta, R. (2018). Risk Management in Food Safety: A Framework for Compliance. *Food Protection Trends*, *38*(4), 297–305.

Hahn, D., Munir, A., & Behzadan, V. (2021). Security and privacy issues in intelligent transportation systems: Classification and challenges. *IEEE Intelligent Transportation Systems Magazine*, *13*(1), 181–196. DOI: 10.1109/MITS.2019.2898973

Haleem, A., Javaid, M., Qadri, M. A., & Suman, R. (2022). Understanding the role of digital technologies in education: A review. *Sustainable Operations and Computers*, *3*, 275–285. DOI: 10.1016/j.susoc.2022.05.004

Haruna, C., Abdulkareem, S., & Hermawan, A. (2017). Neural networks optimization through genetic algorithm searches: A review. *Applied Mathematics & Information Sciences*, •••, 1543–1564.

Hawkes, C., & Ruel, M. (2006). Value chains for nutrition. Reshaping Agriculture for Nutrition and Health. 73-81.

Hermann, M., Pentek, T., & Otto, B. (2016, January). Design principles for industrie 4.0 scenarios. In *2016 49th Hawaii international conference on system sciences (HICSS)* (pp. 3928-3937). IEEE.

Highsmith, J. (2009). *Agile Project Management: Creating Innovative Products* (2nd ed.). Addison-Wesley Professional.

Hill, T. (2000). *Operations Management: Strategic Context and Managerial Analysis*. Palgrave.

Himmelsbach, V., & Billot, J. (n.d.). *AI in the Real World-Canadian Tire*. https://cdnscaleai.ca. https://www.google.com/url?sa=t&source=web&rct=j&opi=89978449&url=https://cdn.scaleai.ca/uploads/2022/05/scaleai-whitepaper-canadiantire-vf.pdf

Hochreiter, S., & Schmidhuber, J. (1997). Long short-term memory. *Neural Computation*, *9*(8), 1735–1780. DOI: 10.1162/neco.1997.9.8.1735 PMID: 9377276

Hofmann, E., Rüsch, M., & Fuger, A. (2017). Industry 4.0 and the current status as well as future prospects on logistics. *Computers in Industry*, *89*, 23–34. DOI: 10.1016/j.compind.2017.04.002

Holleran, E., Bredahl, M. E., & Zaibet, L. (1999). Private incentives for adopting food safety and quality assurance. *Food Policy*, *24*(6), 669–683. DOI: 10.1016/S0306-9192(99)00071-8

Hopp, W. J., & Spearman, M. L. (1997). *Factory Physics: Foundations of Manufacturing Management*. McGraw-Hill.

Huang, H., Lu, J., & Wang, H. (2021). Using AI and NLP to analyze social media for emerging food safety risks. *Food Control, 123*, 107893.

Infosys, B. P. M. (2024). *Infosys BPM*. https://www.infosysbpm.com/

Innovation, Science and Economic Development Canada, Government of Canada. (2024, May 27). *Eight organizations to join Canada's voluntary AI Code of Conduct*. Canada.ca. https://www.canada.ca/en/innovation-science-economic-development/news/2024/05/eight-organizations-to-join-canadas-voluntary-ai-code-of-conduct.html

International Food Policy Research Institute. (n.d.). Research publications on food security. https://www.ifpri.org

International Telecommunication Union (ITU). (2021). *Measuring digital development: Facts and figures 2021*. ITU.

Irfan, M., Sulehri, N. A., & Manickiam, N. (2024). Digital threads in turbulent times: Unraveling technostress and cleaner production in the food industry. *Frontiers in Robotics and AI, 10*, 1293904. DOI: 10.3389/frobt.2023.1293904 PMID: 38273894

IRPAAI. (2024). *Institute for Robotic Process Automation & Artificial Intelligence*. https://irpaai.com/

Ise, T., & Oba, Y. (2019). Forecasting climatic trends using neural networks: An experimental study using global historical data. *Frontiers in Robotics and AI, 6*, 32. DOI: 10.3389/frobt.2019.00032 PMID: 33501048

Ivančić, L., Suša Vugec, D., & Vukšić, V. (2019). Robotic process automation: Systematic literature review. In *Organizational Transformation and Managing Innovation in the Fourth Industrial Revolution*. Springer., DOI: 10.1007/978-3-030-30429-4_19

Ivanov, D., & Dolgui, A. (2021). A digital supply chain twin for managing the disruption risks and resilience in the era of Industry 4.0. *Production Planning and Control, 32*(9), 775–788. DOI: 10.1080/09537287.2020.1768450

Ivanov, D., Dolgui, A., & Sokolov, B. (2019). The impact of digital technology and Industry 4.0 on the ripple effect and supply chain risk analytics. *International Journal of Production Research, 57*(3), 829–846. DOI: 10.1080/00207543.2018.1488086

Ivanov, D., Tsipoulanidis, A., & Schönberger, J. (2021). *Global supply chain and operations management: A decision-oriented introduction to the creation of value* (3rd ed.). Springer. DOI: 10.1007/978-3-030-72331-6

Jadhav, P., & Dongare, R. (2023). Robotic process automation (RPA) in business processes: Benefits and challenges. *International Journal of Advanced Research in Science, Communication and Technology, 3*(1). ISSN 2581-9429.

Jahanger, A., Ozturk, I., Chukwuma Onwe, J., Joseph, T. E., & Razib Hossain, M. (2023). Do technology and renewable energy contribute to energy efficiency and carbon neutrality? Evidence from top ten manufacturing countries. *Sustainable Energy Technologies and Assessments*, *56*, 103084. DOI: 10.1016/j.seta.2023.103084

Jain, A., Sarsaiya, S., Gong, Q., Wu, Q., & Shi, J. (2024). Bioresources from biowaste for advancing manufacturing of packaging materials with modern circular economy towards a sustainable approach. *Environment, Development and Sustainability*, •••, 1–34. DOI: 10.1007/s10668-024-05734-8

Jain, V., Balakrishnan, A., Beeram, D., Najana, M., & Chintale, P. (2024). Leveraging artificial intelligence for enhancing regulatory compliance in the financial sector. *International Journal of Computer Trends and Technology*, *72*(5), 124–140. DOI: 10.14445/22312803/IJCTT-V72I5P116

Januszewski, A., Kujawski, J., & Buchalska-Sugajska, N. (2021). Benefits of and obstacles to RPA implementation in accounting firms. *Procedia Computer Science*, *192*, 4672–4680. DOI: 10.1016/j.procs.2021.09.245

Jardine, A. K., Lin, D., & Banjevic, D. (2006). A review on machinery diagnostics and prognostics implementing condition-based maintenance. *Mechanical Systems and Signal Processing*, *20*(7), 1483–1510. DOI: 10.1016/j.ymssp.2005.09.012

Javaid, M., Haleem, A., Singh, R. P., & Suman, R. (2022). Artificial intelligence applications for industry 4.0: A literature-based study. *J Ind Integr Manag*, *7*(1), 83–111. DOI: 10.1142/S2424862221300040

Jena, M. C., Mishra, S. K., & Moharana, H. S. (2020). Application of Industry 4.0 to enhance sustainable manufacturing. *Environmental Progress & Sustainable Energy*, *39*(1), 13360. DOI: 10.1002/ep.13360

Jenkins, A. (2024, November 3). Warehouse Automation Explained: Trends, Types & Best Practices. *Oracle Netsuite*. December 17, 2024, https://www.netsuite.com/portal/resource/articles/inventory-management/warehouse-automation.shtml

Jeske, T., Sitter, M., & Schilcher, U. (2019). Supply chain 4.0: Concept, challenges and implementation. *Procedia Manufacturing*, *39*, 745–754.

Jha, S. K., Bilalovic, J., Jha, A., Patel, N., & Zhang, H. (2017). Renewable energy: Present research and future scope of artificial intelligence. *Renewable & Sustainable Energy Reviews*, *77*, 297–317. DOI: 10.1016/j.rser.2017.04.018

Jiang, B., Lee, H., & Zeng, D. (2020). Supply chain resilience: A multi-method approach to analyze supply chain disruptions. *International Journal of Production Research, 58*(10), 3180–3197.

Jin, X., Shi, X., Wang, J., & Zhu, L. (2020). Application of artificial intelligence in food safety: A review. *Trends in Food Science & Technology, 99*, 94–103.

Joel, O., Oyewole, A., Odunaiya, O., & Soyombo, O. (2024). Leveraging Artificial Intelligence for Enhanced Supply Chain Optimization: A Comprehensive Review of Current Practices and Future Potentials. *International Journal of Management & Entrepreneurship Research., 6*(3), 707–721. DOI: 10.51594/ijmer.v6i3.882

Johnson, M., & Lee, S. (2023). AI-Driven Analytics for Food Safety Compliance. *Journal of Food Safety*.

Jopson, D. (2021). How Amazon uses artificial intelligence to optimize its inventory management. *Harvard Business Review*.

Jopson, M. (2021). Amazon's use of artificial intelligence to optimize inventory management and reduce holding costs. *Journal of Business Research, 125*, 29–38.

Jordan, M. I., & Mitchell, T. M. (2015). Machine learning: Trends, perspectives, and prospects. *Science, 349*(6245), 255–260. DOI: 10.1126/science.aaa8415 PMID: 26185243

Jungwirth, D., & Haluza, D. (2023). Artificial intelligence and the sustainable development goals: An exploratory study in the context of the society domain. *Journal of Software Engineering and Applications, 16*(4), 91–112. DOI: 10.4236/jsea.2023.164006

Kadow, C., Hall, D. M., & Ulbrich, U. (2020). Artificial intelligence reconstructs missing climate information. *Nature Geoscience, 13*(6), 408–413. DOI: 10.1038/s41561-020-0582-5

Kagermann, H., Helbig, J., Hellinger, A., & Wahlster, W. (2013). *Recommendations for implementing the strategic initiative INDUSTRIE 4.0: Securing the future of German manufacturing industry; final report of the Industrie 4.0 Working Group*. Forschungsunion.

Kaisler, S., Armour, F., Espinosa, J. A., & Money, W. (2013). Big data: Issues and challenges moving forward. In *2013 46th Hawaii International Conference on System Sciences* (pp. 995-1004). IEEE.

Kajla, P., Chaudhary, V., Dewan, A., Bangar, S. P., Ramniwas, S., Rustagi, S., & Pandiselvam, R. (2024). Seaweed-based biopolymers for food packaging: A sustainable approach for a cleaner tomorrow. *International Journal of Biological Macromolecules, 274,* 133166. DOI: 10.1016/j.ijbiomac.2024.133166 PMID: 38908645

Kamble, S. S., Gunasekaran, A., & Arha, H. (2020). Understanding the Blockchain technology adoption in supply chains-Indian context. *International Journal of Production Research, 58*(7), 2100–2123.

Kamble, S. S., Gunasekaran, A., & Gawankar, S. A. (2020). Achieving sustainable performance in a data-driven agriculture supply chain: A review for research and applications. *International Journal of Production Economics, 219,* 179–194. DOI: 10.1016/j.ijpe.2019.05.022

Kamble, S. S., Gunasekaran, A., & Gawankar, S. A. (2021). Achieving sustainable performance in a data-driven agriculture supply chain: A review for research and applications. *Sustainable Production and Consumption, 24,* 83–98.

Kamilaris, A., & Prenafeta-Boldú, F. X. (2018). Deep learning in agriculture: A survey. *Computers and Electronics in Agriculture, 147,* 70–90. DOI: 10.1016/j.compag.2018.02.016

Kanjilal, R., Saenz, J. E., & Uysal, I. (2025). Large-scale data-driven uniformity analysis and sensory prediction of commercial banana ripening process. *Postharvest Biology and Technology, 219,* 113203. Advance online publication. DOI: 10.1016/j.postharvbio.2024.113203

Kankanamge, N. D., Lee, E., & Zhang, L. (2022). Predictive food safety risk analysis using machine learning: A systematic review. *Food Control, 132,* 108543.

Kant, S., & Adula, M. (2024). Mediated by AI-Based Generative Re-Enforcement Learning and Work Attitude: Are Intrinsic Rewards Transforming Employee Perceived Organizational Support? In *Generative AI for Transformational Management* (pp. 83-108). IGI Global.

Kant, S., Niguse, T., & Adula, M. (2025). Family Business Entrepreneurial Orientation Effect on Improving Entrepreneurial Processes by Mediation of Open Innovation in the Horn of Africa. In *AI, Corporate Social Responsibility, and Marketing in Modern Organizations* (pp. 1-30). IGI Global Scientific Publishing.

Kant, S., & Adula, M. (2024). AI Learning and Work Attitude Mediation Between Reward and Organizational Support in Ethiopia. In *Generative AI for Transformational Management* (pp. 109–136). IGI Global. DOI: 10.4018/979-8-3693-5578-7.ch005

Kaplan, A., & Haenlein, M. (2019). Siri, Siri, in my hand: Who's the fairest in the land? On the interpretations, illustrations, and implications of artificial intelligence. *Business Horizons*, *62*(1), 15–25. DOI: 10.1016/j.bushor.2018.08.004

Karthiayani, A., & Raman, R. (2024). IoT-Based Smart Pharmacies for Optimizing Stock Management with Long Short-Term Memory Model. In S. B., A. R., K. S.K., S. K.M., S. A.V., J. S., S. N., C. A., & G. R. (Eds.), *2024 11th International Conference on Reliability, Infocom Technologies and Optimization (Trends and Future Directions), ICRITO 2024*. Institute of Electrical and Electronics Engineers Inc. DOI: 10.1109/ICRITO61523.2024.10522450

Khan, A., Khalid, W., Faisal, S., & Suliman Fadl Elseed, E. (2023). The future: Artificial intelligence's profound impact on accounting and society—Anticipating job market dynamics and evolving skill requirements for accountants. *Seybold Report*, *18*(11), 1596–1622.

Khandeparkar, A. S., Paul, R., Sridhar, A., Lakshmaiah, V. V., & Nagella, P. (2024). Eco-friendly innovations in food packaging: A sustainable revolution. *Sustainable Chemistry and Pharmacy*, *39*, 101579. DOI: 10.1016/j.scp.2024.101579

Khatri, V., & Brown, C. V. (2010). Designing data governance. *Communications of the ACM*, *53*(1), 148–152. DOI: 10.1145/1629175.1629210

Kingma, D. P., & Welling, M. (2013). Auto-encoding variational bayes. *arXiv preprint arXiv:1312.6114*.

Kollia, I., Stevenson, J., & Kollias, S. (2021). Ai-enabled efficient and safe food supply chain. *Electronics (Basel)*, *10*(11), 1–22. DOI: 10.3390/electronics10111223

Kotter, J. P. (2012). *Leading change*. Harvard Business Press.

KPMG. (2024). *KPMG*. https://assets.kpmg.com/

Kritzinger, W., Karner, M., Traar, G., Henjes, J., & Sihn, W. (2018). Digital twin in manufacturing: A categorical literature review and classification. *IFAC-PapersOnLine*, *51*(11), 1016–1022. DOI: 10.1016/j.ifacol.2018.08.474

Krizhevsky, A., Sutskever, I., & Hinton, G. E. (2012). Imagenet classification with deep convolutional neural networks. *Advances in Neural Information Processing Systems*, •••, 25.

Krupitzer, C. (2024). Generative artificial intelligence in the agri-food value chain - overview, potential, and research challenges. *Frontiers in Food Science and Technology*, *4*, 1473357. Advance online publication. DOI: 10.3389/frfst.2024.1473357

Kshetri, N. (2018). Blockchain's roles in meeting key supply chain management objectives. *International Journal of Information Management, 39*, 80–89. DOI: 10.1016/j.ijinfomgt.2017.12.005

Kshetri, N., Bhusal, C. S., Kumar, D., & Chapagain, D. (2023). SugarChain: Blockchain technology meets Agriculture — The case study and analysis of the Indian sugarcane farming. In arXiv (Cornell University). Cornell University. https://doi.org/DOI: 10.48550/arXiv.2301

Kulkarni, N., & Bansal, S. (2023). Finance business process automation using RPA. *Journal of Economics & Management Research, 4*(3), 1–5. DOI: 10.47363/JESMR/2022(4)220

Kumar, A., & Budhwar, N. (2020). Artificial Intelligence and Traceability in Food Supply Chain: A Framework for Future Research. *The Journal of Supply Chain Management, 15*(3), 245–262.

Kumari, N., & Pandey, S. (2023). Application of artificial intelligence in environmental sustainability and climate change. In *Visualization techniques for climate change with machine learning and artificial intelligence* (pp. 293–316). Elsevier. DOI: 10.1016/B978-0-323-99714-0.00018-2

Kumar, P., & Singh, A. (2019). Systems Theory and Food Safety: An Interdisciplinary Approach. *International Journal of Food Sciences, 54*(1), 55–62.

Kumar, S., Ahlawat, S., & Garg, A. (2021). The impact of AI and machine learning in supply chain management. *The Journal of Supply Chain Management, 57*(2), 15–30.

Kumar, V., Rao, P., & Shenoy, N. (2023). Transparency in smart supply chains: IoT-enabled solutions. *Supply Chain Review, 9*(1), 12–18.

Kunduru, K. R., Hogerat, R., Ghosal, K., Shaheen-Mualim, M., & Farah, S. (2023). Renewable polyol-based biodegradable polyesters as greener plastics for industrial applications. *Chemical Engineering Journal, 459*, 141211. DOI: 10.1016/j.cej.2022.141211

Kuusisto, M. (2024). Artificial Intelligence Solutions for the Recycling and Utilization of Biomaterials for Industrial Applications.

Ladva, V., Shukla, M., & Vaghela, C. (2024). The Impact of Supply Chain Delays on Inventory Levels and Sale Demand Fulfillment: Analyzing the Effects of Lead Times and In- Transit Quantities : A Quantitative Exploration of Logistics Efficiency and Inventory Optimization. *Engineering. Technology and Applied Science Research, 14*(4), 15700–15710. DOI: 10.48084/etasr.7834

Lasi, H., Fettke, P., Kemper, H. G., Feld, T., & Hoffmann, M. (2014). Industry 4.0. *Business & Information Systems Engineering*, *6*(4), 239–242. DOI: 10.1007/s12599-014-0334-4

LeCun, Y., Bengio, Y., & Hinton, G. (2015). Deep learning. *Nature*, *521*(7553), 436–444. DOI: 10.1038/nature14539 PMID: 26017442

Lee, H., & Thompson, D. (2022). Global Food Safety and Supply Chain Management. International Journal of Food Safety.

Lee, J., Bagheri, B., & Kao, H. A. (2015). A cyber-physical systems architecture for Industry 4.0-based manufacturing systems. *Manufacturing Letters*, *3*, 18–23. DOI: 10.1016/j.mfglet.2014.12.001

Lee, J., & Yoo, H. J. (2021). An overview of energy-efficient hardware accelerators for on-device deep-neural-network training. *IEEE Open Journal of the Solid-State Circuits Society*, *1*, 115–128. DOI: 10.1109/OJSSCS.2021.3119554

Lee, S., Lee, S., & Lee, Y. (2021). Artificial intelligence in supply chain management: Applications, challenges, and opportunities. *Computers & Industrial Engineering*, *157*, 107274.

Lee, Y. W., & Strong, D. M. (2003). Knowing-why about data processes and data quality. *Journal of Management Information Systems*, *20*(3), 13–39. DOI: 10.1080/07421222.2003.11045775

Legg, S., & Hutter, M. (2007). Universal intelligence: A definition of machine intelligence. *Minds and Machines*, *17*(4), 391–444. DOI: 10.1007/s11023-007-9079-x

Liao, Y., Rocha Loures, E., & Deschamps, F. (2018). Industrial internet of things: A systematic literature review and insights. *IEEE Internet of Things Journal*, *5*(6), 4515–4525. DOI: 10.1109/JIOT.2018.2834151

Light Guide Systems. (June 2024). *Smart warehousing technologies: A comprehensive guide*. Retrieved from https://www.lightguidesys.com/resource-center/blog/smart-warehousing-technologies-a-comprehensive-guide/

Li, J., Huang, Q., & Lin, W. (2022). AI-powered predictive analytics for food safety regulation. *Food Control*, *130*, 108302.

Li, J., Wang, S., & Song, S. (2021). An exploration into the mechanism of blockchain application in food supply chain traceability. *Journal of Cleaner Production*, *290*, 125921.

Li, L., Li, B., & Yu, Z. (2018). Demand forecasting using generative adversarial networks. *International Journal of Production Economics*, *203*, 12–23.

Li, N., Liu, D., & Boadu, F. (2023). The impact of digital supply chain capabilities on enterprise sustainable competitive performance: An ambidextrous view. *Industrial Management & Data Systems, 123*(6), 1670–1689. DOI: 10.1108/IMDS-11-2022-0699

Li, Q., Zhang, X., & Liu, Y. (2020). Machine learning-based detection of foodborne pathogens: A review. *Foodborne Pathogens and Disease, 17*(4), 216–227.

Liu, S., Song, X., & Hu, Z. (2019). Machine learning applications in predictive quality control of food products. *Journal of Food Quality, 2019*, 1–11.

Liu, Y., Zhang, J., & Wang, H. (2023). Adapting to global trade policy shifts: A logistics perspective. *Global Trade Journal, 15*(2), 89–105.

Lodhi, S. K., Gill, A. Y., & Hussain, H. K. (2024). Green Innovations: Artificial Intelligence and Sustainable Materials in Production. *BULLET: Jurnal Multidisiplin Ilmu, 3*(4), 492–507.

Lo, W., Yang, C. M., Zhang, Q., & Li, M. (2024). Increased Productivity and Reduced Waste with Robotic Process Automation and Generative AI-Powered IoE Services. *Journal of Web Engineering, 23*(1), 53–87. DOI: 10.13052/jwe1540-9589.2313

Luthra, S., Garg, D., & Haleem, A. (2016). The impacts of critical success factors for implementing green supply chain management towards sustainability: An empirical investigation of Indian automobile industry. *Journal of Cleaner Production, 121*, 142–158. DOI: 10.1016/j.jclepro.2016.01.095

Machado, C. G., Winroth, M. P., & Ribeiro da Silva, E. H. D. (2020). Sustainable manufacturing in Industry 4.0: An emerging research agenda. *International Journal of Production Research, 58*(5), 1462–1484. DOI: 10.1080/00207543.2019.1652777

Madakam, S., Holmukhe, R., & Jaiswal, D. (2019). The future digital workforce: Robotic process automation (RPA). *Journal of Information Systems and Technology Management, 16*(e201916005). Advance online publication. DOI: 10.4301/S1807-1775201916001

Madanaguli, A., Sjödin, D., Parida, V., & Mikalef, P. (2024). Artificial intelligence capabilities for circular business models: Research synthesis and future agenda. *Technological Forecasting and Social Change, 200*, 123189. DOI: 10.1016/j.techfore.2023.123189

Madusanka, C., Udayanga, D., Nilmini, R., Rajapaksha, S., Hewawasam, C., Manamgoda, D., & Vasco-Correa, J. (2024). A review of recent advances in fungal mycelium based composites. *Discover Materials, 4*(1), 13. DOI: 10.1007/s43939-024-00084-8

Mahmoudi, A., Adl, M., & Barbalho, S. M. (2022). The role of IoT and AI in food safety. *Food Control, 134*, 108691.

Maind, S. B., & Wankar, P. (2014). Research paper on basic of artificial neural network. *International Journal on Recent and Innovation Trends in Computing and Communication*, *2*(1), 96–100.

Makridakis, S. (2017). The forthcoming Artificial Intelligence (AI) revolution: Its impact on society and firms. *Futures*, *90*, 46–60. DOI: 10.1016/j.futures.2017.03.006

Mandvikar, S. (2023). Indexing robotic process automation products. *International Journal of Computer Trends and Technology*, *71*(4), 52–56. DOI: 10.14445/22312803/IJCTT-V71I8P108

Manning, L., & Soon, J. M. (2016). Food safety, food fraud, and food defense: A fast-evolving literature. *Current Opinion in Food Science*, *10*, 114–120. PMID: 26934423

Manzoor, A., Khan, S., Dar, A. H., Pandey, V. K., Shams, R., Ahmad, S., & Pandiselvam, R. (2023). Recent insights into green antimicrobial packaging towards food safety reinforcement: A review. *Journal of Food Safety*, *43*(4), e13046. DOI: 10.1111/jfs.13046

Mastilović, J., Kukolj, D., Kevrešan, Ž., Ostojić, G., Kovač, R., Đerić, M., & Samek, D. U. (2023). Emerging Perspectives of Blockchains in Food Supply Chain Traceability Based on Patent Analysis. In Foods (Vol. 12, Issue 5, p. 1036). Multidisciplinary Digital Publishing Institute. DOI: 10.3390/foods12051036

McKinsey & Company. (2018–2023). *McKinsey & Company*. https://www.mckinsey.com/

McKinsey & Company. (2022, June 22). *Decarbonizing logistics: Charting the path ahead*. https://www.mckinsey.com/capabilities/operations/our-insights/decarbonizing-logistics-charting-the-path-ahead

McKinsey Global Institute. (2017). *Jobs lost, jobs gained: Workforce transitions in a time of automation*. McKinsey & Company.

McQuarrie. J. (2022). "Canadian Tire". *The Canadian Encyclopedia, Historica Canada*. www.thecanadianencyclopedia.ca/en/article/canadian-tire-corporation-limited. Accessed 19 December 2024.

Medium. (2024). *Medium*. https://medium.com/

Mhatre, M. S., Dongre, M., & Thakur, P. (2015). A review paper on artificial neural network: A prediction techniques. *International Journal of Scientific and Engineering Research*, •••, 6.

Miller, R., et al. (2023). Future Innovations in AI for Supply Chain Resilience. Global Logistics Magazine.

Min, H. (2019). Artificial intelligence in supply chain management: Theory and applications. *International Journal of Logistics Research and Applications*, *22*(3), 267–275.

Ministry of Agriculture & Farmers Welfare, Government of India. (n.d.). Home. https://agricoop.nic.in

Mishra, A., Dutta, P., Jayasankar, S., Jain, P., & Mathiyazhagan, K. (2023). A review of reverse logistics and closed-loop supply chains in the perspective of circular economy. *Benchmarking*, *30*(3), 975–1020. DOI: 10.1108/BIJ-11-2021-0669

Misra, A., Sharma, S., & Srivastava, A. (2019). AI in supply chain and inventory management. *The Journal of Supply Chain Management*, *55*(1), 70–80.

Misra, S., Goh, M., & Ding, H. (2019). AI-driven inventory management systems: Automation and optimization techniques. *International Journal of Production Research*, *57*(13), 4093–4111.

Mittelstadt, B. D., Allo, P., Taddeo, M., Wachter, S., & Floridi, L. (2016). The ethics of algorithms: Mapping the debate. *Big Data & Society*, *3*(2), 2053951716679679. DOI: 10.1177/2053951716679679

Mohamed, S., & Frank, L. (2024). Improved efficiency and reduced costs: The role of robotic process automation (RPA) in automating price adjustments.

Monczka, R. M., Handfield, R. B., Giunipero, L. C., & Patterson, J. L. (2015). *Purchasing and supply chain management*. Cengage Learning.

Moreira, S., Mamede, H. S., & Santos, A. (2023). Process automation using RPA – A literature review. *Procedia Computer Science*, *219*, 244–254. DOI: 10.1016/j.procs.2023.01.287

Morgan, J. P. (2020–2021). *J.P. Morgan*. https://www.jpmorgan.com/

Mortimore, S., & Wallace, C. (2013). *HACCP: A Practical Approach*. Springer. DOI: 10.1007/978-1-4614-5028-3

Moubray, J. (1997). *Reliability-centered maintenance*. Industrial Press.

Mrówczyńska, M., Sztubecka, M., Skiba, M., Bazan-Krzywoszańska, A., & Bejga, P. (2019). The use of artificial intelligence as a tool supporting sustainable development local policy. *Sustainability (Basel), 11*(15), 4199. DOI: 10.3390/su11154199

Müller, S., Meyer, B., & Lindemann, U. (2020). Automation in supply chain management: Challenges and opportunities. *International Journal of Operations & Production Management, 40*(6), 637–657.

Nahmias, S. (2013). *Production and Operations Analysis* (7th ed.). Waveland Press.

National Food Security Mission. (n.d.). Home. https://www.nfsm.gov.in

Nawaz, R., Akhtar, R., Shahid, M. A., Qureshi, I. M., & Mahmood, M. H. (2021). Machine learning based false data injection in smart grid. *International Journal of Electrical Power & Energy Systems, 130*, 106819. DOI: 10.1016/j.ijepes.2021.106819

Nayak, A., Patnaik, A., Satpathy, I., Tripathy, S., & Patnaik, B. C. M. (2024). Role of Industrial Internet of Things (IIoT) Enabled Smart Warehousing for Sustainable Perishable Food Supply Chains. In *Digital Technology Enabled Circular Economy: Models for Environmental and Resource Sustainability*. CRC Press., DOI: 10.1201/9781003349877-9

Nayal, K., Raut, R. D., Kumar, M., Paul, S. K., & Narkhede, B. E. (2024). Role of Artificial Intelligence Capability in the Interrelation Between Manufacturing Strategies and Operational Resilience. *Global Journal of Flexible Systems Managment, 25*(1), 137–162. DOI: 10.1007/s40171-023-00367-8

Netland, T. (2023, August 11). *IKEA's Journey With Warehouse Drones*. Better Operations. https://better-operations.com/2023/08/11/ikeas-journey-with-warehouse-drones/

Noe, R. A., Hollenbeck, J. R., Gerhart, B., & Wright, P. M. (2017). *Human resource management: Gaining a competitive advantage* (10th ed.). McGraw-Hill Education.

OECD. (2019). *Artificial intelligence in society*. OECD Publishing.

Ohno, T. (1988). *Toyota Production System: Beyond Large-Scale Production*. Productivity Press.

Olabi, A. G., Abdelghafar, A. A., Maghrabie, H. M., Sayed, E. T., Rezk, H., Radi, M. A., Obaideen, K., & Abdelkareem, M. A. (2023). Application of artificial intelligence for prediction, optimization, and control of thermal energy storage systems. *Thermal Science and Engineering Progress, 39*, 101730. DOI: 10.1016/j.tsep.2023.101730

Onyeaka, H., Akinsemolu, A., Miri, T., Nnaji, N. D., Emeka, C., Tamasiga, P., Pang, G., & Al-sharify, Z. (2024). Advancing food security: The role of machine learning in pathogen detection. *Applied Food Research*, *4*(2), 100532. Advance online publication. DOI: 10.1016/j.afres.2024.100532

Osman, C.-C. (2019). Robotic process automation: Lessons learned from case studies. *Informações Econômicas*, *23*(3), 66–71. DOI: 10.12948/issn14531305/23.4.2019.06

Oyeniyi, L., Ugochukwu, C., & Mhlongo, N.Lawrence Damilare OyeniyiChinonye Esther UgochukwuNoluthando Zamanjomane Mhlongo. (2024). Robotic process automation in routine accounting tasks: A review and efficiency analysis. *World Journal of Advanced Research and Reviews*, *22*(2), 695–711. DOI: 10.30574/wjarr.2024.22.1.1156

Pandey, D. K., & Mishra, R. (2024). Towards sustainable agriculture: Harnessing AI for global food security. *Artificial Intelligence in Agriculture*, *12*, 72–84. DOI: 10.1016/j.aiia.2024.04.003

Panigrahi, R. R., Shrivastava, A. K., Qureshi, K. M., Mewada, B. G., Alghamdi, S. Y., Almakayeel, N., Almuflih, A. S., & Qureshi, M. R. N. (2023). AI Chatbot adoption in SMEs for sustainable manufacturing supply chain performance: A mediational research in an emerging country. *Sustainability (Basel)*, *15*(18), 13743. DOI: 10.3390/su151813743

Paprocki, M. (2024). DESIGNING AND DEVELOPING SUSTAINABLE PRODUCTS. Design and Assessment of Sustainable Products: A Conceptual and Practical Framework, 38.

Partel, V., Charan Kakarla, S., & Ampatzidis, Y. (2019). Development and evaluation of a low-cost and smart technology for precision weed management utilizing artificial intelligence. *Computers and Electronics in Agriculture*, *157*, 339–350. DOI: 10.1016/j.compag.2018.12.048

Pasupuleti, V., Thuraka, B., Kodete, C. S., & Malisetty, S. (2024). Enhancing Supply Chain Agility and Sustainability through Machine Learning: Optimization Techniques for Logistics and Inventory Management. *Logistics (Basel)*, *8*(3), 73. Advance online publication. DOI: 10.3390/logistics8030073

Patrício, L., Costa, L., Varela, L., & Avila, P. (2023). Sustainable implementation of robotic process automation based on a multi-objective mathematical model. *Sustainability (Basel)*, *15*(18), 15045. DOI: 10.3390/su152015045

Paula, S. K., & Azeema, A. (2011). An artificial neural network model for optimization of finished goods inventory. *International Journal of Industrial Engineering Computations*, 431-438.

Pearson, T., May, D., Leontidis, G., Swainson, M., Brewer, D., Bidaut, L., & Zisserman, A. (2019). Deep learning for environmental GPR data processing. *IEEE Geoscience and Remote Sensing Letters*, *16*(9), 1385–1389.

Perera, T., Hettige, S. P. T., Jayatilake, S. A. P., Abeygunawardhana, W., & Bandara, H. M. N. D. (2020). Blockchain-Based Product Traceability Beyond Proof of Existence. In *2022 Moratuwa Engineering Research Conference (MERCon)* (p. 1). DOI: 10.1109/MERCon50084.2020.9185207

Perumalsamy, R., & Natarajan, J. (2010). Predictive analytics using genetic algorithm for efficient supply chain inventory optimization. *2010 2nd International Conference on Computing, Communication and Networking Technologies, ICCCNT 2010*. DOI: 10.1109/ICCCNT.2010.5591607

Pesovski, I., Santos, R., Henriques, R., & Trajkovik, V. (2024). Generative ai for customizable learning experiences. *Sustainability (Basel)*, *16*(7), 3034. DOI: 10.3390/su16073034

Peterson, C., & James, R. A. (1988). Neural networks and NP-complete optimization problems: A performance study on the graph bisection problem. *Complex Systems*, •••, 59–89.

Pfleeger, C. P., Pfleeger, S. L., & Margulies, J. (2015). *Security in computing* (5th ed.). Prentice Hall Press.

Piotrowski, N. (2024). Machine learning approach to packaging compatibility testing in the new product development process. *Journal of Intelligent Manufacturing*, *35*(3), 963–975. DOI: 10.1007/s10845-023-02090-8

Porter, M. E. (1985). *Competitive Advantage: Creating and Sustaining Superior Performance*. Free Press.

Pradhan Mantri Kisan SAMPADA Yojana. (n.d.). Home. https://www.pmksy.gov.in

Prashar, D., Jha, N., Jha, S., Lee, Y.-J., & Joshi, G. P. (2020). Blockchain-Based Traceability and Visibility for Agricultural Products: A Decentralized Way of Ensuring Food Safety in India. In Sustainability (Vol. 12, Issue 8, p. 3497). Multidisciplinary Digital Publishing Institute. DOI: 10.3390/su12083497

Provost, F., & Fawcett, T. (2013). *Data Science for Business: What you need to know about data mining and data-analytic thinking*. O'Reilly Media, Inc.

Punia, S., Singh, S. P., & Madaan, J. K. (2020). From predictive to prescriptive analytics: A data-driven multi-item newsvendor model. *Decision Support Systems*, *136*, 113340. Advance online publication. DOI: 10.1016/j.dss.2020.113340

Puvanasvaran, R., Nor, K. M., & Liew, L. K. (2020). Artificial intelligence in warehouse management: An exploration of its impact on operational efficiency. *Journal of Intelligent Manufacturing*, *31*(2), 499–511.

PwC Germany. (2019). *PwC*. https://www.pwc.de/

Queiroz, M. M., Fosso Wamba, S., Chiappetta Jabbour, C. J., Lopes de Sousa Jabbour, A. B., & Machado, M. C. (2022). Adoption of Industry 4.0 technologies by organizations: A maturity levels perspective. *Annals of Operations Research*, •••, 1–27. PMID: 36217321

Radford, A., Narasimhan, K., Salimans, T., & Sutskever, I. (2018). *Improving language understanding by generative pre-training*. OpenAI.

Radhakrishnan, P., Prasad, V., & Gopalan, M. (2009). Optimizing inventory using genetic algorithm for efficient supply chain management. *Journal of Computational Science*, *5*(3), 233–241. DOI: 10.3844/jcssp.2009.233.241

Radhika, K., Mohammed, S., Tadepalli, S. K., & Puvvula, K. (2024). Food supply chain management by leveraging AI, IoT, and blockchain technologies. In *The Future of Agriculture: IoT, AI and Blockchain Technology for Sustainable Farming*. Bentham Science Publishers., DOI: 10.2174/9789815274349124010010

Raffel, C., Shazeer, N., Roberts, A., Lee, K., Narang, S., Matena, M., & Liu, P. J. (2020). Exploring the limits of transfer learning with a unified text-to-text transformer. *Journal of Machine Learning Research*, *21*(140), 1–67.

Ramesh, A., Pavlov, M., Goh, G., Gray, S., Voss, C., Radford, A., & Sutskever, I. (2021). Zero-shot text-to-image generation. In *International Conference on Machine Learning* (pp. 8821-8831). PMLR.

Ramírez, A. P., & Labadie, N. (2017). Stochastic inventory control and distribution of blood products. *Industrial Engineering and Operations Management*, 420-430.

Rane, N. (2023a). ChatGPT and Similar Generative Artificial Intelligence (AI) for Smart Industry: role, challenges and opportunities for industry 4.0, industry 5.0 and society 5.0. Challenges and Opportunities for Industry, 4.

Rane, N., Choudhary, S., & Rane, J. (2024b). Contribution of chatgpt and similar generative artificial intelligence for enhanced climate change mitigation strategies. Available at *SSRN* 4681720. DOI: 10.2139/ssrn.4681720

Rani, A., Negi, S., Fan, C., Lam, S. S., Kim, H., & Pan, S. Y. (2024). Revitalizing plastic wastes employing bio-circular-green economy principles for carbon neutrality. *Journal of Hazardous Materials, 472*, 134394. DOI: 10.1016/j.jhazmat.2024.134394 PMID: 38703690

Reddy, A. D., Gupta, S., Kumar, D., & Kashyap, S. (2024). AI-enabled food quality and safety assurance. In *Agriculture 4.0: Smart Farming with IoT and Artificial Intelligence*. CRC Press., https://www.scopus.com/inward/record.uri?eid=2-s2.0-85205659966&partnerID=40&md5=b3221c30a7f765e037a14d9eb5e96432 DOI: 10.1201/9781003570219-14

Redman, T. C. (2013). *Data driven: Profiting from your most important business asset*. Harvard Business Press.

Ribeiro, J., Lima, R., Eckhardt, T., & Paiva, S. (2021). Robotic process automation and artificial intelligence in Industry 4.0: A literature review. *Procedia Computer Science, 181*, 51–58. DOI: 10.1016/j.procs.2021.01.104

Richards, G. (2017). *Warehouse management: A complete guide to improving efficiency and minimizing costs in the modern warehouse* (3rd ed.). Kogan Page Publishers.

Richey, R.G, Glenn, R., Chowdhury, S., Davis-Sramek, B., Giannakis, M., & Dwivedi, Y. K. (2023). Artificial Intelligence in Logistics and Supply Chain Management: A Primer and Roadmap for Research. *Journal of Business Logistics, 44*(4), 532–549. DOI: 10.1111/jbl.12364

Rimel, A., Grungier, P., Gamache, M., Gendreau, M., & Rousseau, L.-M. (2021). Supervised learning and tree search for the real-time storage allocation in robotic mobile fulfillment systems. *arXiv preprint arXiv:2106.02450*. Retrieved from https://arxiv.org/abs/2106.02450

Rodrigo, R. (2023). Impact of robotic process automation (RPA) implementation in banking sector of South Asian countries. DOI: 10.13140/RG.2.2.20867.27680

Rodríguez Guzmán, M. M. (2024). Using generative AI tools to improve marketing communication: opportunities, challenges and best practices.

Rogers, D. S., & Tibben-Lembke, R. S. (1999). *Going backwards: Reverse logistics trends and practices*. Reverse Logistics Executive Council.

Rojas-Reyes, J. J., Rivera-Cadavid, L., & Peña-Orozco, D. L. (2024). Disruptions in the food supply chain: A literature review. *Heliyon, 10*(14), e34730. DOI: 10.1016/j.heliyon.2024.e34730 PMID: 39149054

Román, A. R., & Menéndez, M. R. (2024). The Future of Retail: Harnessing Generative AI for Disruptive Innovation and Sector Transformation. In *Reshaping Marketing Science in Wholesaling and Retailing*. IGI Global., DOI: 10.4018/979-8-3693-6145-0.ch013

Rombach, R., Blattmann, A., Lorenz, D., Esser, P., & Ommer, B. (2022). High-resolution image synthesis with latent diffusion models. In *Proceedings of the IEEE/CVF Conference on Computer Vision and Pattern Recognition* (pp. 10684-10695). DOI: 10.1109/CVPR52688.2022.01042

Roszkowska, P. (2020). Fintech in financial reporting and audit for fraud prevention and safeguarding equity investments. *Journal of Accounting & Organizational Change, Advance online publication.*

Ruan, J., Shi, Y., Han, D., & Wen, Q. (2019). Deep learning for supply chain demand forecasting: Methods, challenges, and directions. *Computers & Industrial Engineering*, *127*, 225–234.

Russell, S. J., & Norvig, P. (2021). *Artificial intelligence: A modern approach* (4th ed.). Pearson.

Rustia, D. J. A., Lin, C. E., Chung, J.-Y., Zhuang, Y.-J., Hsu, J.-C., & Lin, T.-T. (2020). Application of an image and environmental sensor network for automated greenhouse insect pest monitoring. *Journal of Asia-Pacific Entomology*, *23*(1), 17–28. DOI: 10.1016/j.aspen.2019.11.006

Rutschi, C., & Dibbern, J. (2020). Towards a framework of implementing software robots: Transforming human-executed routines into machines. *The Data Base for Advances in Information Systems*, *51*(2), 104–128. DOI: 10.1145/3380799.3380808

Saheb, T., Dehghani, M., & Saheb, T. (2022). Artificial intelligence for sustainable energy: A contextual topic modeling and content analysis. *Sustainable Computing : Informatics and Systems*, *35*, 100699. DOI: 10.1016/j.suscom.2022.100699

Santos, F., Pereira, R., & Braga de Vasconcelos, J. (2019). Toward robotic process automation implementation: An end-to-end perspective. *Business Process Management Journal*, *26*(2), 405–420. Advance online publication. DOI: 10.1108/BPMJ-12-2018-0380

Sarker, M. N. I., Wu, M., & Liu, X. (2022). Resilient supply chain management through AI-powered risk analysis in the post-COVID era. *International Journal of Logistics Research and Applications*, *25*(3), 475–492.

Schein, E. H. (2010). *Organizational culture and leadership* (Vol. 2). John Wiley & Sons.

Schneider, K., van der Vorst, J. G., & Garcia-Martinez, M. (2021). The impact of digitalization on food supply chains: Insights from the European Union. *Sustainability, 13*(16), 9150.

Schuh, G., Anderl, R., Gausemeier, J., ten Hompel, M., & Wahlster, W. (Eds.). (2017). *Industrie 4.0 Maturity Index. Managing the Digital Transformation of Companies (acatech STUDY)*. Herbert Utz Verlag.

Schwabacher, M. (2005). A survey of data-driven prognostics. In AIAA Infotech@ Aerospace Conference (p. 7002). DOI: 10.2514/6.2005-7002

Scott, J. (2021, March 16). *NuPort Robotics receives $1 million from Ontario for an autonomous trucking project with Canadian Tire*. Canadian Startup News & Tech Innovation. https://betakit.com/nuport-robotics-receives-1-million-from-ontario-for-autonomous-trucking-project-with-canadian-tire/

Sedkaoui, S., & Benaichouba, R. (2024). Generative AI as a transformative force for innovation: A review of opportunities, applications and challenges. *European Journal of Innovation Management*. Advance online publication. DOI: 10.1108/EJIM-02-2024-0129

Senanayake, T. E., Mahanama, T. V., & Prasadika, J. (2024). Strategizing Inventory Write-Off Process in Sri Lankan Apparel Sector: A Systematic Literature Review. *ICARC 2024 - 4th International Conference on Advanced Research in Computing: Smart and Innovative Trends in Next Generation Computing Technologies*, 334 – 339. DOI: 10.1109/ICARC61713.2024.10499743

Sengar, S. S., Hasan, A., Kumar, S., & Carroll, F. (2024, August 14). Bin, Kumar, S., & Carroll, F. (2024). Generative Artificial Intelligence: A Systematic Review and Applications. *Multimedia Tools and Applications*. Advance online publication. DOI: 10.1007/s11042-024-20016-1

Shafique, M. N., Adeel, U., & Rashid, A. (2024). The Synergy Between Industry 5.0 and Circular Economy for Sustainable Performance in the Chinese Manufacturing Industry. *Sustainability (Basel), 16*(22), 9952. DOI: 10.3390/su16229952

Shahin, M., Chen, F. F., Maghanaki, M., Firouzranjbar, S., & Hosseinzadeh, A. (2024). Evaluating the fidelity of statistical forecasting and predictive intelligence by utilizing a stochastic dataset. *International Journal of Advanced Manufacturing Technology*. Advance online publication. DOI: 10.1007/s00170-024-14505-8

Sharma, M., Luthra, S., Joshi, S., & Kumar, A. (2020). Developing a framework for enhancing survivability of sustainable supply chains during and post-COVID-19 pandemic. *International Journal of Logistics Management, 31*(4), 869–885.

Sheffi, Y. (2005). *The Resilient Enterprise: Overcoming Vulnerability for Competitive Advantage*. MIT Press.

Shi, J., Jain, R., Duan, R., & Ramani, K. (2023). Understanding Generative AI in Art: An Interview Study with Artists on G-AI from an HCI Perspective. arXiv preprint arXiv:2310.13149.

Shi, W., Cao, J., Zhang, Q., Li, Y., & Xu, L. (2016). Edge computing: Vision and challenges. *IEEE Internet of Things Journal*, *3*(5), 637–646. DOI: 10.1109/JIOT.2016.2579198

Shrivastava, S., & Pal, S. N. (2019). A framework for next generation agricultural marketing system in Indian context. *2019 5th IEEE International WIE Conference on Electrical and Computer Engineering, WIECON-ECE 2019 - Proceedings*. DOI: 10.1109/WIECON-ECE48653.2019.9019983

Shriyam, S., Palkar, P., & Srivastava, A. (2023). On fulfilling the exigent need for automating and modernizing logistics infrastructure in India: Enabling AI-based integration, digitalization, and smart automation of industrial parks and robotic warehouses. *arXiv preprint arXiv:2310.01077*. Retrieved from https://arxiv.org/abs/2310.01077

Shukla, S. S., Jain, M., Jain, A., Bajpai, S., Mishra, A. K., & Pramanik, S. (2024). The Circular Economy: Changing the Future of India. In Green Transition Impacts on the Economy, Society, and Environment (pp. 149-165). IGI Global.

Silver, E. A., Pyke, D. F., & Peterson, R. (1998). *Inventory management and production planning and scheduling* (3rd ed.). John Wiley & Sons.

Silver, E. A., Pyke, D. F., & Thomas, D. J. (2017). *Inventory and Production Management in Supply Chains* (4th ed.). CRC Press.

Singh, A., Patel, M., & Verma, R. (2021). Upskilling logistics professionals for smart warehousing. *Human Resource Development Quarterly*, *32*(1), 78–95.

Singh, P., Pandey, V. K., Singh, R., Negi, P., Maurya, S. N., & Rustagi, S. (2024). Substantial Enhancement of Overall Efficiency and Effectiveness of the Pasteurization and Packaging Process Using Artificial Intelligence in the Food Industry. *Food and Bioprocess Technology*, •••, 1–16.

Smith, A. (2022). AI in Food Supply Chains: Enhancing Resilience. Food Tech Journal.

Smith, J., Johnson, R. D., & Williams, C. (2021). Navigating food safety regulations with artificial intelligence: An application of natural language processing in compliance management. *Journal of Food Protection*, *84*(3), 451–460.

Sonck-Rautio, K., Lahtinen, T., & Tynkkynen, N. (2024). Consumer meaning-making of packaging functions for sustainable food packaging–Insights from qualitative research in Finland. *Current Research in Environmental Sustainability*, *7*, 100259. DOI: 10.1016/j.crsust.2024.100259

Šperka, R., & Halaška, M. (2023). The performance assessment framework (PPA-FR) for RPA implementation in a loan application process using process mining. *Information Systems and e-Business Management*, *21*(2), 277–321. DOI: 10.1007/s10257-022-00602-2

Spitalleri, A., Kavasidis, I., Cartelli, V., Mineo, R., Rundo, F., Palazzo, S., Spampinato, C., & Giordano, D. (2023). BioTrak: A Blockchain-based Platform for Food Chain Logistics Traceability. In arXiv (Cornell University). Cornell University. https://doi.org//arxiv.2304.09601DOI: 10.48550

Stadtler, H. (2015). Supply chain management: An overview. In Stadtler, H., Kilger, C., & Meyr, H. (Eds.), *Supply chain management and advanced planning: Concepts, models, software, and case studies* (pp. 9–36). Springer. DOI: 10.1007/978-3-642-55309-7_1

Stobbe, R. (2024, August). *The state of AI Regulation in Canada (2024)*. Field Law. https://www.fieldlaw.com/News-Views-Events/237102/The-State-of-AI-Regulation-in-Canada-2024)

Stock, T., & Seliger, G. (2016). Opportunities of sustainable manufacturing in industry 4.0. *Procedia CIRP*, *40*, 536–541. DOI: 10.1016/j.procir.2016.01.129

Stouffer, K., Pillitteri, V., Lightman, S., Abrams, M., & Hahn, A. (2015). *Guide to industrial control systems (ICS) security*. National Institute of Standards and Technology. DOI: 10.6028/NIST.SP.800-82r2

Strijbosch, L. W. G., & Moors, J. J. A. (1999). Simple expressions for safety factors in inventory control. *Tilburg. Ekonometria*, •••, 1–25.

Suali, A. S., Srai, J. S., & Tsolakis, N. (2024). The role of digital platforms in e-commerce food supply chain resilience under exogenous disruptions. *Supply Chain Management*, *29*(3), 573–601. DOI: 10.1108/SCM-02-2023-0064

Sumathi, N., & Jeyalakshmi, A. (2022). Implementing robotic process automation to improve operational efficiency in the health-care sector. *International Journal of Health Sciences*, *6*(S1), 9666–9672. Advance online publication. DOI: 10.53730/ijhs.v6nS1.7256

Sun, Y., Gao, C., Li, J., Wang, R., & Liu, J. (2019b). Quantifying the effects of urban form on land surface temperature in subtropical high-density urban areas using machine learning. *Remote Sensing (Basel)*, *11*(8), 959. DOI: 10.3390/rs11080959

Tambe, P., Cappelli, P., & Yakubovich, V. (2019). Artificial intelligence in human resources management: An agenda for research and practice. *California Management Review*, *61*(4), 5–26. DOI: 10.1177/0008125619867910

Tang, C. S. (2006). Perspectives in supply chain risk management. *International Journal of Production Economics*, *103*(2), 451–488. DOI: 10.1016/j.ijpe.2005.12.006

Tao, J., Li (2016). Design Optimization of Amazon Robotics. Automation. Control and Intelligent Systems, 4(2), 48-52.

Tao, F., Cheng, J., Qi, Q., Zhang, M., Zhang, H., & Sui, F. (2018). Digital twin-driven prognostics and health management for complex equipment. *CIRP Annals*, *67*(1), 169–172. DOI: 10.1016/j.cirp.2018.04.055

Tao, F., Zhang, H., Liu, A., & Nee, A. Y. C. (2018). Digital twin in industry: State-of-the-art. *IEEE Transactions on Industrial Informatics*, *15*(4), 2405–2415. DOI: 10.1109/TII.2018.2873186

Thekkethil, M., Shukla, V., Beena, F., & Chopra, A. (2021). Robotic process automation in banking and finance sector for loan processing and fraud detection. In *2021 6th International Conference on Recent Innovations in Technology and Engineering (ICRITO)* (pp. 1–6). IEEE. DOI: 10.1109/ICRITO51393.2021.9596076

Thompson, B. (2024, January 25). *Amazon Grows To More Than 750,000 Robots*. Global Trade News. https://incodocs.com/blog/amazon-robot-growth/

Thomson Reuters. (2022). *Thomson Reuters*. Retrieved May 18, 2024, from https://www.thomsonreuters.com/

Tian, F. (2017). A supply chain traceability system for food safety based on HACCP, blockchain & Internet of Things. *IEEE International Conference on Service Systems and Service Management*, 1-6.

Tian, Z., Shi, X., & Hong, S.-M. (2021). Exploring data-driven building energy-efficient design of envelopes based on their quantified impacts. *Journal of Building Engineering*, *42*, 103018. DOI: 10.1016/j.jobe.2021.103018

Tidd, J., & Bessant, J. (2009). *Innovation and Entrepreneurship: The Lean Startup Approach to Disruptive Innovation*. Wiley.

Tiwari, S., Wee, H. M., & Daryanto, Y. (2018). Big data analytics in supply chain management: A state-of-the-art literature review. *Computers & Operations Research, 98*, 212–227.

Toneguzzi, M. (2024, May 14). *Canadian Tire launches AI shopping assistants and humanoid robots to enhance customer service and operations [interview]*. Retail Insider. https://retail-insider.com/retail-insider/2024/05/canadian-tire-launches-ai-shopping-assistant-and-humanoid-robots-to-enhance-customer-service-and-operations-interview/

Treacy, S., Adyanthaya, A., Kearny, C., Anand, J., O'Sullivan, K., & Xu, Y. (2023). From hype to reality: Navigating the challenges of RPA implementation. In *Proceedings of the 18th European Conference on Innovation and Entrepreneurship* (pp. 875–882). DOI: 10.34190/ecie.18.2.1721

Tse, Y. K., & Tan, K. H. (2020). Managing food supply chain risks: The interplay of demand and supply management practices. *Transportation Research Part E, Logistics and Transportation Review, 136*, 101887.

Tufano, P., & Giustiniano, L. (2020). AI in the Supply Chain: Optimizing Order Quantities and Timing. *The Journal of Supply Chain Management, 56*(4), 45–62.

Tukamuhabwa, B. R., Stevenson, M., Busby, J., & Zorzini, M. (2015). Supply chain resilience: Definition, review and theoretical foundations for future research. *International Journal of Production Research, 53*(18), 5592–5623. DOI: 10.1080/00207543.2015.1037934

Tungsten Automation. (2023). *Tungsten Automation*. Retrieved May 21, 2024, from https://www.tungstenautomation.com/

Tzounis, A., Katsoulas, N., & Fountas, S. (2017). Robotics in agriculture. A review. *Agronomy for Sustainable Development, 37*(3), 21.

UiPath. (2021, 2022, 2023). *UiPath*. Retrieved May 20, 2024, from https://www.uipath.com/

United Nations. (2015). *Transforming our world: The 2030 Agenda for Sustainable Development*. United Nations.

Using AI for smarter demand forecasting. IKEA global. (2021, May 27). https://www.ikea.com/global/en/stories/design/using-artificial-intelligence-for-smarter-demand-forecasting-210527/

Van Geest, M., Tekinerdogan, B., & Catal, C. (2021). Design of a reference architecture for developing smart warehouses in industry 4.0. *Computers in Industry*, *124*, 103343. Advance online publication. DOI: 10.1016/j.compind.2020.103343

Vaswani, A., Shazeer, N., Parmar, N., Uszkoreit, J., Jones, L., Gomez, A. N., & Polosukhin, I. (2017). Attention is all you need. *Advances in Neural Information Processing Systems*, •••, 30.

Vegesna, V. V., Prabhakaran, S. M., & Whig, P. (2024). AI for food safety: Leveraging artificial intelligence to ensure a safe and reliable food supply. In *Nutrition Controversies and Advances in Autoimmune Disease*. IGI Global., DOI: 10.4018/979-8-3693-5528-2.ch019

Vicente, D., Proença, D. N., & Morais, P. V. (2023). The role of bacterial polyhydroalkanoate (PHA) in a sustainable future: A review on the biological diversity. *International Journal of Environmental Research and Public Health*, *20*(4), 2959. DOI: 10.3390/ijerph20042959 PMID: 36833658

Vijai, C., Suriyalakshmi, S., & Elayaraja, M. (2020). The future of robotic process automation (RPA) in the banking sector for better customer experience. *Shanlax International Journal of Commerce*, *8*(3), 61–65. DOI: 10.34293/commerce.v8i2.1709

Villar, A., & Khan, N. (2021). Robotic process automation in banking industry: A case study on Deutsche Bank. *Journal of Banking and Financial Technology*, *5*(1). Advance online publication. DOI: 10.1007/s42786-021-00030-9

Vladimir, M., & Tasho, T. (2011). A programme implementation of several inventory control algorithms. *Bulgarian Academy of Sciences*, 64-78.

Waller, M. A., & Fawcett, S. E. (2013). Data science, predictive analytics, and big data: A revolution that will transform supply chain design and management. *Journal of Business Logistics*, *34*(2), 77–84. DOI: 10.1111/jbl.12010

Wamba, S. F., Akter, S., & Bhattacharya, M. (2020). Artificial Intelligence applications and their implications in warehousing and supply chain management. *Journal of Business Research*, *120*, 38–48.

Wamba, S. F., Akter, S., & Ren, S. (2020). Artificial intelligence in supply chain management: A survey and research directions. *The Journal of Supply Chain Management*, *56*(4), 29–42.

Wamba-Taguimdje, S.-L., Fosso Wamba, S., Kala Kamdjoug, J. R., & Tchatchouang Wanko, C. E. (2020). Influence of artificial intelligence (AI) on firm performance: The business value of AI-based transformation projects. *Business Process Management Journal*, *26*(7), 1893–1924. DOI: 10.1108/BPMJ-10-2019-0411

Wang, G., Gunasekaran, A., Ngai, E. W., & Papadopoulos, T. (2016). Big data analytics in logistics and supply chain management: Certain investigations for research and applications. *International Journal of Production Economics*, *176*, 98–110. DOI: 10.1016/j.ijpe.2016.03.014

Wang, S., Ghadge, A., & Aktas, E. (2024). Digital transformation in food supply chains: An implementation framework. *Supply Chain Management*, *29*(2), 328–350. DOI: 10.1108/SCM-09-2023-0463

Wang, X., Lin, X., & Zhang, Q. (2021). Integrating blockchain and Internet of Things for food traceability: Research trends, challenges, and future prospects. *Journal of Food Engineering*, *292*, 110369.

Wang, Y., Zhang, Q., & Li, Z. (2020). A machine learning-based approach for optimizing food supply chain operations. *Journal of Cleaner Production*, *275*, 123482.

Warehousing, S. (August 2024). *The role of real-time data in enhancing inventory turnover rates*. Retrieved from https://www.smartwarehousing.com/blog/role-of-real-time-data-in-enhancing-inventory-turnover-rates

Westerman, G., Calméjane, C., Ferraris, P., & Bonnet, D. (2014). The Digital Advantage: How Digital Leaders Outperform Their Peers in Every Industry. *MIT Sloan Management Review*.

Whitmore, A., Agarwal, A., & Da Xu, L. (2015). The Internet of Things—A survey of topics and trends. *Information Systems Frontiers*, *17*(2), 261–274. DOI: 10.1007/s10796-014-9489-2

World Bank. (2021). *World Development Report 2021: Data for Better Lives*. World Bank.

World Economic Forum (WEF). (2020). *The Future of Jobs Report 2020*. World Economic Forum.

World Health Organization. (2020).. . *Food Safety (Tokyo, Japan)*.

Wu, H., Liu, J., & Liang, B. (2024). AI-Driven Supply Chain Transformation in Industry 5.0: Enhancing Resilience and Sustainability. *Journal of the Knowledge Economy*, *16*(1), 1–43. DOI: 10.1007/s13132-024-01999-6

Wunderlich, S. M., & Martinez, N. M. (2018). Conserving natural resources through food loss reduction: Production and consumption stages of the food supply chain. *International Soil and Water Conservation Research*, *6*(4), 331–339. DOI: 10.1016/j.iswcr.2018.06.002

Wu, Y., & Pagell, M. (2011). Balancing supply chain efficiency and flexibility: A multi-case study analysis. *International Journal of Operations & Production Management*, *31*(5), 566–591.

Xiaonan, Z., Lai, P.-L., Wang, X., & Ma, F. (2023). Evolution, Challenges, and Opportunities of Transportation Methods in the Last-Mile Delivery Process. *Systems*, *11*(10), 509. DOI: 10.3390/systems11100509

Xu, L. D., Xu, E. L., & Li, L. (2018). Industry 4.0: State of the art and future trends. *International Journal of Production Research*, *56*(8), 2941–2962. DOI: 10.1080/00207543.2018.1444806

Yakoubi, S. (2024). Sustainable Revolution: AI-Driven Enhancements for Composite Polymer Processing and Optimization in Intelligent Food Packaging. *Food and Bioprocess Technology*, •••, 1–26.

Yampolskiy, R. V., & Fox, J. (2012). Artificial general intelligence and the human mental model. In *Singularity hypotheses* (pp. 129–145). Springer. DOI: 10.1007/978-3-642-32560-1_7

Yao, F., Li, C., Nekipelov, D., Wang, H., & Xu, H. (2024). Human vs. Generative AI in Content Creation Competition: Symbiosis or Conflict? arXiv preprint arXiv:2402.15467.

Yao, M. (2024). RPA technology enables highly automated development of corporate financial accounting processes. *Applied Mathematics and Nonlinear Sciences*, *9*(1), 1–13. DOI: 10.2478/amns-2024-0541

Yaroson, E. V., Chowdhury, S., Mangla, S. K., & Dey, P. K. (2024). Unearthing the interplay between organisational resources, knowledge and industry 4.0 analytical decision support tools to achieve sustainability and supply chain wellbeing. *Annals of Operations Research*, *342*(2), 1–48. DOI: 10.1007/s10479-024-05845-5

Yuan, X., Wang, Y., Yang, C., & Gui, W. (2020). Stacked isomorphic autoencoder based soft analyzer and its application to sulfur recovery unit. *Information Sciences*, *534*, 72–84. DOI: 10.1016/j.ins.2020.03.018

Yue, W., Yao, Y., Su, M., Rong, Q., & Xu, C. (2023). Identifying distributions of urban ecosystem health based on Latin-hypercube sampling and multi-criteria decision analysis framework. *Ecological Indicators*, *147*, 109957. DOI: 10.1016/j.ecolind.2023.109957

Yue, X.-L., & Gao, Q.-X. (2018). Contributions of natural systems and human activity to greenhouse gas emissions. *Advances in Climate Change Research*, *9*(4), 243–252. DOI: 10.1016/j.accre.2018.12.003

Zahraee, S. M., Khalaji Assadi, M., & Saidur, R. (2016). Application of artificial intelligence methods for hybrid energy system optimization. *Renewable & Sustainable Energy Reviews*, *66*, 617–630. DOI: 10.1016/j.rser.2016.08.028

Zhang, S., Roller, S., Goyal, N., Artetxe, M., Chen, M., Chen, S., . . . Zettlemoyer, L. (2022). OPT: Open Pre-trained Transformer Language Models. *arXiv preprint arXiv:2205.01068.*

Zhang, Q., Ullah, A., Ashraf, S., & Abdullah, M. (2024). Synergistic Impact of Internet of Things and Big-Data-Driven Supply Chain on Sustainable Firm Performance. *Sustainability (Basel)*, *16*(13), 5717. DOI: 10.3390/su16135717

Zhao, S., Chen, H., Liu, R., & Li, G. (2023). Deep Learning for Food Quality and Safety: A Review. *Comprehensive Reviews in Food Science and Food Safety*, *22*(1), 531–556.

Zhao, Y., Zhan, Y., & Liu, G. (2019). IoT and AI technologies for food safety surveillance: A review. *Sensors (Basel)*, *19*(19), 4185. PMID: 31561597

Zheng, G., Haq, M. Z. U., Huo, B., Zhang, Y., & Yue, X. (2024). Leveraging intellectual capital for building a supply chain circular economy system: A knowledge-based view. *International Journal of Production Economics*, *272*, 109225. DOI: 10.1016/j.ijpe.2024.109225

Zhou, X., & Li, F. (2022). Sustainability challenges in automated warehousing. *Journal of Environmental Management*, *14*(3), 89–97.

Zong, Z., & Guan, Y. (2024). AI-Driven Intelligent Data Analytics and Predictive Analysis in Industry 4.0: Transforming Knowledge, Innovation, and Efficiency. *Journal of the Knowledge Economy*, *16*(1), 1–40. DOI: 10.1007/s13132-024-02001-z

About The Contributors

Geetika Madaan is an Assistant Professor at the Research Centre of Marwadi University (NAAC A+, Gujarat). She holds a Master's degree in Business Administration with a specialization in Finance (Honors) from Apeejay School of Management, Jalandhar, Punjab, and earned her Doctorate in Business Management from Chandigarh University, Punjab. With over nine years of academic teaching experience and six years in the corporate sector, Dr. Madaan is a recognized expert in Technology, Business Management, and International Business. She has an impressive portfolio of more than 45 research publications, including contributions to one Q1 and two Q2 journals, and has edited three books. Her research contributions are well-regarded in both national and international academic circles.

Amrinder Singh is currently holding a position as a professor in Jain(Deemed to be a university student in Bangalore). He has a Master's in Business Administration and a Doctorate in Business Studies from Punjabi University, Patiala. With more than 15 years of teaching experience, he is a personality to reckon with in marketing and research. He has more than 30 research papers published to his credit in various National and International journals of repute. He has published one international book on CRM in General Insurance. He received the Outstanding Researcher Award from the Punjab Institute of Management and Technology. He attended a summer exchange program at the University Of Derby, UK, in Aug 2018 and organized a joint workshop on Strategic management and Marketing strategies. Dr.Singh has received the award for being the best resource person for National and International collaborations. During his tenure in the last organization, he received a letter of appreciation from CT University, Ludhiana, for his overall good performance during his job in 2020. He is a member of academic and administrative bodies at various universities and institutions. He is a research guide nationally and internationally with Liverpool John Moores University.

Bhupinder Pal Singh Chahal is an Adjunct Marketing Professor at Yorkville University in New Westminster, BC, Canada. Additionally, the Chartered Association of Business Schools in London has awarded him the Certified Management and Business Educator (CMBE) designation. With a distinguished career spanning over 15 years in the realm of management education, Dr. Chahal boasts an impressive background as the Professor & Head of Department at Apex Institute of Technology (AIT-Management) within Chandigarh University, a QS Asia-ranked institution, where he dedicated 11 years of his professional journey. Dr. Chahal's expertise lies in Marketing and International Business, and he has provided mentorship for research work across diverse topics within these domains, contributing to papers, case studies, chapters in books, and publications in international and national refereed journals, including those listed in Scopus, ABDC, and UGC. His research contributions have been showcased in conference proceedings at multiple international and national conferences, as well as in edited books. In addition to mentoring Ph.D. research scholars on topics such as knowledge management, financial inclusion, online and social media marketing, international business, and entrepreneurship development, Dr. Chahal has successfully supervised the completion of five Ph.D. research projects under his guidance. Utilizing participatory and innovative teaching methodologies, Dr. Chahal imparts comprehensive knowledge to students, aligning with the course curriculum. Dr. Chahal's wealth of experience extends to conducting training and capacity-building programs at the national level. His innate expertise lies in team building, survey implementation, data analysis, competency enhancement, and strategic planning. With two international patents to his credit, Dr. Chahal has demonstrated exceptional accomplishments in the realms of research and innovation. Over his 18-year career (3 years Ph.D. research scholar), with a predominant focus on academia for over 15 years, Dr. Chahal has held pivotal decision-making positions in various committees, teams, and clubs dedicated to transforming institutions into Centres of Excellence. As a dedicated educator, he has a natural flair for teaching and evaluating learners, employing special assessment techniques that challenge and enhance the learning experience. Dr. Chahal remains committed to staying updated with the latest business management, marketing, and knowledge management developments.

Arokiaraj David is currently an associate professor at SBS Swiss Business School, RAK, UAE. He has written many book chapters and articles for reputed journals. He holds a doctorate from Pondicherry Central University. He is Skilled in Analytical Skills, Curriculum Development, Employee Relations, Soft Skills, E-Learning, and Start-up mentoring. His recent research publications include Sl.no. Patents/ Journals/ Books/ Books Chapters in Nos. Number of Papers Indexed in Scopus Journal 25 Number of papers published in ABDC Journal 08 Number of

Patents 08 Number of Books Published 04 Number of Book Chapters 16 Number of papers published in Springer 3 Number of articles in Web of Science (WoS) 05 Total number of published 75

Gurmeet Singh, With over 27 years of academic teaching experience, Professor Gurmeet Singh is a renowned scholar and academic leader. As the current Deputy Vice-Chancellor, Research and Innovation and Head of the School of Business & Management and Director of MBA and DBA Programmes at the University of the South Pacific, he has contributed over 185 articles to prominent international journals and conferences. He has a distinguished record of service, having served on the editorial boards of numerous top-ranked journals, secured significant research funding, and been recognized with various awards for his contributions to academia. Professor Singh's primary research expertise lies in the domain of Marketing. He delves into areas such as Consumer Behavior, Electronic Marketing, and the innovative intersection of Services Marketing with Artificial Intelligence. Furthermore, he possesses extensive knowledge in supply chain management and has dedicated his research efforts to SMEs (Small and medium-sized enterprises). The transformative impact of the digital era has led Dr. Singh to explore Digital Governance & Learning, with a focus on E-Governance and E-Learning platforms. His keen interest in the rapidly evolving field of Information Systems emphasizes System Development, Business Informatics, and Digital Transformation strategies. Additionally, Dr. Singh has cultivated a research niche in Tourism, studying its economic, sociocultural, and environmental implications in today's global context.

***

Harleen Aggarwal has been teaching Business Management courses at the University level for 25 years. She has to her credit the publication of 11 books under the titles of 'Business Environment' and 'Strategic Management' for various university-level courses. Harleen has also presented and published research papers in India and Canada as well.

Neeraj Anand is a Professor and Head of Department at Chitkara Business School, Chitkara University, Punjab India. He was the Head of Department at University of Petroleum and Energy Studies, India. He was the Associate Dean Research during 2017. He received professional memberships in Indian Rail Transport Institute; IIM Bangalore; Energy Institute, London and IGET India, to name a few. His research interests include supply chain management and research methodology.

Madhu Arora is presently working at Symbiosis Institute of Computer Studies and Research, Symbiosis International (Deemed University). She completed her PhD (Management) from University of Petroleum and Energy Studies Dehradun, India. She holds a Post-graduate degree in Business Administration with specialization in Operations Management from IGNOU, India along with Masters in Personnel Management from Pune University India. Her research interests include cold chain management, consumer adoption and business. She has more than 20 years experience in teaching.

Muralidhar L B is a professionally qualified dental surgeon. He holds Ph.D. from Canara Bank School of Management Studies, Bangalore University after his MBA from Christ University. He has 18 years of total teaching experience spanning across dental education, healthcare management and general management and 4 years of corporate experience. He has worked as research associate in Institute for Social and Economic Change, Bangalore in a project entitled as "Management By and For Common Man: An Exploration of the Concept and Direction for Future"post Ph.D. He had secured 28th rank in S.S.L.C examination for Karnataka State in 1993. His 21 patents have been published and 2 design patents have been granted. He has 32 research publications to his credit out of which 6 are Scopus-indexed and 5 are ABDC indexed. He has co-authored 5 books. He is currently working as Assistant Professor in Department of Business Management Studies, School of Commerce, Jain Deemed to be University, Bangalore.

R. Baghel is working as Associate professor, department of mathematics, Poornima university. He has 15 years of experience in academics and research. He has published more than 20 research papers in high indexing journals.

RituDahiya is a dedicated academician and faculty member in the Department of Chemistry at Chhotu Ram Arya College, located in Sonepat, Haryana. With a passion for chemistry and education, Dr. Ritu has made significant contributions to the academic community. Her work focuses on fostering a deep understanding of chemistry while also mentoring students to excel in their studies and careers. Dr. Ritu's commitment to teaching and research continues to inspire the next generation of scientists at her institution.

G. Sofia serves as an Associate Professor in the Department of Computer Science at Lady Doak College, Madurai. She holds an M.C.A. from Manonmaniam Sundaranar University, Tirunelveli, an M.Phil. in Computer Science from Mother Teresa Women's University, Kodaikanal, and a Ph.D. in Computer Science from Bharathiar University, Coimbatore. She has actively contributed to the academic community by presenting numerous research papers at national and international conferences. She has published articles in reputed journals and authored book chapters in her areas of expertise. She has served as a resource person for various workshops and faculty development programmes, where she shares her knowledge and insights with academic peers and professionals. Her outstanding contributions to teaching and research have earned her accolades such as the Katie Wilcox Teacher of the Year Award and the Best Paper Award at the Virtual Conference on Smart Computing and Advanced Communication. She holds academic memberships in various institutions, serving as a subject expert and member of Board of Studies meetings, member of doctoral committees, and external expert for academic council meeting. Her areas of specialization include Image Processing, Data Science, and Virtual Reality.

Swapna H R holds two Ph.D degrees in Management from CMR University, Bangalore and second PhD with Rayalaseema University, Kurnool. She is an MBA in Finance specialization from VTU University. She has over 15 years of teaching experience in the area of Finance, Analytics, Statistics and Human Resource Management and 4 years industry experience. Her research expertise has been in the area of Statistics, Behavioural Finance, Employee Engagement, Artificial Intelligence, Entrepreneurship, research on millennials, customer engagement, social media and also in health sectors. She has published over 100 research articles in National and International Journals, where 46 articles have been published in Scopus, Web of science (WoS), Emerald Publications, Springer, ABDC and has presented various papers in National and International Conferences including in USA, Malaysia, etc. She was awarded as a best paper presenter in many National Conferences. She has authored Nine text books in the field of Statistics, Entrepreneurship, Business analytics and Event management. Among which Business Analytics was awarded the "Golden Book Award of 2022". She has been also awarded as 'The Best Researcher Award' on 5th September 2022 from Research Education Solution which is a body of MSME, GOI. Another feather for her hat is Patents with 9 Patents published, 2 Grants and 4 patents filed. She is been editor for many Scopus indexed journals of IGI Global publishing house, Public Administration and Policy', Emerald Publishing house and pubmed indexed medical journals. She is an active member of ENACTUS, International Social entrepreneurship forum. She has always been in front line in contributing for social services like organising Blood Donation Camp in association with Rotary Club and ttk, presently she is associated with Ministry of education for One on One Covid-19 drive for building emotional support among the society and the recent one was environmental awareness drive saying "One Earth".

Mohsina Hayat is an accomplished academic professional with over 23 years of teaching experience in the fields of Economics and Finance. Currently serving as an Assistant Professor and Finance Coordinator at the School of Commerce, Jain (Deemed-to-be University), she has previously held permanent faculty positions at Arya College of Management and Technology, Rajasthan Technical University, and Akhtari Quershi Girls Degree College. Her extensive teaching portfolio also includes guest faculty roles at prestigious institutions such as Jamia Millia Islamia and Aligarh Muslim University. Dr. Hayat holds a PhD, M.A., and MBA (First Class) and has consistently demonstrated her dedication to higher education through her numerous publications, including 30 peer-reviewed research papers, 12 books, and 6 book chapters. She is actively engaged in academic and research endeavors, having presented papers at 23 conferences and participated in 31 workshops and faculty development programs. In addition to her teaching career, Dr. Hayat is a member of the Indian Accounting Association (NCR Chapter) and serves as a reviewer for the International Journal of Economics, Finance and Management Sciences. Her innovative contributions to academia include two patent applications and ongoing research guidance for PhD candidates at Jain University. Known for her strong leadership and organizational skills, Dr. Hayat is dedicated to fostering student success through the development of self-reliance, research interests, and effective communication. She is highly regarded for her ability to create and implement programs that not only enhance academic learning but also boost the morale and leadership capabilities of students.

Lakshman.K is currently working as Associate Professor & Head, Assistant Dean- Student Affairs, Department of management, JAIN (Deemed-to-be-University) -CMS, SBMJEC, Bangalore. He is an organized and energetic Academician with fourteen plus years of academic experience and Three plus years of industry experience with strong academic background completed his Doctorate in Philosophy PhD from Annamalai University, Chidambaram, Tamilnadu Thesis Tiled on "E-Banking services on Senior citizen's Buying behaviour "evaluated by Professor from Aurora University, Illinois, United State of America. He is contributed with various research papers published in reputed Scopus Index Journal, Web of science, UGC Care list Journal and Peer reviewed journal. Also presented various papers at National and International Conferences also received more than six times best paper awards for his work.He is been associated & invited as resource person, Keynote speaker, session chair, BOE member, BOS Chair Person member, Resource person, International speaker, Editorial board member with various National & International Educational organizations & Institutions.

Nethravathi K, She has a professional experience of 16 years in Education and a 1-year industry. She has a Ph.D. from CMR University. She has worked in various capacities such as HOD of Management and PG Coordinator and attended many workshops and Faculty Development Programs, seminars, and orientation programs. She has worked in various capacities as NAAC Coordinator, BBA Coordinator, Guide, and Mentor for three years. She has presented more than 35+ research articles at various National and International conferences. Out of which few were awarded the best research paper. She has traveled to various states to participate in workshops, conferences, and paper presentations. She has published many research articles in ISSN and ISBN, UGC care listed, Scopus indexed, ABDC, and published 3 Indian Patents and also has authored the textbook Accountant in Business. She has organized many offline and online FDPs and SDPs and is well appreciated as a resource person on various topics such as research methodologies, finance, and taxation. She is currently working in the Department of Business Management Studies, and guiding research scholars in the School of Commerce, Jain Deemed to be University.

Sheeba K P., M. Sc, PhD (pursuing). She is having a academic Experience of 5 years and 8 years in health care industries like hospitals, health insurance organizations and TPA companies both in abroad and in India. She is a post graduate in Hospital Administration from Rajiv Gandhi University of Health Sciences, currently pursuing her PhD in management. She is a member of associate ship in general insurance from Insurance Institute of India (III). She has attended many workshops and Faculty Development Programs, seminars, and orientation programs. . She has presented more than 3 research articles at National and International conferences. She has published few research articles in UGC care listed journals. She is currently working in a Department of Business Management Studies Assistant Professor, and guiding research scholars in School of Commerce, Jain Deemed to be University.

Bhavana S. Karmore working as Assistant Professor and Dean at BCA and MCA Department of G. H. Raisoni University, Amravati for U.G. and P. G. Department . Total Years of Experience 15+ . She publish more than 20+ Papers in International Journals . She work as International Expert at International Conferences. She worked as a Editorial Board Members in 5+ International Journals and Reviewer in 10+ International Journals and International Conferences . Project Guided 10+ PG students and 10+ UG Students . She publish more than 10+ books of Electronics and Computer Engg. subjects. 3 Design patent, 2 copyright .

Arun Kumar, a Ph.D. scholar affiliated with the Department of Horticulture at the School of Agriculture, Lovely Professional University, Phagwara, Punjab-144411, India, holds a Master's degree from the same institution, specializing in Fruit Science. His research interests encompass production technology, postharvest technology, and organic farming. With a repertoire comprising ten publications, including review papers, research papers, book chapters, and articles, Mr. Kumar has garnered recognition in his field. Notably, he has been honored with the Best Researcher Award and Emerging Scientist Award by Vigyan Varta Magazine (2022) for his substantial contributions, notably his thesis on the "Effect of nano silicon dioxide coating on the quality of winter guava." Mr. Kumar has participated in several training workshops, including Precision Agriculture (by ICAR), Agricultural Extension (MANAGE), Climate Smart Viticultural Technologies (ICAR-NRC Grapes), Artificial Intelligence in Livestock Sector (MANAGE).

Pawan Kumar, PhD, is a prominent academic and researcher with over 20 years of expertise in business studies. A professor at Graphic Era Deemed to be University, Dehradun, India, he has supervised six PhD scholars and published over 60 research papers, including 40 in SCOPUS/ Web of Science, and 14 books, including eight indexed in Scopus. His works, featured by prominent publishers like Emerald, IGI Global, Nova Science, and Cambridge Scholars, have gained recognition. Dr. Kumar earned his PhD in Commerce from Kurukshetra University in 2007 and holds an M.Phil. and M.Com. Degrees. Specializing in finance, banking, and securities markets, he has authored books on green finance, AI, and fintech. His innovative research includes five patents and has served as a guest editor for Scopus-indexed journals. In 2022, he received the prestigious Researcher Award from Chandigarh University. Dr. Kumar's impactful work on sustainable finance, blockchain, and fintech is globally recognized. His active participation in high-impact conferences and faculty development programs reflects his commitment to advancing research, teaching, and innovation in commerce and management.

Maksud A.Madraswale working as an Associate Professor, Department of Management, Presidency University, Bengaluru. For few years he served at Oman and other Arab Countries. To his credit he published five articles in Scopus Indexed journals.

Parinita has a doctorate in marketing with teaching experience of 23 years. She is a thesis supervisor for Liverpool Business School. She has authored 2 books on 'Consumer Behaviour' caselets.

Latha Manickam, I hold a Doctorate in Commerce and possess dual Master's degrees in Commerce and Information Technology. With over 20 years of experience in higher education, I have served in various academic and administrative capacities, including teaching, curriculum development, and programme coordination across institutions in India, Malaysia, and Singapore. I have previously served as the MBA Open and Distance Learning (ODL) Programme Coordinator, contributing to the design and delivery of flexible learning experiences tailored to the needs of working professionals and international learners. Currently, I serve as the Ph.D. Programme Coordinator at MAHSA University, Malaysia. Throughout my career, I have demonstrated a strong commitment to academic excellence, student mentorship, and cross-border educational collaboration. My diverse academic background and extensive leadership experience contribute significantly to postgraduate programme management and scholarly advancement in the field of commerce and information systems.

T.C. Manjunath is a Professor in the Department of Computer Science and Engineering at Rajarajeswari College of Engineering, located on Mysore Road, Bangalore, India. He specializes in IoT Cybersecurity and Blockchain Technology and currently serves as the Dean of Research & Development (R&D) at the institution. With extensive expertise in these cutting-edge areas of technology, Dr. Manjunath contributes significantly to academic research and the development of innovative solutions in his fields of interest. His work continues to shape advancements in the cybersecurity and blockchain domains.

Dinesh.N MCom, MHRM, MLM, Ph.D, Dlitt. Associate Professor OB and HRM Cms Business School Jain Deemed to be University 10+ years of Experience in industry and 18 + years of Experience in teaching dinesh.rao@cms.ac.in A result-oriented professional with a strong background of Twenty Eight years. Facilitated excellent support services in Teaching, Technical & Administration. Excelled at multi-tasking activities. Team worker, building kind relations with colleagues/peers & superiors. Dr. Dinesh.N is a doctorate in Management (Human Resource), having academic and industrial blend as a unique quality, 10 years of various industry and 18 years of teaching the Management subjects authored ten books. Published Twenty papers in International and National Journals and made Fifteen conference presentations, and attended more than 50 Faculty Development workshops and PhD Supervisor guiding ten research scholars. Member of International Journal of Creative Research Thoughts (IJCRT), Member of various universities as Board of Examiner and Board of Studies.

P. Selvakumar is currently an Associate Professor in the Department of Science and Humanities at Nehru Institute of Technology, Coimbatore, Tamil Nadu, India. Since 2023, he has served as the Department IQAC Coordinator at the institute. Prior to joining Nehru Institute of Technology, he held positions as an Assistant Professor in the Department of Science and Humanities various engineering colleges in Coimbatore, Tamil Nadu, India, spanning from 2011 to 2023. Dr.P. Selvakumar's academic and research interests lie primarily in the field of phytochemistry. He has made significant contributions to his field, with an impressive record that includes 30 journal publications, 12 patents, and 20 conference papers. Additionally, he has organized 5 seminars and conferences, further demonstrating his commitment to academic and scientific exchange. Throughout his career, Dr. P.Selvakumar has been recognized for his dedication and excellence in teaching, having been honored with several awards, including the Best Teacher Awards. He holds a prominent position in the academic community, contributing actively to research.

P. Sudheer is currently working as an Assistant Professor in the Department of Computer Science and Engineering (CSE) with a focus on Artificial Intelligence and Machine Learning (AI & ML) at CVR College of Engineering, Hyderabad. With a keen interest in Image Processing, Deep Learning, and Machine Learning, he is dedicated to advancing research and teaching in these dynamic fields.

R.Bhuvaneswari working as an Assistant Professor of English at Dr.Mahalingam College of Engineering and Technology, Pollachi. To her credit she completed 10 NPTEL / SWAYAM and published few articles in Scopus indexed journals.

R. Velmurugan presently working as an Associate Professor in Commerce, Karpagam Academy of Higher Education, Coimbatore. He started his career as a Project Fellow under UGC Sponsored Major Research Project at NGM College, Pollachi. He has 18 years of teaching experience has 15 years of research experience. To his credit he produced 22 Ph.D. Scholars and 5 M.Phil. Scholars. At present 1 scholar submitter his thesis, 6 Ph.D. Scholars are under his guidance. Moreover, to his credit he published 163 articles in various International Journals, of which 54 articles are published in Scopus Indexed Journals and 15 articles in Web of Science and 76 articles in edited volumes. He also participates and presented 27 papers in various International Seminar / Conference and 54 National Seminar / Conference. He acted as an external examiner for various M.Phil. and Ph.D. Viva voce examinations. He acted as a Vice Chancellor nominee for Doctoral Committee Expert at Bharathiar University, Coimbatore and Periyar University, Salem. He acted as a Chairperson at various National Level Conferences. He also acted as an Expert for Syllabus Scrutiny Committee, Question Paper Scrutiny Committee and Expert members for Board of Studies at various reputed Colleges. Considering his meritorious contribution, he received awards like Best Paper Presenter, Best Research Supervisor and Best Faculty Awards from various research foundations and councils. He too acted as a resource person for 27 Guest Lectures, 30 Workshops and 10 faculty development programmes.

Kamal Kumar Rajagopalan is an experienced academician and finance professional, currently serving as an Assistant Professor at Uttaranchal Institute of

Management, Uttaranchal University, Dehradun, India. With expertise in finance, accounting, business analytics, and strategic management, he integrates analytical tools into finance curricula for data-driven decision-making. He holds a Ph.D. and MBA in Finance from ICFAI, along with ACMA certification. His industry experience spans roles in accounting, auditing, and financial management. A published researcher in Scopus and UGC Care journals, he is proficient in Python, R, Power BI, and SAP. Dr. Rajagopalan is a member of ICAI.

Manickam Ramasamy, A chartered engineer registered with Engineering Council United Kingdom. Currently working in UCSI University Malaysia. Vast experience in working with different ehtnic groups gained through a career spanning 34 years working in countries like Inidia, Malaysia and Singapore.

Himanshu, a Ph.D scholar with a specialization in Fruit Science at Lovely Professional University, Phagwara, Punjab. With a deep focus on horticultural research, he is particularly interested in addressing critical issues in fruit crop management, specializing in the study of fruit drop mitigation and physiochemical properties of fruits. He has a strong academic foundation in horticulture, having completed his master's degree in Horticulture (Fruit Science) at Lovely Professional University, Phagwara, Punjab and graduated from Maharishi Markandeshwar University, Ambala, Haryana. His areas of interest include post-harvest technology, precision agriculture, hydroponics, nursery management, and biotechnology. He is particularly focused on optimizing post-harvest quality, advancing soilless cultivation methods, and integrating biotechnological approaches to enhance crop resilience and productivity. He has made significant contributions to the academic community, with 12 publications to his name, including research articles, review papers, and book chapters published in well-regarded NAAS and Scopus-indexed journals. His research outputs highlight his commitment to advancing horticultural science and contributing practical insights for improving crop quality and resilience. Beyond academia, he actively engages in professional development, attending workshops, conferences, and training in areas like mushroom cultivation, Hi-tech seedling production and nanotechnology, further enriching his expertise. His dedication to research, combined with hands-on learning and participation in various scientific gatherings, positions him as a promising young researcher with a strong foundation in horticulture and fruit science.

B. Jonathan Samuel is a highly accomplished professional with over 17 years of experience in the banking sector and 8 years in the logistics industry. He currently serves as a Chief Manager at a leading private sector bank in India. His diverse professional journey has equipped him with deep expertise in banking operations, logistics management, and legal compliance. He holds a Bachelor of Commerce (B.Com.), an MBA, and a Bachelor's degree in Law, equipping him with a strong foundation in finance, management, and legal affairs. His expertise spans a wide range of areas, including labour law, banking-related legal matters, and corporate governance. He is well-versed in handling complex legal issues and providing strategic legal support within corporate and financial environments. He is particularly well-regarded for his ability to navigate complex labour issues and ensure regulatory compliance in dynamic business environments. In addition to his legal and industry experience, he has also completed a Higher Diploma in Software Engineering, enabling him to bridge the gap between technology, law, and finance. This multidisciplinary background allows him to approach problems with a unique, integrated perspective.

K. Sasikala is a faculty member in the Department of Information Technology at R P Sarathy Institute of Technology, Salem. With a strong academic background and dedication to teaching, she contributes to shaping the future of students in the field of Information Technology.

G S Sharma is working as Associate professor, department of mathematics in Vivekananda Global University, Jaipur. He has 7+ years of experience in academics and research. He has the specialization in operations research, inventory control and supply chain management.

MohitSharma is a research scholar at the Institute of Management Studies and Research, Maharshi Dayanand University, Rohtak, Haryana, India. His research focuses on the dynamic relationship between brand image and purchase intentions, particularly in the context of brand extensions within the FMCG sector. He has co-authored several scholarly works and continues to contribute to the academic community. Mohit can be reached at mohitattri.rs.imsar@mdurohtak.ac.in

Jatinder Singh has been serving as an Professor in Lovely Professional University, Jalandhar (India) since, 2013. His area of research is production of seasonal fruits and their upto-date preservation by using of different post-harvest technologies. He has published around 100 publications along with 25 book chapters with the national/international publishers. He has won many honours/prizes like First in poster presentation (2021), best academician award (2023) and Vigyan Varta all rounder award (2022). He has guided 26 students of postgraduation and three students of P.hD. Most of the students are/had worked on post-harvest technology. He got published six patents through Govt of India and ten copyright. He has worked in various government procurement organizations such as Punjab Agro Food Grains Corp. Ltd. He also has experience in private organizations such as Genesis seeds,

Krishna seeds. He also has worked internationally in East Land Agro Pvt Ltd. As a Project Manager in West Africa. He also excels in teaching experience such as he contributed in colleges like SGTB Khalsa College, Sri Anandpur Sahib and Miri Piri Khalsa College Bhadour, Barnala. He has attended various trainings such as National training on production technology of Mushrooms, Vermicomposting, Grain storage, Entrepreneurship development programme, 15 days training on novel technologies and various workshops also

Vikanksha is a Ph.D. scholar at Department of Horticulture, School of Agriculture, Lovely Professional University, Phagwara, Punjab-144411, India. She has completed her masters from, Lovely Professional University, in the discipline of Fruit Science. Her research interests are production technology, postharvest technology and organic farming. She has authored twelve publications that include review papers, research papers, book chapters and articles. She has been awarded with Best researcher award and Emerging scientist award by Vigyan Varta Magazine (2022) for her contribution and recognition in field with thesis "Effect of nanoencapsulation coating on quality and shelf-life of guava". She has attended various training/workshops such as one month training on Precision agriculture (by ICAR), Agricultural extension: An exciting profession for the future (MANAGE), Innovations in Agriculture Extension (MANAGE& SAMETI), Climate Smart Viticultural Technologies for Sustainable Quality Grape Production (ICAR-NRC Grapes), Prospects and applications of artificial intelligence in livestock sector (MANAGE), Labour Codes for Agricultural Workers (V.V. Giri National Labour Institute, NOIDA & MANAGE), Drone: A boom for agriculture (CSJM University Kanpur and agrimeet foundation), Drone marathon (DFSRDA- NAHEP, VNMKV, Parbhani & Just Agriculture Education Group), Hydroponics (Hydrogreen), Sustainable agriculture (MANAGE).

Ravi Thirumalaisamy working as an Assistant Professor, Department of Business and Economics, Modern College of Business and Science. To his credit, he published more than 10 articles in Scopus indexed journals.

Shraddha N. Zanjat currently serves as an Assistant Professor in the EEE Department at the School of Engineering and Technology (SOET), Sandip University, Nashik. With 10 years of teaching experience and 2 years of industry exposure, she brings a wealth of knowledge to her role. She holds a post-graduation degree, a Master of Technology in Electronics and Communication, from RTMN University. Additionally, she is pursuing her PhD at Sandip University, Nashik. Prof. Shraddha N. Zanjat has presented ten conference papers at both national and international levels and has published five papers in esteemed international journals such as IEEE, UGC, and others. Furthermore, she has obtained five Indian design patents. She is also a member of two technical association bodies. Prof. Shraddha N. Zanjat has registered four copyrights under her name and has published two book chapters in Scopus-indexed publications. Additionally, she has authored more than 50 books with international publications and has received royalties of about 20k.

Index

A

B

C

D

E

F

G

H

I

L

M

O

P

Q

R

S

W

www.ingramcontent.com/pod-product-compliance
Lightning Source LLC
LaVergne TN
LVHW080346011026
849161LV00009B/2211